Launched Bridges

Prestressed Concrete Bridges Built on the Ground and Launched into Their Final Position

Marco Rosignoli

ASCE PRESS
1801 Alexander Bell Drive
Reston, VA 20191-4400

Abstract: *Launched Bridges* is a book designed to merge the static, organizational, and economic aspects of construction techniques for the launch of monolithic bridges, as these aspects are related to each other and must be considered as a whole in the design of such type of structures. The book is geared toward building contractors, technical and design engineers, and owners, as it can benefit all involved with bridge building.

Library of Congress Cataloging-in-Publication Data

Rosignoli, Marco.
Launched bridges: prestressed concrete bridges built on the ground and launched into their final position/Marco Rosignoli.
p. cm.
Includes bibliographical references and index.
ISBN 0-7844-0314-7
1. Bridges, Concrete–Design and construction. 2. Prestressed concrete construction. I. Title.
TG340.R62 1998
624'.2–dc21 98-4585
CIP

Library of Congress Catalog Card No.: 98-4585 ISBN 0-7844-0314-7
Manufactured in the United States of America.

Before introducing this book, I would like to thank my wife Carla, whose dedication and tenacity made it possible, my children Chiara and Luca, my brother Dario who made the CAD drawings, and Maria Grazia Bruschi, who checked my English translation.

ABOUT THE AUTHOR

Marco Rosignoli received his degree in structural engineering with Honors from the University of Ancona in 1981.

He worked for large Italian builders for almost twenty years. As site director, then project manager, and, for eight years, at the head of the bridge department of the seventh Italian general contractor, he managed design and construction of a large number of bridges; among them, the most important launched bridges in Italy.

He is currently an independent consulting engineer practicing in Italy and abroad.

His activities include preliminary study, budgeting, bidding, and structural design, as well as the design and construction supervision of equipment, site organization, and technical assistance.

He has contributed to the diffusion and evolution of these construction methods by publishing many technical articles.

The author can be contacted at fax number +39 (0) 521 252875, by the Internet, E-mail mrosign@tin.it, or by mail: Marco Rosignoli, Via Ravà 15, 43100 Parma, Italy.

CONTENTS

PREFACE

In the last decades, the progressive growth of the environmental conscience led communities to rediscover the aspiration to build beautiful and efficient public works that characterized the whole history of mankind, an ethic abandoned only at the end of the Victorian age for the aesthetic poverty considered indispensable in the dark years of recession and of the first world war. At the same time, the availability of new calculating means and of new structural materials permitted a progressive evolution of the prestressed concrete bridge construction techniques toward the prefabrication of always larger and heavier elements. More recently, new management techniques and commercial strategies singled out the intrinsic quality of the final product and its good adaptability as a primary objective.

All these elements produced substantial changes in the structural typologies adopted in medium span prestressed concrete bridges, and in particular made the continuous box girder preferable over the simply supported precast I beams. Similarly, the rise of labor cost and the reduction of the average length of bridges stimulated research on new box girder construction techniques that combine low labor consumption and low investments, which could be amortized in more jobs. From here came the rediscovery of techniques of prehistoric intuition, if one thinks of the trunk of a tree, which involve the handling of monolithic elements previously prepared on the ground.

Clearly, the handling of large and heavy structures involves many problems, and designers have to find a delicate balance between unusual structural requirements, technological constraints, and economic implications. This balance requires good communication between the designer and the contractor, and therefore humbleness and teamwork to refine the design, a task which is anything but light. On the one hand, this requires that the builder comprehend the implications of the design choices and adopt faithfully the decided procedures. On the other, it imposes that the designer consider it his duty to optimize even the organizational aspects which move the builder toward the expected economic result.

As a matter of fact, it is also because of the difficulty in reaching this balance that many contractors operating with these technologies have in-house design teams which develop and promote research and innovation, using the feedback of data and experiences from the field, and put themselves in direct communication with the owner even in the most advanced projects. For contractors who operate in the international market, dynamic innovation and export are two basic and tightly interconnected requirements, as the ability to export can be measured by the ability to propose competitive alternatives, which in turn derives only from cultivating creative innovation.

In these last years, the general contraction of the public works market and the resulting tighter competition led contractors to reduce overhead expenses, and with them research, and to seek a systematic reuse of available equipment, which in turn is prevented by innovative ideas involving new investments. But, in spite of that, innovation is an inevitable requirement for international activity of high-tech countries, which cannot compete with emerging countries in labor cost, and therefore must exercise know-how to propose economic and technologically advanced alternatives in which savings occur on amortization and consumption of materials. And since some emerging countries have already reached high technological levels, the most evolved nations, to preserve their export capacity, cannot but continue the research of new ideas, which will be accepted abroad only after having found a sufficiently consolidated field of application in the domestic market.

In this context, the aim of this book is to merge the static, organizational, and economic aspects of construction techniques for the launch of monolithic bridges, as these aspects are related to each other and must be considered as a whole in the design of such structures. This book was written by a project manager who spent many years leading work groups composed of designers and field engineers, verifying each time that the broad view can result in a winning approach to high-tech structures. The book is addressed to building contractors, technical and design-office engineers, and owners, to help each one complete his partial knowledge of the subject.

In Chapter 1 the general aspects and the main differences between three construction techniques (the "continuous casting" with incremental launching, the "balanced casting" with symmetrical launching or rotation, and the casting on a general centering followed by side translation) are detailed. The bridges built by incremental launching are described in Chapters 2 through 9. Chapter 10 is dedicated to the bridges built by balanced casting followed by rotation, monolithic front launch, or side translation.

In Chapter 2, after a comparison with the alternative techniques, the main implications of incremental launch on the cross section, the static scheme, and the plan and profile layout of such bridges are analyzed. Chapter 3 describes the construction methods for bridge segments, and Chapter 4 the launching techniques and equipment needed. In Chapter 5, the static stresses deriving from the handling of the continuous beam are examined. Specific design methods allow builders to optimize both the support equipment (launching nose, cable-stayed schemes, temporary piers) and the continuous beam itself, whose load effects are analyzed by means of different representations of the stress state. Chapter 6 examines additional effects (thermal gradients, time-dependent phenomena, and differential settlement of bearings) by means of specific design methods, as well as the techniques of verifying the design assumptions by means of field monitoring and data recording. Chapter 7 describes launch and final prestressing, and Chapter 8, the load conditions of the temporary and the final piers. Finally, Chapter 9 illustrates the present trends and the ongoing research regarding incremental launching, ranging from the use of alternative materials to the prestressed composite sections.

Chapter 1

INTRODUCTION

BIRTH OF INCREMENTAL LAUNCHING

The first prestressed concrete bridge was built by Freyssinet in 1941. From the end of the second world war onwards, prestressed concrete bridge construction developed quickly, thanks to the pioneers of the new technology: Guyon, Freyssinet, Leonhardt, Magnel, Morandi, Mörsch, and Ross.

In the past, the structural design of almost all these new bridges was limited to quasi-isostatic schemes, since analysis and dimensioning of indeterminate systems were not practical using the only calculating means available then: the slide rule. For many years, the basic criterion in the design of prestressed concrete bridges has been to assure the possibility of simple static analysis. So, multi-span bridges built on falseworks used the simple support scheme, the first bridges built by free cantilevering in the fifties were hinged in mid-span, and even the first decks built with self launching centering, in the sixties, were simply supported or articulated with Gerber joints at about the zero-moment points.

From the sixties onwards, the extraordinary progress of computers, the technological advances in prestressing, and the improved knowledge of materials rapidly extended the possibilities in design and, above all, analysis of prestressed concrete. In their turn, these new possibilities stimulated technological innovations and the definition of new construction methods, in spite of their structural implications.

In the creative thinking of these fifty years, many construction techniques were developed in order to guarantee adequate margins of competitiveness of prestressed concrete versus steel constructions in spite of the different evolution of the cost of materials and, above all, of labor. New construction procedures took advantage of the most recent technological advances both in the design of new equipment and in the application of techniques already tested. And so, if the idea of the monolithic handling of bridge decks derives in itself from Paleolithic technology, its recent application to prestressed concrete structures has been permitted by the ability to adapt launching techniques already proven in the field of steel, and by the commercial availability of innovative materials such as Teflon.

For many decades, the lighter weight of steel structures had permitted their front launch by means of winches and lubricated wooden slides or steel rollers[122], with a friction that, although considerable, did not induce inadmissible stresses in the piers or require excessively expensive launching equipment. The high general efficiency of the steel deck (ratio of service load to total design load) and the ability of the material to work both in tension and in compression permitted easy absorption of the temporary launching stresses produced by dead load, avoiding need for oversizing with respect to service loads. The relative simplicity of the technique attracted the interest of bridge owners and operators, and showed many advantages compared with in-situ assembly. The fitting and joining of segments was performed on the ground instead of in the air, without affecting the areas below the bridge and with a greater safety and simplicity (figure 1.1).

Figure 1.1 - Incremental launch of the Schrotetal Bridge steel girder (author).

At the same time, compared with traditional hoisting techniques, higher costs resulted from the operational slow-down caused by the availability of only one working yard, and as a consequence the adoption of this technique was, and is, generally limited to decks very high above the ground or in inaccessible areas.

To the eyes of the prestressed concrete deck pioneers, some of these factors were less critical, some others even very promising. In fact, the construction times and the yard organization for prestressed concrete bridges are very different from those for steel superstructures, and the cost of labor and equipment is so high that every possible alternative has to be carefully examined. The monolithic launch of decks built on the ground near their final position promised interesting savings in both equipment and labor, but actual chances were limited by the much higher weight of concrete and by its low tensile strength.

These obstacles were gradually overcome by:

- the advances in prestressing technology, which lightens the deck, makes it elastic and less subject to cracking, permits joining subsequent segments, and allows it to be introduced or removed according to necessity;
- the commercial availability of computer programs which facilitate continuous beam analysis of the prestressed concrete in the subsequent support configurations assumed during launch;
- the introduction of the steel-Teflon contact along sliding surfaces in order to reduce friction;
- the evolution of construction equipment to the technological level necessary to build and to move enormous masses with due precision.

Nothing of this was available in 1959, the year which saw a first attempt to move individual concrete precast segments for the highway bridge over the Ager River[19], composed of four spans of *3 m*, *85 m*, *85 m*, and *36 m* respectively (figure 1.2).

Figure 1.2 - Ager Bridge precast segments (courtesy Leonhardt[19]).

The deck was built on a timber scaffolding which supported precast segments about *10 m*-long, built on one bank and transported one by one into the final position, for the full length of the bridge. The *0.5 m*-thick joints between adjacent segments were then cast in-situ, and prestressing was obtained by means of long external tendons placed inside the box cell and tensioned from the two deck ends. The segments were moved to their final position by sliding them above a wooden rail by means of wooden skids lubricated with engine oil. The friction coefficient reached the value of *0.25* in this way, but without negative effects due to the low weight of segments, which were moved one by one.

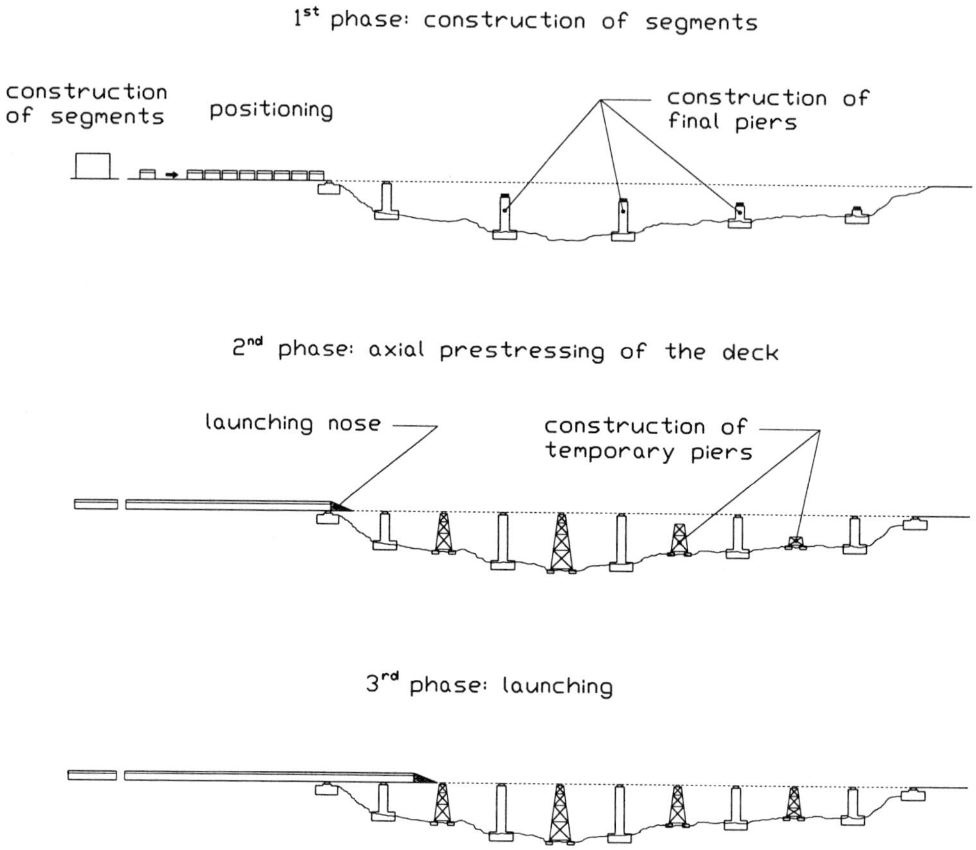

Figure 1.3 - Construction scheme of the Rio Caroni Bridge.

This procedure, in spite of its cleverness, demonstrated how it would have been simpler to avoid the scaffolding at all and to launch the whole deck entirely precast behind an abutment directly above the piers, utilizing its own flexural stiffness. This concept was applied not much later for the Rio Caroni Bridge[20], in Venezuela, designed also by Leonhardt in 1961, and built in 1962-63. The scaffolding was eliminated, and the deck was built behind an abutment in its whole length by joining *9.2 m*-long segments built in a casting yard and advanced to the assembly position by sliding above a wooden rail, with *0.4 m* cast in-situ joints.

Upon completion of the assembly (figure 1.3) the deck was prestressed using several rectilinear external tendons distributed inside the box cell so as to obtain a centroidal force. Their anchorages were arranged in the front end diaphragm of the deck, while at the opposite end they were deviated around a semi-cylindrical concrete block, whose jacking to a distance of *2.8 m* from the deck back end produced a prestressing force of *50 MN* in only one operation.

Figure 1.4 - Launch of the Rio Caroni Bridge (courtesy Leonhardt[20]).

The deck, which at the end of launch would have covered two *48 m* side spans and four *96 m* interior ones as a continuous beam, was launched with the help of a *17 m* launching nose over the permanent piers and over one temporary truss pier in each span, to reduce the launching span to *48 m* (figure 1.4). Two prestressing jacks anchored to the back abutment and pulling strand cables fastened to the sides of the deck produced the horizontal force necessary to launch the deck. The deck was moved above integral launch bearings, and at the end of each *0.96 m* stroke it had to be lifted to relocate the bearings (figure 1.5).

At the end of launch, the axial prestressing would not have been sufficient for the design loads, so the external tendons were made eccentric by shifting vertically their mid-span and support deviation points by means of jacks. At the same time, the back deviation semi-cylinder was released, and at the end of tendon realignment it reached a definitive distance of *0.20 m* from the deck. Tendons were finally drawn near the webs, fastened to them, and protected by means of pre-pack concrete (figure 1.6).

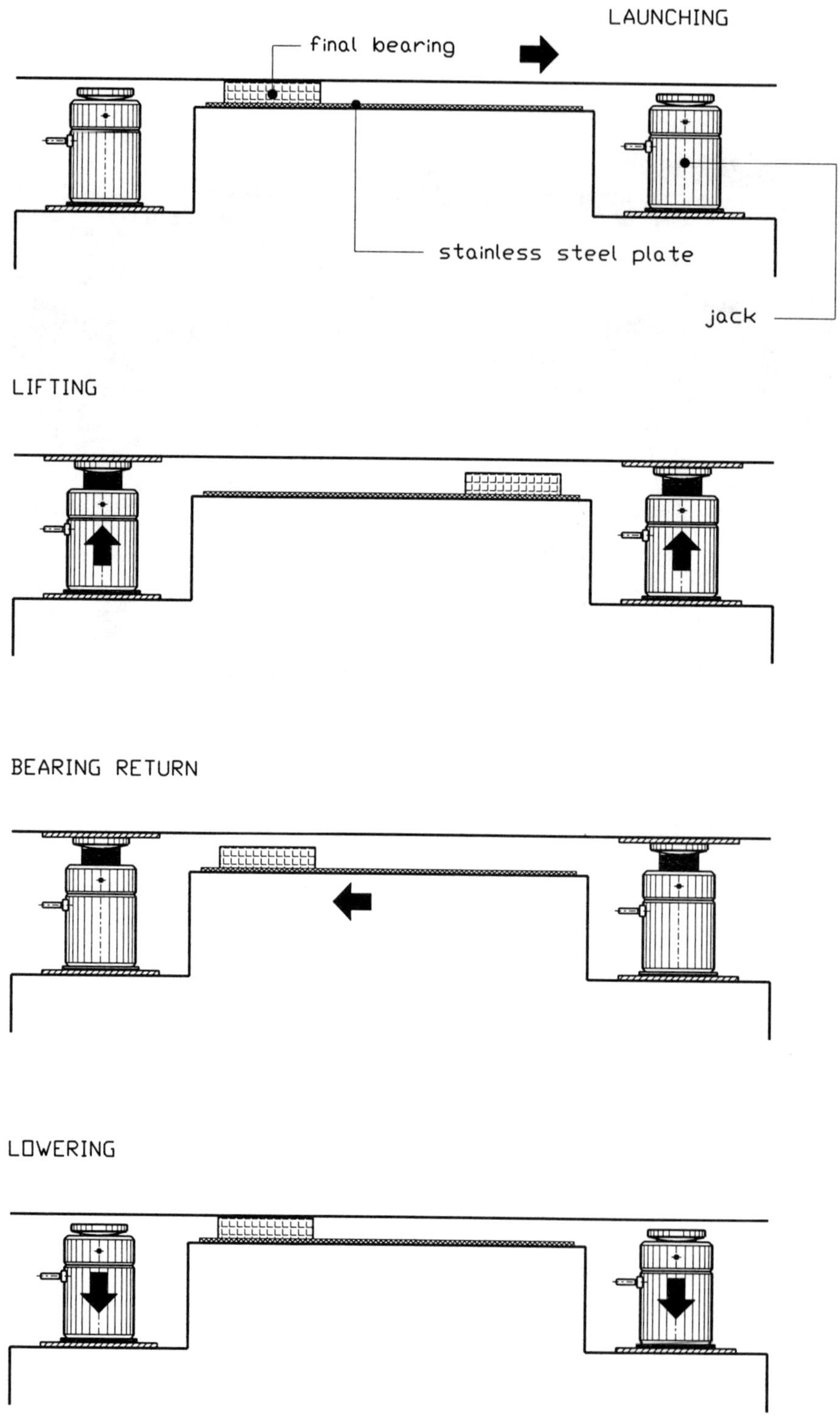

Figure 1.5 - Discontinuous launch with mobile bearings.

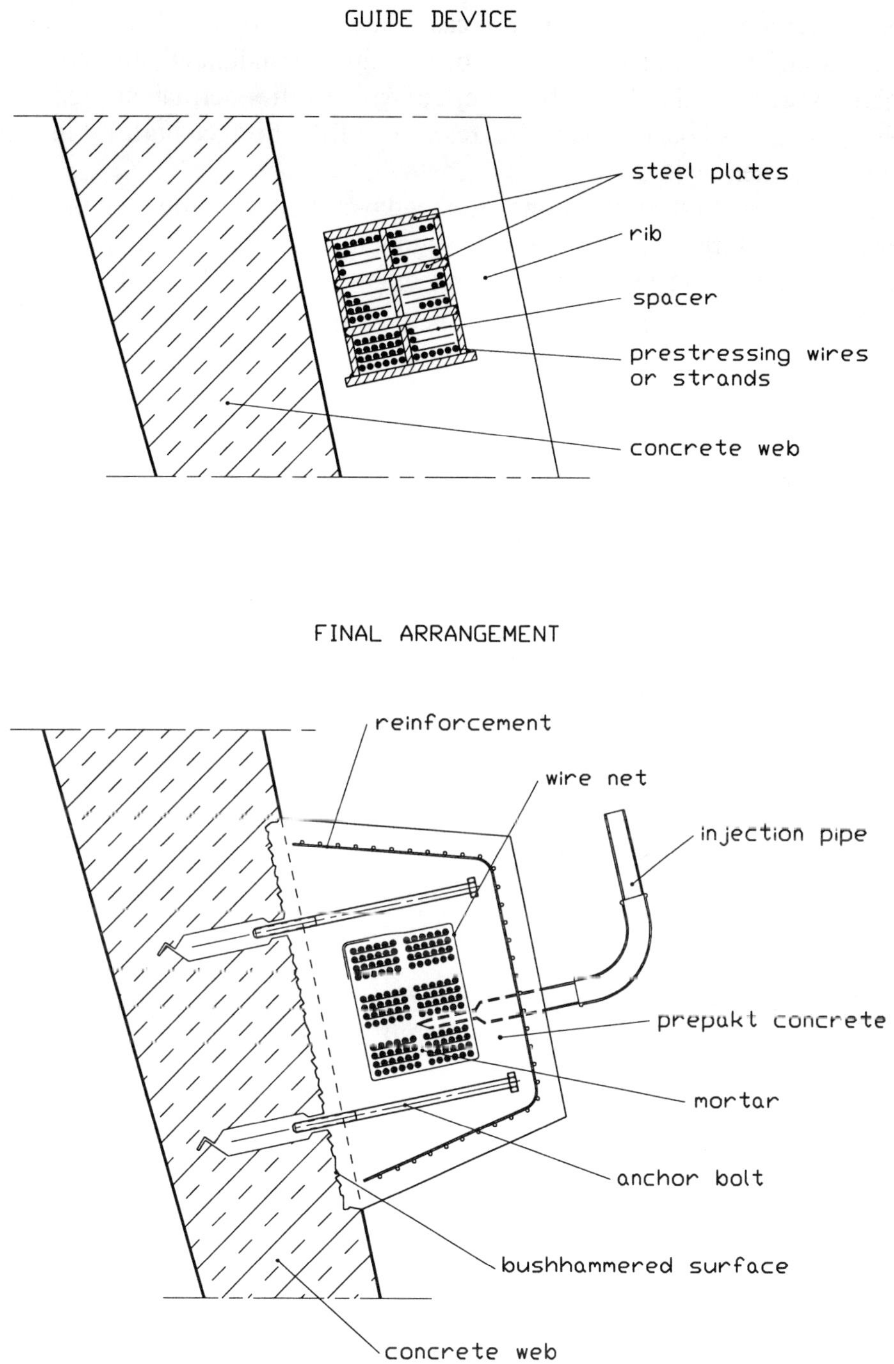

Figure 1.6 - Final arrangement of prestressing tendons in the Rio Caroni Bridge.

Although conceptually ingenious, the shifting of stressed tendons was such a complex and expensive operation (it took longer than the whole deck launch) that it could be convenient only in very special cases, and only for small corrections of the tendon position. This stimulated the idea of introducing undulated tendons at the end of launch to correct the low efficiency of a specific permanent axial launch prestressing, and this concept characterizes most of the bridges built by incremental launching since then.

Moreover, these first applications of the dawning technology emphasized the uncompetitive features of the precast segment assembly. It requires a great placement precision and long joint casting operations, which result in high labor costs.

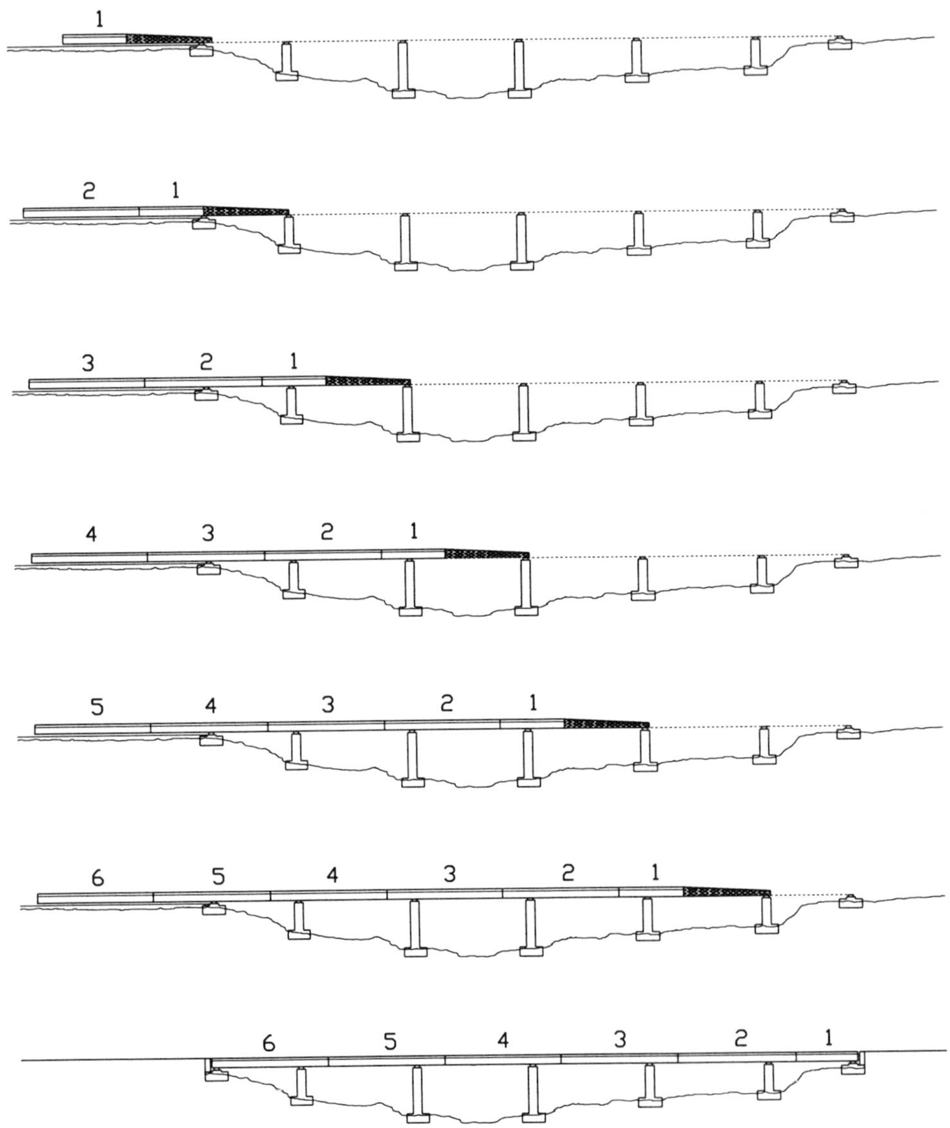

Figure 1.7 - Incremental launch.

That was how, in 1965, the incremental launching method (construction of cast in-situ deck segments alternated with the launch of the whole deck section) was applied for the first time in the construction of a highway bridge near Kufstein[188]. Once the back segment concrete has reached the necessary strength and launching prestressing has been introduced, the whole deck is launched forward a length equal to the last segment. This gives free access to the formwork, and another segment can be built in contact with the back one, and so on until the completion of the deck (figure 1.7). From a certain point of view, the construction procedure simulates a horizontal "slip form," with the difference that the formwork is fixed and the deck itself moves with respect to it in a process which can be likened to a cadenced extrusion.

In this first experience of incremental launching, the use of a specific launch prestressing was avoided[108] by reducing launching spans with temporary piers and by accepting edge cracking, temporarily controlled by reinforcement and permanently closed by parabolic prestressing (figure 1.8).

Figure 1.8 - Launch of the third box girder of the Kufstein Bridge: rectangular piers are temporary (courtesy Leonhardt[108]).

In subsequent years, many innovations permitted refinement of construction and handling techniques. Sliding is currently facilitated by inserting light neoprene-Teflon plates (from which derives the neo-flon acronym) between the deck and fixed steel launching bearings, thus eliminating the need for deck lifting. The front cantilever support devices range from steel or concrete launching noses to cable-stayed schemes. Thrust devices have evolved toward compact friction units which guarantee wide safety margins and smooth travel thanks to the joined use of high-pressure hydraulic and digital controls. At the same time, fast continuous beam solver algorithms permit easy evaluation of the stresses and deformations of the deck in each launching phase, and monitoring of their envelopes.

Figure 1.9 - Skye Bridge (reprinted with permission, Dywidag Systems International).

The technological advances of incremental launching led to standardization of production, and in turn permitted experimentation with new types of monolithic deck handling. So, though incremental launch still represents the most common handling technique, and its evolution has influenced the development of others, the symmetrical approach and span rotation have won a role in critical applications, and in many cases they represent the least expensive and most elegant solutions.

The maturation of these construction techniques was ensured by the current historical moment, in which the extreme competitivity of the market dictates, whatever the dimensions of the bridge may be, the triumph of the least expensive technology. So, although launch techniques have been refined for small or medium-size bridges (big bridges permit single-use amortization of even very expensive equipment) the search for the greatest cost reduction lent them such a cost-effectiveness as to extend their adoption even to large bridges. The result is that these techniques are now applied to small bridges, where low labor cost can be reached without major investments, and to large bridges, where even lower labor costs are obtained by means of the mass production permitted by higher investments, which in turn can be amortized in more bridges if equipment is adaptable.

In the last two decades, the launch techniques have known ingenious applications, which range from the first prestressed composite highway overpasses to beams supported by arches, curved decks, cable-stayed bridges launched symmetrically from the two banks, or built along one bank and then turned into their final position, or built launching the deck as a continuous beam above temporary piers then suspended once it reached the final position.

Among the many railway bridges, decks launched above arches, or more modest, but always innovating, decks launched as a continuous beam and then divided into several shorter beams separated by expansion joints, as well as simple support schemes, have all brought their individual, precious contribution to the evolution of these construction methods.

LAUNCHED BRIDGE TECHNOLOGIES

Compared with traditional construction techniques, the monolithic handling of cast in-situ prestressed concrete decks is surely among the most revolutionary, from many points of view:

- the standard equipment is simple, being limited to a formwork supported from the ground, a launching device and, only in front launching, special devices to reduce the launching stresses in the front spans;
- the formwork, batching plant, and steel cage assembly yard are placed in a radius of between *50 m* and *100 m*, so the stocking and handling of materials by means of a tower crane are considerably simplified;
- the deck is built in a fixed, sheltered location. Each operation, from the steel cage assembly to the concrete casting or prestressing, is simpler and safer than on a falsework or at the tip of a cantilever, and can be organized in parallel rather than in series;
- a high level of quality control is indispensable in each construction stage, since mistakes or irregularities produce launching difficulties and therefore higher costs. The construction of well-built and durable works is therefore in the best interest of the contractor;
- the construction of the deck does not require placement of equipment such as footbridges or falseworks between the piers, with clear advantages when the

bridge spans navigable rivers, towns, highways, railways, and inaccessible places, or presents reduced clearances (figure 1.9);

- unlike free cantilevering, it is possible and convenient to build long deck segments (in incremental launching, even more than *30 m* long) and this reduces the number of construction joints, weak points of the structure;
- compared to the free cantilevering method, and assuming the same duration for casting of each segment, launching results in faster construction and requires only one casting bed instead of two or more (four or six for symmetry reasons).

From a geometrical point of view, the launch of the deck from the construction point to the final position can take place along four different directions (figure 1.10), eventually combined between them: two translations along the horizontal axes of the $x-y$ plane, a translation along the vertical axis z, and a rotation about z.

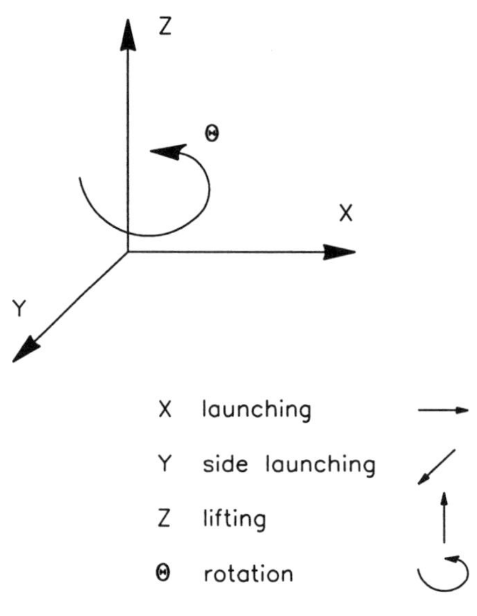

Figure 1.10 - Launch types (the deck is built along axis x).

Placing the x-axis along the deck, the translation along x is called front launching, the translation along y is side launching, the one along z is lifting, and rotation takes place about z.

These four different handling types are nowadays markedly specialized:

- front launch can be incremental or monolithic. Incremental launching is used for continuous beams composed of a number of bays with similar spans, and permits building bridges even of large dimensions, longer than one kilometer. Monolithic launch is used for bridges composed of two or three spans, even with long spans in the case of cable-stayed schemes, and launching can occur from one bank toward the other one, or symmetrically from both banks with mid-span closure;

- rotation is adopted both in the three bay continuous beam with the central span double the side ones (with rotation of the two deck halves on the opposite banks and mid-span closure) and in asymmetrical schemes. It allows use of varying depth cross sections and cable-stayed schemes, and therefore can reach longer spans;
- side launching has a marginal role, and is limited to small decks and to roofs for hangars or industrial sheds;
- finally, lifting is not analyzed in this book because of its extreme rarity and the absence of particular problems.

The different handling types require different yard layouts and construction sequences, and can be distinguished in two schemes. The first one, "continuous casting," is the typical scheme of incremental launching, which involves the construction of the deck in a fixed formwork and a sequence alternating segment casting and progressive launching. The second one, "balanced casting," is the typical scheme of monolithic launching, in which the deck is built progressively in a fixed position by means of one or more mobile formworks, and handling takes place only once the whole deck is finished and prestressing is introduced.

"Continuous Casting" and Incremental Launch

"Continuous casting" technology is generally adopted in medium-length bridges composed of many spans of similar length, as an alternative to fixed falseworks or self-launching centering.

Figure 1.11 - "Continuous casting" yard of the Serio River Bridge (author).

The deck is built in a yard placed behind an abutment, in consecutive segments generally as long as a span or its half, to reach the highest degree of repetition in the construction of the support and mid-span segments. The casting of each segment is followed by the launch of the whole deck (figure 1.11), giving access to the formwork to build the next segment, match-cast against the previous one (figure 1.12).

Figure 1.12 - Launch gives access to the formwork (author).

In case of short decks, the use of long segments is uneconomical, since their small number is insufficient both to obtain a good repetition of operations and to amortize the large investments. Therefore, each segment is subdivided into several sub-segments built in a free sequence along a continuous support formed by two reinforced concrete beams. This scheme allows the use of small modular formworks sliding along the beams. Obviously, the "free sequence" is only apparent, since the segments containing the launch prestressing anchorages shall be cast first to allow for their longer curing before stressing, and thc segments subject to uniform compression can be cast at the end. In addition, steel cage assembly and concrete pouring should not occur on two adjacent segments.

During launch, the deck acts as a continuous beam, whose number of supports increases progressively because of the presence of the new piers. As the temporary stresses, produced by the deck overhanging beyond the front pier until it reaches the next pier, are incompatible with a correct general dimensioning, they must be reduced by means of cantilever support or lightening devices. In case of long or very different spans, it is also convenient to use temporary piers.

The overload of the deck front zone is not limited to negative bending moment and shear force. Even the positive moment is higher than in the back zone of the deck, since the moment diagram of the continuous beam with spans of equal length and constant unit load reaches higher values in the end spans. In bridges built in their final position, the end spans are generally shorter in order to limit these overstresses. On the contrary, during launch the deck has to overhang the full interior spans.

The adoption of support or lightening devices for the cantilever permits design of the deck for the load effects in its back zone, and absorption of the higher stresses in the front zone by means of structural stiffening or local strengthening of launch prestressing.

Longitudinal prestressing is composed of several tendon groups introduced at different times:

- rectilinear launch tendons distributed in such a way to produce in each cross section of the deck the axial force necessary to cover, partially or totally, the tensile stresses produced by the bending moment envelope. These tendons are tensioned before launch and are joined longitudinally to each other by couplers or overlapping;
- internal parabolic tendons or external polygonal ones tensioned after launch, extending for one or more spans, and generally joined by overlapping;
- eventual "cap tendons" above supports and "span-tendons" in mid-span, which complement the action of the parabolic tendons and increase the efficiency of the total prestress.

Especially in wide decks, longitudinal prestressing is combined with transverse prestressing of the top slab.

"Balanced Casting" and Front Launch or Rotation

In most cases, balanced casting and monolithic launching are adopted in three span symmetrical bridges with side spans shorter than the central one (figure 1.13).

The two halves of the bridge are built on the opposite banks, along the obstacle to overpass in case of rotation (figure 1.14) or along the final layout of the deck in case of symmetrical approach. The deck is built in segments, symmetrically from the central segment (which in case of rotation is placed directly above the final interior pier, pivot of the movement, while in case of symmetrical approach is in a temporary position) as in the conventional free cantilevering, but the formwork is supported on the ground instead of being suspended from the deck, and can be single (figure 1.15), or paired.

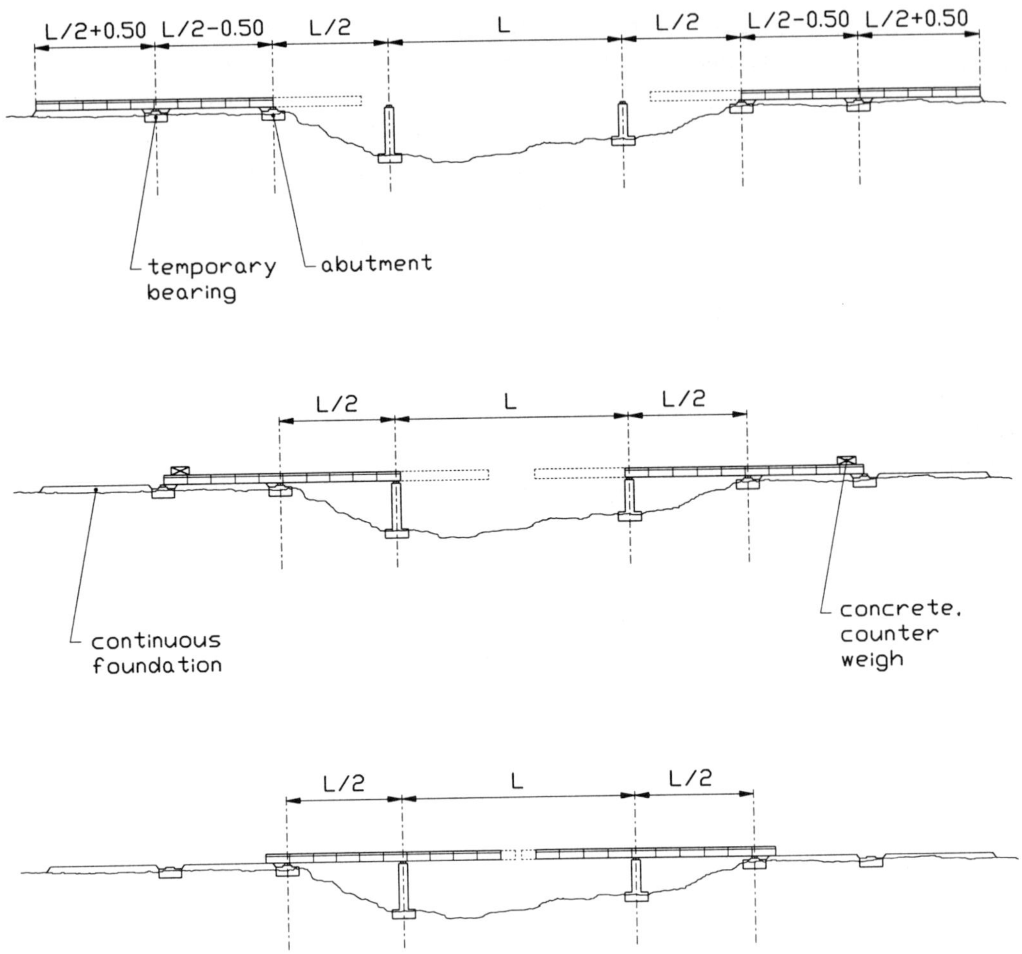

Figure 1.13 - Balanced casting and monolithic launch.

The symmetrical decks require the organization of one yard on each bank. The equipment is so simple and construction sequence so fast that the transfer of the whole equipment from one bank to the other is rather frequent, and the construction of the two half-decks occurs in two subsequent stages.

Launching occurs generally after the completion of the second half-deck, and the cast in-situ mid-span closure segment creates the final continuous beam static scheme. Appropriate temporary bearings support the deck during launch and guarantee adequate safety from tilting.

Rotation does not induce temporary stresses in the deck, apart from small torsion and dynamic effects in large cable-stayed bridges.

Front launching can be obtained either on movable bearings, which avoid temporary stresses in the deck but require stiff launching rails, or on fixed bearings, which produce temporary stresses similar to the ones of incrementally launched bridges, but require small temporary foundations.

The problem of longitudinal tilting is more complex in rotation, since the pivot coincides with the interior pier, which must be placed relatively distant from the obstacle to guarantee an adequate clearance to the deck falsework. This can create lack of balance between the interior cantilever and the external balance span, which can be solved by lightening the cantilever using lightweight concrete or HPC, or by permanently or temporarily ballasting the external span.

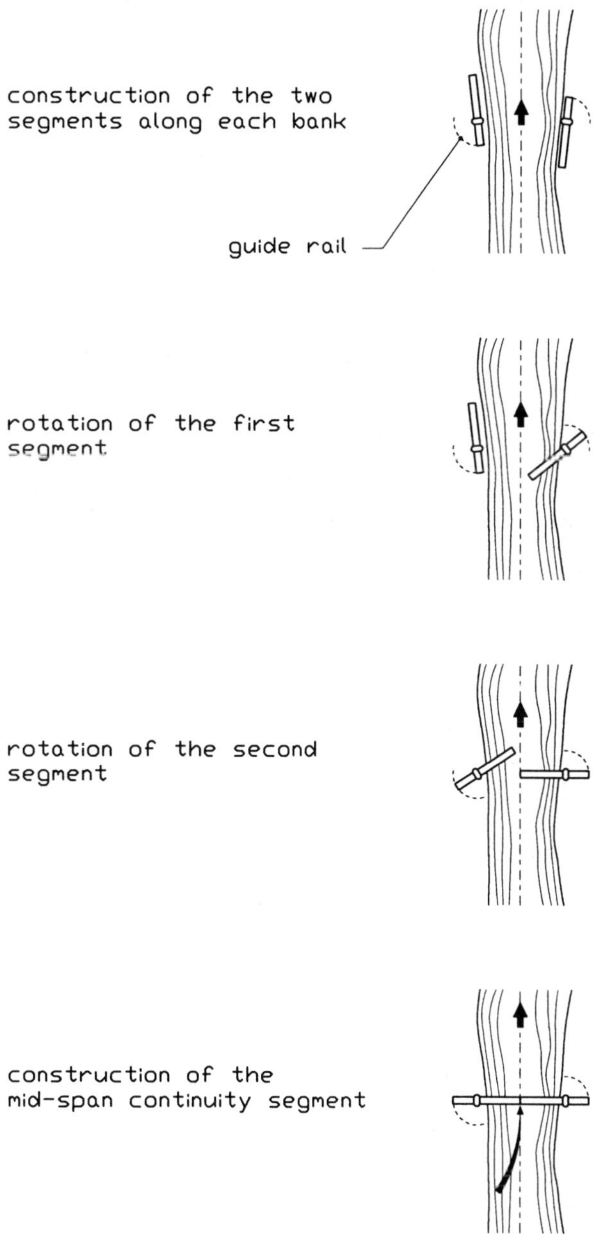

Figure 1.14 - Rotation with closure at mid-span.

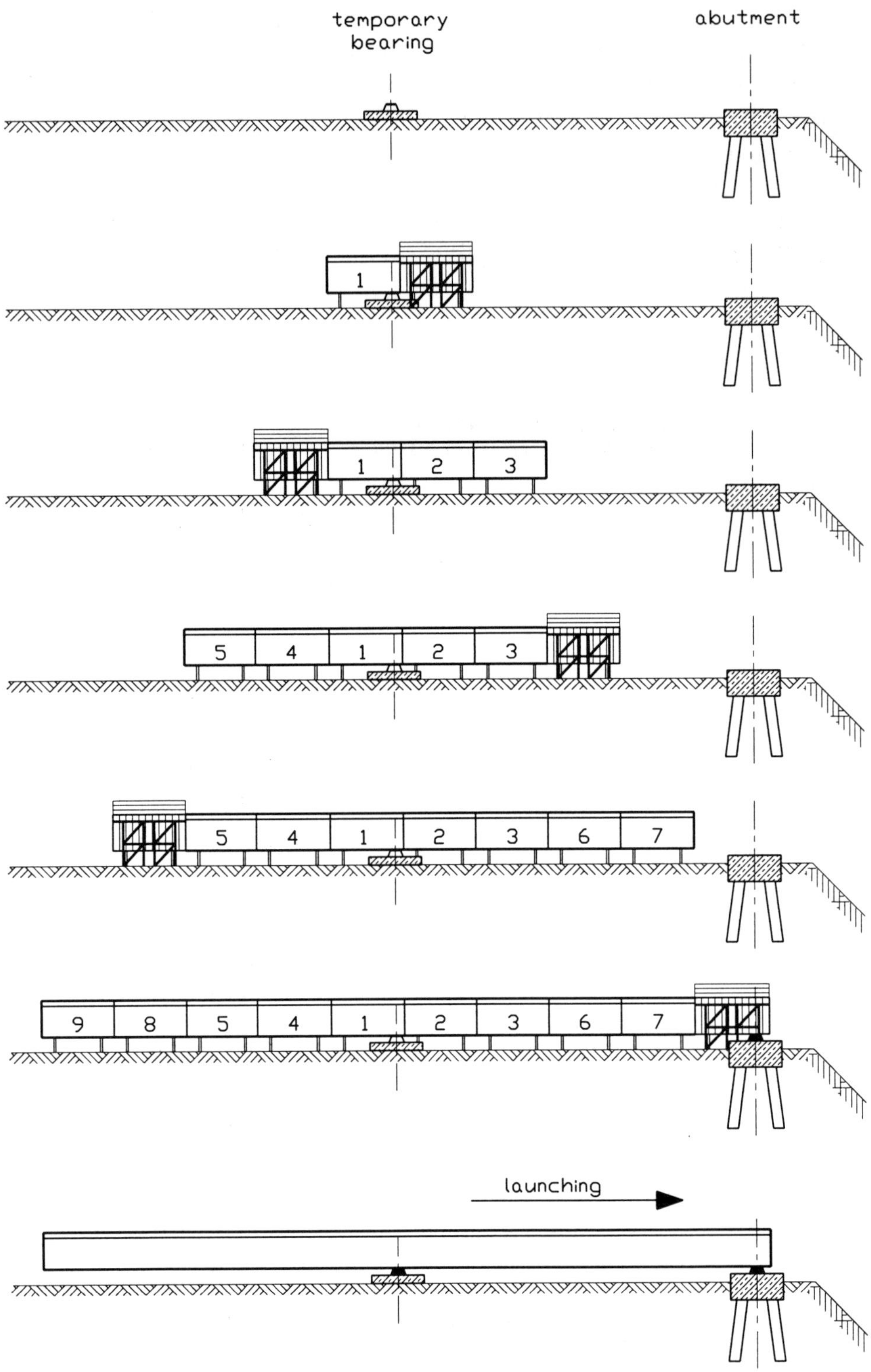

Figure 1.15 - Balanced casting with only one formwork.

Longitudinal prestressing is accomplished by three groups of tendons:

- in each half-deck, conventional cantilever tendons placed in the top slab, descending or not in the webs at their anchorages, and tensioned before launch;
- continuity tendons placed in the bottom slab of the central span and tensioned only after curing of the mid-span segment;
- additional undulated tendons necessary to resist service loads.

Balanced casting technology has been adopted in the construction of many viaducts, and particularly in cable-stayed bridges, due to the minimal influence of temporary construction phases on the deck dimensioning. Notable, among others, are the Milan-Certosa Bridge, built by symmetrical approach, and the Ben-Ahid Bridge[85] (figure 1.16), rotated about the tower with lightweight concrete used for the interior span (*168 m* long) and normal weight concrete for the counterweight back span (*128 m* long).

Figure 1.16 - Ben-Ahin Bridge, built on closed temporary supports, cable-stayed and then rotated (courtesy Greisch[85]).

Side Launch

Construction by side launching involves the construction of the deck on one side of the final position and its transfer by side translation (figure 1.17).

Unlike all the other launch techniques, in side launching the position of the deck supports is immediately the final one, and the static scheme is the same both in launching phase and in service.

The purposes of this technique are generally two: to use the same falsework several times, and to place the construction yard outside the final location of the bridge.

The first case is justified only in parallel decks built on such an expensive falsework (very high above the ground, very long, placed above operating ways or supported on indirect foundations in settling soils) that rebuilding it to face construction in a traditional way is not competitive. The second case is justified when it is necessary to substitute railway and highway bridges still in service, to minimize the interference between the falsework and their regular service.

The disadvantages of this technique are limited to the need to build the falsework, and to the transfer of high loads along sliding rails dimensioned for the same loads as the final bearings, which can be restrictive in case of soils which require indirect foundations.

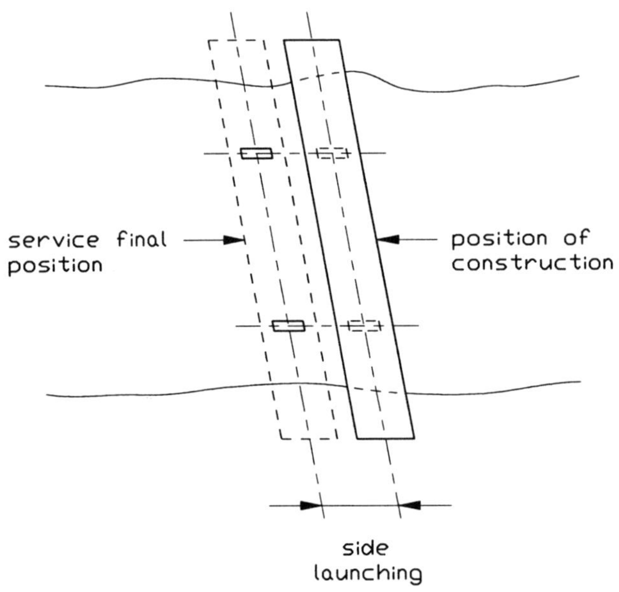

Figure 1.17 - Side launch.

Symbols

The symbols listed below have been used in the text; specific passages have required further symbols, described in their respective paragraphs.

a	distance between web centers on the top slab axis
a_f	base of the fold in a folded plate web
A	area of the cross section
A_a	area of the prestressing anchorage
A_b	area of the bottom slab or bottom plate, area of the braces for the temporary pier
A_j	area of the nose-deck joint cross section
A_p	area of each pillar of the temporary pier
A_t	area of the top slab or top plate
A_w	area of each web
b	distance between web centers on the bottom slab axis
b_b	width of the bottom plate of each girder of the launching nose
b_f	step of the folds in a folded plate web
b_{nf}	width of the neo-flon plate
b_t	width of the top plate of each girder of the launching nose
B	total width of the top slab
B_b	total width of the bottom slab
c	web sloped depth
C_d	inertial term of the BEF equation
C_f	friction coefficient
C_i	I-constant
C_t	torsional stiffness
d	theoretical span of the side cantilever of the cross section
d_b	distance of the bottom slab axis from the centroid
d_t	distance of the top slab axis from the centroid
D_x	longitudinal flexural stiffness of a cross sectional plate
D_z	transverse flexural stiffness of a unit length of web
e	eccentricity
e_a	eccentricity of the additional prestress in the front zone of the deck
e_d	downward eccentricity of the resulting force
e_M	efficiency of material
e_u	upward eccentricity of the resulting force
E	elastic modulus
E_b	elastic modulus of the braces of the temporary pier
E_n	elastic modulus of the launching nose
E_x	axial deformation modulus of a folded steel plate
EI	flexural stiffness of the superstructure

E_nI_n flexural stiffness of the launching nose
f ultimate tensile stress
f' ultimate compressive stress
f_{bc} ultimate biaxial tension of the concrete
f'_{bc} ultimate biaxial compression of the concrete
f_c ultimate axial tension of the concrete
f'_c ultimate axial compression of the concrete
f_y yield stress
F thrust, friction, or prestressing force
F_a additional prestress in the front zone of the deck
F_b prestressing force in the bottom slab
F_h hemisymmetrical force
F_{min} minimum launch prestressing
F_s force in the stays of the cable-stayed front scheme
F_t prestressing force in the top slab
F_{und} undulated prestressing force
G shear modulus
h theoretical depth of the cross section (distance between the center lines of the slabs or the flanges)
h_c average thickness of concrete
h_f depth of the fold in a folded plate web
h_p height of the pier, net height of the web panel
h_t height of the tower in the cable-stayed front scheme
H total depth of the cross section (distance between the edges)
H_c depth of the cross sectional central core
H_m depth of the cross section at mid-span
H_s depth of the cross section at support
I moment of inertia about the centroid of the cross section
I_j moment of inertia of the nose-deck joint cross section
I_n moment of inertia of the launching nose
k BEF modulus of foundation
K generic constant
l length of the span
l_{cr} critical length of the front cantilever
l_n length of the launching nose
l_{nf} length of the neo-flon plate
l_0 effective length
l_s length of the side span of the continuous beam
m inverse $\left(1\ \upsilon\right)$ of Poisson's ratio
M bending moment
M_B^* negative moment in the first support section due to cantilevered masses

M_B^{EOL} negative moment in the first support section at the end of launch
N compression force
N_E Euler's critical load
p uniformly distributed service load
p_h hemisymmetrical component of the unit service load
p_p distributed upward load produced by undulated prestressing
p_{perm} uniformly distributed superimposed dead load
P_a load due to the fall of the formwork hung on the cantilever
P_{c+w} construction loads and relieving wind action
P_{cr} critical load
q uniformly distributed dead load
q_n uniformly distributed dead load of the launching nose
Q_{max} maximum dead load of the cantilever
Q_{min} minimum dead load of the cantilever
r radius of gyration of the cross section
R_h hemisymmetrical component of the support reaction
R_I I-support reaction
R_m average component of the support reaction
$\{S\}_j$ state array in the section j
$\{S_R\}_J$ reduced state array of the support section J
t time
t_b thickness of the bottom slab or of the launching nose bottom plate
t_t thickness of the top slab or of the launching nose top plate
t_w web thickness
T torsional moment, temperature
$[T]_j^k$ transfer matrix between the cross sections j and k
$[T_R]_J^K$ reduced transfer matrix between the support sections J and K
V shear force
W vertical misplacement of the launching saddles
W_h hemisymmetrical component of the support misplacement
W_m average support misplacement
z_l distance of the centroid from the lower edge
z_u distance of the centroid from the upper edge
α angle, coefficient of thermal expansion, dimensionless progression of launch
α_{cr} critical dimensionless progression of launch
β angle
Δ difference between two values
$\Delta\sigma_p(t)$ loss of prestress at time t
ε orthogonal strain
ϕ friction angle

φ	flexional rotation
γ	specific weight
η	vertical deflection
η_d	distortional component of the cross section deformation
η_t	torsional component of the cross section deformation
λ	BEF characteristic length, slenderness ratio
λ_s	BEF characteristic length for the calculation of the distortional edge stresses
λ_w	development ratio of the folded web
ν	Poisson's Ratio
ϑ	torsional rotation
ρ_g	general efficiency of the deck
ρ_s	structural efficiency of the cross section
ρ_t	geometrical efficiency of the prestressing tendon
σ	orthogonal stress
σ_c	allowable compressive stress (positive)
σ_{crL}	local critical orthogonal stress
σ_{oct}	octaedral orthogonal stress
σ_p	tensile stress in the prestressing tendon
σ_{p0}	initial tension in the tendon
σ_t	allowable tensile stress (negative)
σ_u	ultimate compressive stress
τ	tangential stress
τ_{crG}	general critical tangential stress
τ_{crL}	local critical tangential stress
τ_{oct}	octaedral tangential stress
τ_{Tim}	Timoshenko's critical tangential stress
τ_u	ultimate tangential stress
ψ	reduction coefficient of the launching saddle breakaway force

Chapter 2

INTRODUCTION TO INCREMENTAL LAUNCHING

STATIC SCHEME

In incremental launch, the deck is built as a continuous beam. This static scheme can be maintained permanently or modified once the deck has reached its final position.

Since temporary construction stresses require the structural continuity of the deck, the possibilities to improve its static performance in these temporary stages are limited to the stress control in the front spans and to the launching span optimization, which can be made uniform or reduced by inserting temporary piers.

On the contrary, once the deck reaches the final position the static scheme can be modified in many ways:

- in case of very long decks it can be convenient to reduce the distance between expansion joints by dividing the deck into several shorter continuous beams, or even into simply supported spans;
- in case of variable curvature in the alignment and/or profile that makes front launch impossible from only one abutment, the deck can be launched from both the opposite abutments, and the central closure increases the final degree of redundancy;
- the deck can be permanently framed to the launch supports, to obtain indeterminate arches or portal schemes;
- the deck, launched over temporary piers, can be cable-stayed to a tower.

With the progress of material technology and calculating methods, the continuous beam static scheme has gained a huge advantage over the statically determinate schemes, since it reduces the amount of material and guarantees a better control of the structural deformations and of fatigue phenomena. Compared with both the simply supported beam and the Gerber schemes, its better behavior under seismic loads is well known, since it avoids internal hinges that interrupt the beam continuity in the horizontal plane. The elimination of bearings and intermediate joints permits important reductions of maintenance costs as well. Finally, secondary stresses produced by differential settlement of bearings are reduced by creep, and anyway, if excessive, can be quickly reduced by adequate shimming.

Therefore, it is generally advantageous to maintain in the final configuration the continuous beam static scheme utilized during construction of incrementally launched bridges, at least when their length is compatible with the expansion and contraction produced by thermal and time-dependent phenomena.

After launching, the substitution of the launching saddles with the final bearings allows arrangement of either fixed or sliding bearings on each pier, and therefore achievement of two final longitudinal static schemes:

- fixed bearing at one end of the continuous beam and sliding bearings at all the other supports, which permits concentration at one abutment of all horizontal forces, but which produces high average sliding of bearings and requires a large expansion joint at the sliding end;
- fixed bearing at an internal pier and sliding bearings at all the other supports, which requires a short and stiff central pier on which the horizontal actions can be concentrated, but permits reduction in the movement of bearings and expansion joints.

Particular design requirements (very long bridges, high thermal differentials, slender piers) can dictate deck division into several consecutive continuous beams, or even simply supported spans. As these solutions modify the deck behavior under horizontal loads, they are frequent in railway bridges.

In France, for instance, in the 49.16 Lot of "TGV Atlantique," a deck composed of five consecutive three-bay continuous beams, with spans of *29.2*, *30.6* and *29.2 m* consecutively, was launched as a single *445 m*-long continuous beam. After launching, the temporary prestressing of the final joints was released to permit the deck separation into the five final beams, which were moved towards their final position by jacking the internal expansion joints.

The same technique was adopted in the South-African Olifant's River Bridge[127], composed of twenty-three spans *45 m* long for an overall length of *1035 m*. In this case, the final scheme consisted of two eleven-span continuous beams fixed at the respective abutments and a simply supported central span working as an expansion joint. This scheme permits transfer of the braking forces directly to the abutments, and therefore the use of slender piers. The twenty-three spans were built as a single continuous beam. After launching, the deck was permanently fixed to the launching abutment, and once the launch prestressing of the first internal joint was relieved, the free segment of the deck (a twelve-span continuous beam) was launched forward until the central span was placed in its final position. Once the second joint was opened, the operation was repeated on the remaining eleven-span deck segment, fixing it to the second abutment.

The same solution was adopted in France for the highway Oli Bridge[127], composed of fifteen spans *41 m* long placed with a high longitudinal grade (*5.4%*). The presence of piers *60 m* high dictated two continuous beams fixed at their respective abutments, with a central expansion joint. Since the vectorial component of dead load along the launching surface was higher than the total friction, launching was obtained with a braked downward sliding, with a braking force of *6 MN* instead of the *14 MN* thrust required by the upward launch of the *150 MN* deck weight. The same technique was adopted for the opening of the central expansion joint. Once the upper beam was fixed to its abutment, the lower beam was allowed to slide downward by releasing the temporary prestressing tendons of the central joint.

The Sinntal Bridge Schaippach for the high speed line of the German Railways, composed of eight spans *42.5 m* long, one of *51.5 m*, and one of *30.8 m*, was built by launching as a continuous beam a sequence of single spans temporarily coupled to each other by means of joint prestressing, and disconnected after launching to obtain the design scheme of ten simply supported beams. Another ingenious application of the techniques of temporary connection of expansion joints has permitted the execution of the Basrah Bridge[184] in Iraq, in which the central span of a movable bridge (rotating in plan on a central pier placed inside a wide channel) was launched together with the access spans as a single continuous beam.

The launch static scheme of the deck can be permanently modified not only by dividing the continuous beam into several shorter elements, but also by joining two deck segments into a single continuous beam. This is generally the case for decks with a straight-curved or curved-countercurved layout, which can be obtained by launching two deck segments with constant curvatures from the opposite abutments, and then connecting them by means of a central closure joint. In the last phases of mutual approach, the launching nose is assembled in a raised position to slide above the front-end diaphragm of the deck already placed (figure 2.1).

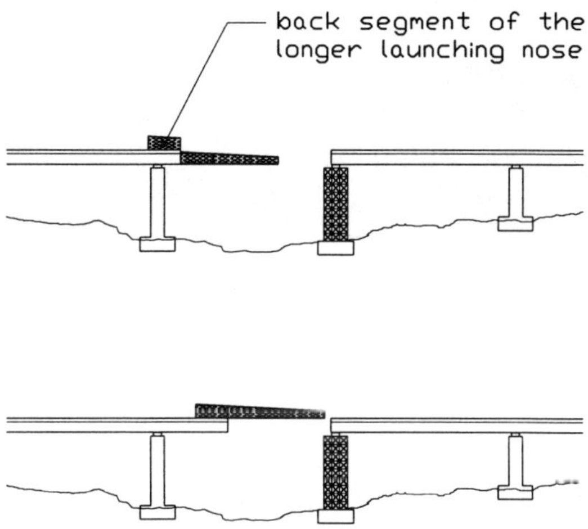

Figure 2.1 - During the final approach of two decks launched from the opposite abutments, the nose is lifted to slide over the deck.

The Schnaittach Bridge in Germany is in construction by launching downhill a *424 m*-long deck segment with a constant radius of curvature of *1000 m* from one abutment, and uphill a *864 m*-long segment composed of a Cornu spiral front segment and a back tangent segment from the second abutment. The spiral segment required the progressive rotation of the formwork during the first launches and the use of transversely movable launching bearings placed on temporary extensions of the final piers (figure 2.2). Upon completion of the launch, the deck segments were connected.

There is also the case, less common, of continuous beams with a central span much longer than the other ones. The deck is launched symmetrically from the two abutments with a mid-span closure, and the temporary support of the two cantilevers

Figure 2.2 - Transversely movable launching bearings in the Schnaittach Bridge (author).

can be obtained by means of temporary piers, or of cable-stayed schemes, as in the Mainflingen Bridge[195].

Some continuous beams launched above arches have represented a brilliant solution to the requirements of short construction time, safety of workers, and cost savings. In this case, the arch must have enough flexural stiffness to resist both the asymmetrical loads and the horizontal forces produced by the deck launch from one abutment toward the other, and can be helped by temporary cable-stays which balance launching loads and brace the higher piers. The Neckarburg Bridge and the Trellins Bridge[139] were built using this technique (figure 2.3). In effect, load asymmetry could be avoided by launching two symmetrical decks at the same time from the opposite abutments toward mid-span, but the presence of symmetrical loads on only some of the piers induces bending stresses in the arch, and the cost of paired launch equipment rarely makes this scheme competitive.

In the Veitshochheim Railway Bridge[110], temporary cable-stays were integrated with counter-weight blocks, and the deck, *1260 m* long and weighing *425 MN*, is one of the highest loads ever launched (figure 2.4).

So far, no rigid frame has been built by incremental launching, although this structural scheme would permit creating arch bridges spans with lower costs. On the

Figure 2.3 - In the Neckarburg Bridge, temporary cable-stays balanced both the horizontal forces transferred by the deck and the load asymmetry (courtesy Leonhardt[188]).

Figure 2.4 - Temporary cable-stays integrated with counter-weight blocks in the Veitschochheim Bridge (courtesy Leonhardt[110]).

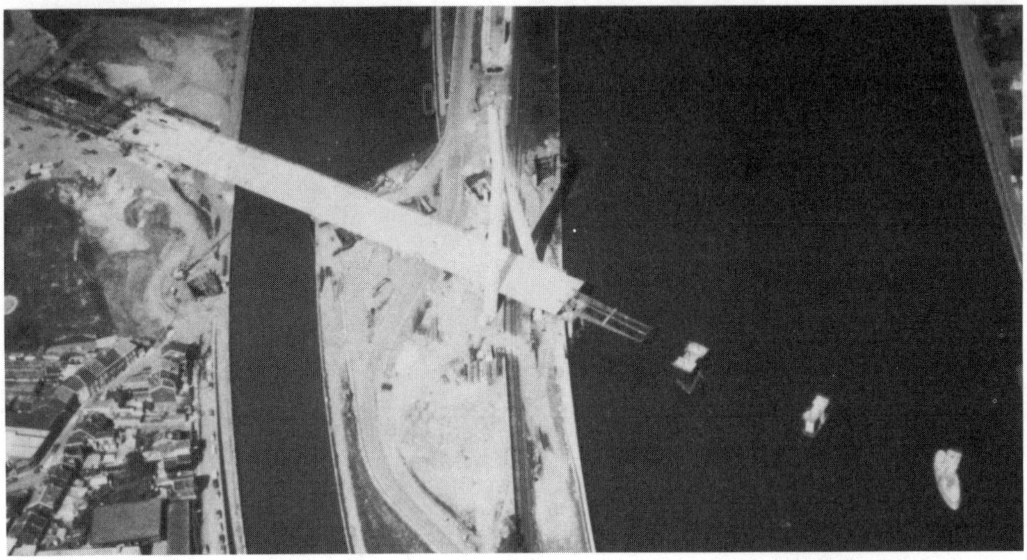

Figure 2.5 - Wandre Bridge, launched over temporary piers and then stayed (reproduced with permission, Greisch).

Figure 2.6 - Palizzi Bridge, launched over temporary piers and then stayed (author).

contrary, many portal steel bridges have been built, and since incremental launch originates from steel construction it is likely that, in the future, launching of concrete rigid frames will be attempted.

Finally, a great elegance characterizes the bridges launched on temporary piers and then permanently cable-stayed to a tower, such as the Wandre Bridge[86] (figure 2.5), and the smaller Palizzi Bridge [174, 176] (figure 2.6).

DECK LAYOUT

During launching, the deck is supported on launching bearings and is guided by anti-drift devices.

Being a continuous beam, each constraint eccentricity (vertical misplacement of launching bearings and horizontal misplacement of guiding devices) with respect to their theoretical position induces secondary stresses in the deck, which can be tolerated only if small and by increasing the reinforcing and/or prestressing steel – that is, with higher costs. Therefore, it is necessary that the deck edges in contact with the launching rails be perfectly aligned with them. This can be guaranteed in each launch position only in case of a common geometry.

For a solid to slide rigidly inside another one, it is necessary that it can be superimposed on itself by translation – that is, its layout must be:

- tangent in plan and tangent or circular in profile;
- circular in plan and tangent in profile;
- curvilinear both in plan and in profile (figure 2.7). In this case, the deck is an arc of circle placed in an inclined plane and its projections on the horizontal plane and the vertical surface which contains its axis are not arcs of a circle but arcs of an ellipse. This should be considered when defining the plan layout of the piers and when placing above them the launching bearings, especially in the case of short radii of curvature;
- a circular helix, very rare and substantially comparable with the above, even in case of the smallest radii of curvature.

Although some types of launching bearings can be adjusted in the transverse direction, and then adapted to varying width cross sections or varying plan curvatures, when designing a launched bridge it is always advantageous to place the varying curvature zones outside the bridge itself. If a transition curve affects the end section of the deck, the lower edge of its cross section, which is the only one subject to the launching constraints, can be designed with a constant curvature and the final superstructure layout can be obtained by locally adapting cantilevers. In fact, transition curves begin with radius variations so small that these corrections are generally acceptable.

Finally, major differences in curvature can be solved by symmetrical launching from opposite abutments, or by placing the formwork in a central position and launching one deck segment on one side and then a second segment on the opposite side, with its different curvature.

As a matter of fact, incremental launching involves some constraints that the designer should take into account from the early stages of design. Technological

Figure 2.7 - Val Restel Bridge, launched with a radius of curvature of only 150 m. Due to the curved profile, the plan layout of the bridge is an arc of ellipse instead of an arc of circle (courtesy Leonhardt[188]).

progress has permitted realizations which were unimaginable just a few years ago, by making the most of equipment and materials. Therefore the technological implications of the different construction methods will continue to influence more and more the design aspects.

DECK CROSS SECTION

Since the debut of this technique, solid or lightened slabs, ribbed slabs (T beams), and box girders have been built by incremental launching.

The present trend is against the use of lightened slabs (despite the economy deriving from their simplicity of execution) and in favor of box girders. These better tolerate launching because of the higher torsional stiffness and higher structural efficiency, rarely present problems of instability or of excessive deformation, permit external prestressing, and support wide cantilevers which lighten the architectural aspect of the deck with pleasantly shaded areas.

Even compared with the ribbed slab, the box girder presents many advantages:

- a higher structural efficiency (5.4), *0.55* to *0.65* compared with an average of *0.40* to *0.45* for ribbed slabs;
- a larger central core, which reduces both launch prestressing and the stress fluctuations generated by live loads;
- a lower position of the centroid, which improves the balance of the section moduli, makes the absorption of negative moments easier (which is determinant in this construction technique) and reduces the irreversible rotations due to time-dependent phenomena;
- a higher torsional stiffness, which permits reduction in the distance between bearings and therefore the dimensions of pier caps, with significant aesthetic improvement, especially in curved bridges;
- the possibility of arranging launch prestressing tendons inside the slabs and keeping the webs available for parabolic tendons;
- the availability of the box cell for external prestressing and for ducts;
- easier inspection;
- comparable consumption of materials.

The use of ribbed slabs is generally limited to small bridges built by symmetrical launching and mid-span closure with main spans of about *25* to *30 m*, and particularly above highways or railways in service. In this case, the slenderness of the deck is about $H/l = 1/17$[28]. A ribbed slab launched over an arch is described in[139]. For preliminary dimensioning the rib thickness can be obtained from the graph of figure 2.8, and the slab thickness is about $B'/18$ with a minimum of *0.20 m*[28].

For spans longer than *30 m*, therefore, box girders are practically always used.

In decks wider than *14 m*, twin box girders (Green Street Bridge) and multi-cell box girders (Wabash River Bridge, *14.1 m* with three webs; Swinburne Bridge, *20.2 m* with four webs) have been used, even though the actual shear force distribution between the webs can deviate from theory, especially in the case of vertical misplacement of launching bearings. A solution could consist in reducing the cross section redundancy during launch by transferring loads between the webs, but this may be, in practice, expensive, complicated, and not very reliable. Therefore, the present trend in incremental launching is to use single-cell box girders even for wider cross sections (Ravensboch Valley Bridge, *17.3 m*; Adour Bridge, *17.8 m*; Drac Bridge at Grenoble and Schnaittach Valley Bridge, *18.4 m*), as the shear force cannot but distribute equally between the webs. The top slab transverse bending can be controlled by transverse prestressing or by stiffeners, and distortion can be analyzed with reliable theories.

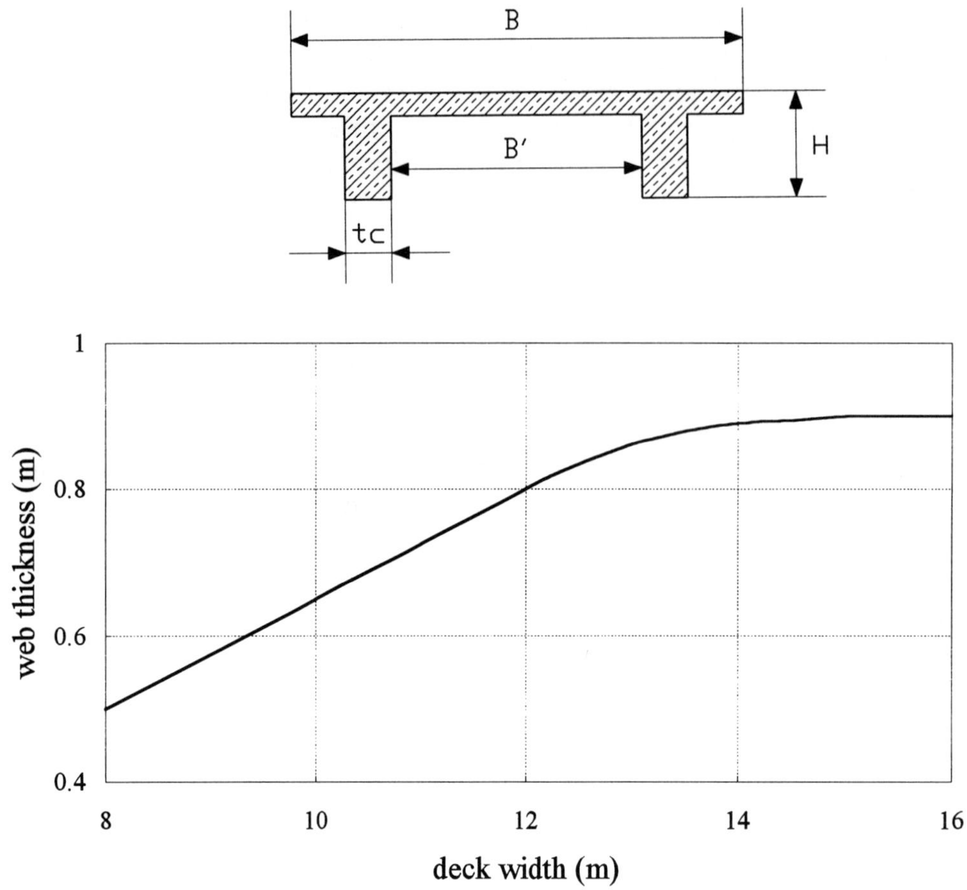

Figure 2.8 - Pre-sizing graph for ribbed slabs.

Higher structural efficiency does not result in higher investments. The cost of the falsework, supported on the ground, is not greatly affected by the weight of the deck (unlike what occurs both with the fixed or self-launching centering and with the free cantilevering method) and even construction is generally simpler, resulting in lower labor costs.

The single-cell cross section permits adaptation without excessive costs of structural dimensions to the local stresses, and, specifically, stiffening of the webs and the bottom slab in proximity of the final bearing sections. The real convenience of these interventions should be evaluated case by case, especially for the webs, as the lower labor costs permitted by a higher structural simplification balance the higher consumption of materials, making stiffening not very competitive. This trend is supported by static considerations too. The web stiffening can be useless (the ultimate shear strength depends mainly on the axial compression, since tension can be fully absorbed by reinforcement) and sometimes even detrimental, since it could produce important internal stresses due to thermal gradients or differential shrinkage.

From the geometrical point of view, incremental launching does not require a cross section with constant external dimensions, but only one that presents two longitudinal directrixes (subject to the already examined plan and profile constraints) along which all the cross sections of the deck can be supported during launching. The transfer of large support reactions imposes that these directrixes be placed in proximity of cross sectional structural nodes which can adequately distribute stresses toward the webs. Their most natural position is under the webs, and this requires the use of constant-depth cross sections.

Moreover, a marked uniformity of the moment of inertia and of the web thickness along the deck is necessary from the static point of view, since during launching all the cross sections are progressively subject to the same stresses. Variable-depth decks would be over-stressed in those sections for which the launch stresses are quite different from the final ones – that is, in the thinner mid-span sections.

Although there are exceptions, incremental launching is then favored in decks with substantially constant cross sections, thus the frequent emphatic reference to extrusion processes.

For longer spans, these constraints can be restrictive and can justify ingenious solutions. In the Danube River Bridge in Ingolstadt (five spans, *54.4*, *60.0*, *75.0*, *115.0* and *74.4 m* long respectively, the last three with varying depth cross section), the deck was launched symmetrically from the opposite abutments with closure in the main span. The varying depth zones were temporarily shimmed with varying thickness precast concrete slabs clamped under the webs to compensate for their haunches and to obtain the two rectilinear launching directrixes.

Variation of the cross-sectional width produces only minor problems. In the B 224 Highway Bridge in Essen, wide launching bearings were used in order to accommodate the neo-flon plates placed under the webs. In the Danube Bridge of Fischerdorf, launched on temporary piers, both launching bearings and friction launchers were designed to allow for their transverse displacement during launch.

Similar devices are shown later, in figures 2.2 and 4.27.

ALTERNATIVE CONSTRUCTION TECHNIQUES

Each construction technique has its advantages and disadvantages, and in the absence of particular constraints which make one solution immediately preferable to the others, the comparative evaluation of several alternatives is always a difficult task.

Comparison in terms of cost of materials, frequent in the bibliography since it can be estimated easily, is based on only one component of the total deck cost, which is more and more influenced by costs involved in the processing of materials (labor, investments, possibility for reuse of equipment in the future, etc.) and by the cost of the accessories of the bridge. Therefore, reasoning only in terms of the cost of materials is misleading: although structural optimization can be reached using a general falsework supported on a rigid foundation, the lower cost of materials permitted by this optimization will normally not balance the huge construction costs.

Although most of these data are part of the cultural luggage of contractors and are not in the public domain, a realistic analysis of the economic implications will almost always permit estimation with sufficient accuracy of the construction cost of a bridge, thus allowing optimization of its design.

Deck construction cost C_c (leaving maintenance costs, life expectancy, and aesthetics out of consideration) can be divided into structural costs C_S and operational costs C_o, which include the costs of bearings, expansion joints, sealing, paving, and railings.

$$C_c = C_s + C_o \qquad (2.1)$$

Structural costs consist of the cost of materials, C_m (concrete, reinforcement, prestressing etc.), and technological costs, C_t, involved in processing of materials (labor, assembly and amortization of equipment, energy and minor costs).

$$C_s = C_m + C_t \qquad (2.2)$$

Once the continuous beam static scheme is chosen, the operational costs, C_o, are practically constant and the different solutions can be compared in terms of structural costs. Reduced material costs will lower overall costs only if they are not obtained through higher technological costs. At the same time, higher material costs will not reduce the competitivity of the solution, if they allow low labor costs and low amortization[165].

Compared to the fixed falseworks, incremental launching reduces labor costs with almost the same investments. Compared to the self-launching centering, it reduces investments with almost the same labor costs. In both cases, therefore, it leads to a reduction of C_t.

But compared with bridges built directly in their final position, for which the static schemes during construction are very similar to the final ones, launched bridges are subject to temporary stresses which increase material consumption, and with it, C_m and the dead load of the deck. For the same service loads, this tends to reduce the efficiency of the cross section and the whole structure: consequently, the cost effectiveness of incremental launching in terms of C_s will be obtained, or improved, by optimizing the cross section properties.

From an organizational point of view, the construction method which comes closest to incremental launching is segmental precasting. Many bridges have been built by joining together precast segments in free or suspended cantilevering (using launching trusses, temporary cable-stayed schemes, etc.) through construction static schemes substantially similar to the service ones. In spite of this, the need for avoiding joint decompression with adequate safety factors increases the cost of materials (especially of prestressing steel), and investments in equipment and labor costs are high as well. Consequently, this construction method applies mainly to longer spans than the ones usually built by full span incremental launching, and therefore to variable depth sections in which construction stresses are absorbed more easily by virtue of the higher depth of the support sections.

Equipment amortization requires decks of a certain length, nor can one forget the problems generated by joints over time, problems that in many administrations have induced such a diffidence toward segmental precasting as to cause underestimation of the advantages involved in the strict qualitative and dimensional control of production, and in the long concrete curing at the time of the segment installation.

As an alternative to prefabrication, spans suited to incremental launching can be built by casting in-situ.

The construction in-situ of a continuous box girder is a rather complex operation, since for bridges with more than two or three spans it is convenient to advance in line by using the same formwork several times. Therefore the deck is built through a sequence of different static schemes which are almost always characterized by complete prestressing and overloads.

Casting can occur either in fixed or in movable centering. Once the length and the number of spans are defined, the factors in the choice between these two types of equipment are essentially economic, since both the labor associated with fixed falseworks and the investments requested by self-launching centering are very considerable.

Figure 2.9 - Lecco Bridge, cast-in-situ on modular trusses (author).

These different construction methods are then applicable to different bridge dimensions, and constitute an alternative respectively to the monolithic launching by front approach or rotation (fixed falseworks) and to incremental launching (self-launching centering).

Starting in the forties, the original wooden falseworks were replaced by modular tubular scaffolding with mechanical joints or by modular braced towers, supporting the formwork directly. In turn, in long spans, modular trusses supported by modular piers (figure 2.9) have replaced these schemes.

Despite the simplification of the support structures and good adaptability, this equipment still involves long assembly times and high labor costs (although less than with ordinary scaffolding) even for decks not very high above the ground. The longest steel girders can reach *30 m*, but with such a low carrying capacity and high deformations that from *20 m* onwards it is preferable to adopt intermediate supports, with additional costs. Assembly duration and number of skilled workers increase noticeably with the height of the deck above the ground (figure 2.10), and the cost for the modular piers increases exponentially, and for the girders linearly.

Thus, the overall cost increases, especially when the girder assembly cannot be amortized on a large number of spans without further changes, as each span variation imposes the reassembly of the girders and their bracing. And as regards the bearing towers, each assembly has its own story, most delicate because of the safety and static implications. They have to limit settlement when loaded and act as a rigidly braced tower, their height must be equal to the pier height, and their plan lay-

Figure 2.10 - Neaples Bridge, cast-in-situ on modular trusses (author).

out must be compatible with the pier footing.

The cost distribution of the superstructure of figure 2.10 (figure 2.11, Italian costs ... notice that it is $C_l = 71\%$ and $C_m = 29\%$) demonstrates that with labor costs higher than *60%* of the overall construction cost this type of equipment cannot stand the progressive increase of labor cost. In industrialized countries the use of fixed falseworks is therefore limited to structures not very high above the ground, with a reduced repetition, and which permit the total obstruction of the lower area.

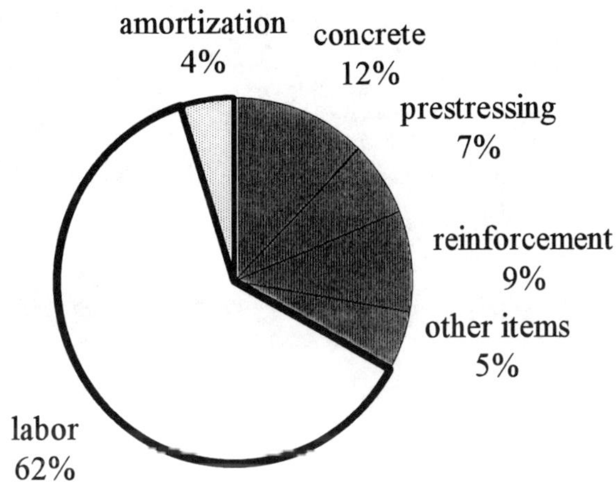

Figure 2.11 - Prime cost distribution for the superstructure of figure 2.10.

A movable falsework is composed of a formwork suspended from (figure 2.12) or supported by a couple of longitudinal beams which allow it to move to the subsequent casting positions. The cost of the equipment and its assembly, its difficult reuse, and the start-up times make the adoption of launching falseworks justified only in the case of many tens of identical spans – that is, in the case of kilometers of superstructure. And even in this case, the influence of their amortization on the prime cost of the superstructure is noticeable (see figure 2.13, with $C_l = 55\%$ and $C_m = 45\%$, showing the construction cost distribution of the deck of figure 1.28, *2300 m* long, where equipment was fully amortized).

With this type of equipment, production is a compromise between casting in-situ and prefabrication. Generally the steel cages are prefabricated and transferred into the formwork by means of gantries which run above the superstructure from a preassembly-yard placed behind the abutment, while formwork and pouring activities are more conventional.

Like fixed formwork, the launching one deflects under the fresh concrete weight and requires a camber in order to obtain a correct final geometry of the deck. Cambers cannot prevent settlements between adjacent segments at the construction

Figure 2.12 - Neaples Bridge movable falsework (author).

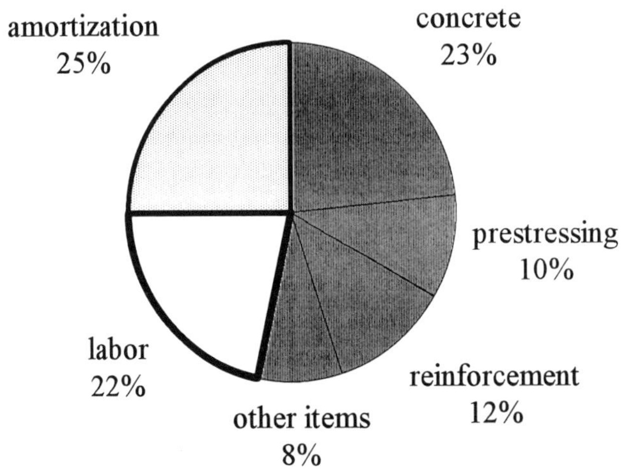

Figure 2.13 - Prime cost distribution of the superstructure of figure 2.12.

joints. However, the consequent secondary stresses are reduced by creep, which is very high in these first curing phases. Moreover, in launching centering, it is necessary to introduce the prestressing as early as possible, as the concrete curing immobilizes equipment and labor, and the joint cross section between two segments (on which the tendon anchorages are almost always arranged) is covered by the casting of the adjacent span. In turn, the introduction of prestressing of the whole after short curing affects the losses of prestress, and the creep rotations due to a prestressing force sized for dead and live loads, and therefore excessive for the dead load only, produce irreversible cambers which are aesthetically harmful and disturb drivers.

Compared with the different alternatives, incremental launching presents most of the advantages of both prefabrication (protected yard, precision, quality control) and casting in-situ (reinforcement passing through match-cast construction joints, short transports), and at the same time it avoids most of their weak points.

LENGTH OF THE SPANS

The first applications of incremental launching trace back to the sixties, when spans between *50 m* and *60 m* represented the upper limit of the constant-depth deck. Above these spans there was no alternative to free cantilevering, in which the higher efficiency of the variable-depth sections was preferable. Constant-depth box girders were mainly used for spans shorter than *40* to *50 m* (although occasionally lengths exceeded *100 m*) since for longer spans both fixed and movable falseworks resulted in very high costs.

Incremental launching has modified this break-even point: statistical studies have shown average spans of about *50 m*, spans of *60* to *65 m* have been reached without particular problems, the adoption of temporary piers has permitted length exceeding *100 m*, and spans of *140 m* may still be competitive.

Thus, full-span incremental launching falls in the transition zone (spans of *40* to *70 m*) between constant- and variable-depth decks, and characterizes the upper limit of the constant-depth structural type. At the same time, for shorter spans, it is a competitive alternative to conventional construction techniques.

In other terms, for longer spans launched bridges can compete with precast or cast-in-situ free cantilevering (e.g., the preliminary study of the Wabash River Bridge[127] has credited a *11%* global economy to incremental launching) while for shorter spans they can be *5* to *10%* less expensive than decks built in fixed or movable falsework.

Since the cost effectiveness of incremental launching is due to the marked repetition of the operations, it requires the highest standardization, and this necessity should be taken into account in the bridge design. The deck segments will be repetitive in dimensions, type of cross section, tendon layout, positioning and sizes of their anchor blisters, and so on, and this suggests that the spans of the continuous beam be equal, or at least arranged on a modular base.

On the contrary, side spans are almost always shorter than the interior ones. From a theoretical point of view, in a semi-infinite continuous beam subject to a constant unit-load, the bending moments at the supports are equal when the side span is[28]:

$$l_s = \sqrt{\frac{2}{3}}\, l \qquad (2.3)$$

where l is the constant interior span. Obviously this does not mean that every continuous beam bridge has to be designed in this way. In the case of incremental launching, however, construction stresses are very high in the deck front zone, which has to overcome the full internal spans. This often requires the adoption of stiffened cross sections and integrative prestressing.

Chapter 3

CONSTRUCTION BY CONTINUOUS CASTING

DECK SEGMENTATION

The organization of production (planning of the yard and the activities, equipment, optimization of labor rotation) depends on the deck length, on the time available for its construction, and on the length of the deck segments compared with the typical span.

The deck length influences the level of industrialization of the yard. Long bridges are composed of a large number of segments and permit amortization of the investments necessary to industrialize the production cycle in order to reduce labor cost. Shorter bridges require a less expensive organization, since it is impossible to reach an adequate operational repetition cycle.

Deck segmentation depends on its length and on the time available for its construction, taking into account the time required for yard assembly, start up, and final disassembly. Clearly this is true only when, as in most cases, the deck segments are match-cast against each other in-situ. In case of prefabrication, times are less restraining, the segment assembly being much faster.

Once defined, the optimum average length of the segment (defined as the deck element built between two subsequent launching stages) should be adapted to the typical span, it being necessary that segments correspond to a span or constitute a whole fraction of it. This adjustment is developed according to well-established criteria:

- in bridges composed of a large number of short, equal spans, segments are generally as long as the whole span in order to reduce the number of construction joints and, with them, the cost of prestressing coupling, and in order to use repeatedly the same internal formwork;
- as span increases, the segments become shorter than the typical span in order to limit the cost of the installation, since the cost of most of the equipment depends linearly on the segment length. In this case, the segments can be one half or one third of the span;
- in short bridges or with external prestressing, for any span length, it can be convenient to use shorter segments, building more segments on a continuous support and launching them as a whole.

Once the number of segments composing the span and their typical length are defined, the position of the construction joints should be verified from the static point of view, in relation to the null points of the service bending moment diagram, and from the construction point of view, as dismantling of the internal formwork always occurs by extraction through the joint section. Therefore, if the diaphragms are cast together with the rest of the section (figure 3.1), the joint can be placed immediately before the diaphragm so that the recovery of most of the formwork can proceed without obstacles. As an alternative, diaphragms can be cast in a second stage, and even after launching is completed, by means of casting pipes crossing the top slab. On the contrary, if the segment is built in two casting phases, the recovery by extraction affects only the top slab internal formwork and the joint can be placed in any position, with the condition that diaphragms must be tapered on their upper part in order to allow the passage of the formwork (figure 3.2).

Finally, if the continuity of launching prestressing is obtained by means of couplers instead of by overlapping, the couplers will reduce considerably the moment of inertia of the joint section, which then shall be placed at about the quarter of the span, in the zone of reduced service moment.

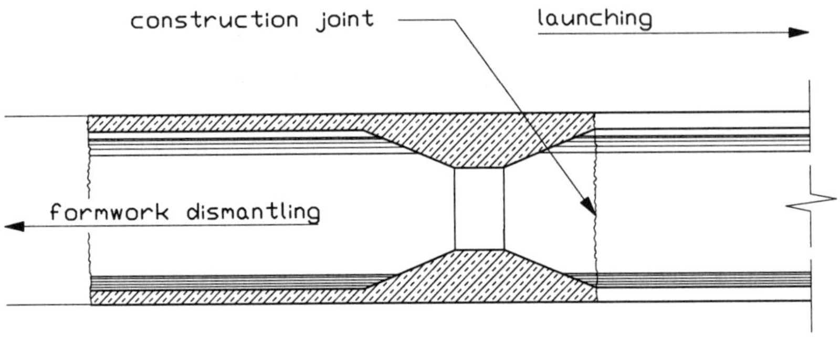

Figure 3.1 - The construction joint between two monolithic segments can be placed immediately before the diaphragm to facilitate the extraction of the internal formwork.

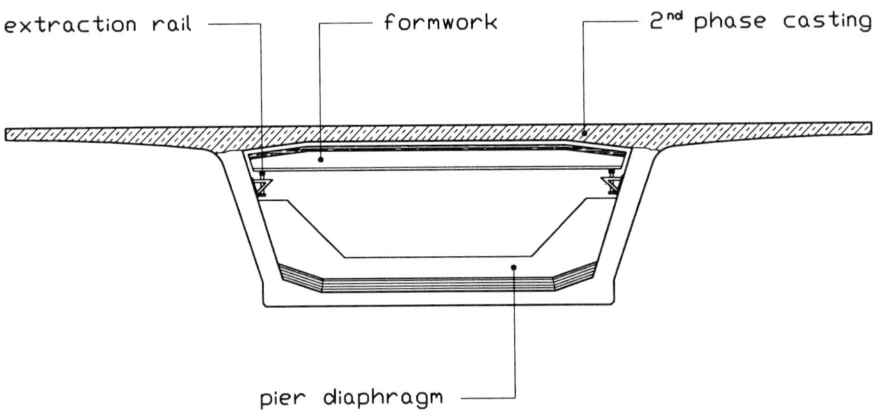

Figure 3.2 - When the segment is built in two casting phases, the support diaphragms are tapered to allow the extraction of the top slab internal formwork.

YARD

In most launched bridges each segment is match-cast in-situ against the previous one, and therefore the whole yard organization is hinged on the formwork. Whatever the subdivision in segments and casting phases may be, in order to build a deck segment in the shortest time it is necessary to avoid concentrating in the formwork activities which could take place elsewhere. Therefore, production is generally organized into parallel rather than serial processes: integral or partial prefabrication of steel cages instead of their assembly in the formwork, continuous instead of single inserts, precast anchor blocks, and so on.

A correct handling scheme is essential, and being a "radial" type organization in a narrow area, the utilization of a tower crane which serves all the activities is decidedly advantageous.

If the number of segments is high and enough space is available, it is convenient to create a preassembly area for most of the loose material (bent steel bars, sheaths, anchorages, inserts) in order to transfer them into the formwork as a whole. Yard industrialization permits reduction in the duration of each activity compared with the conventional execution. Moreover, operations are simplified, the number of workers in the narrow formwork area is reduced to a minimum, their full occupation is permitted by parallel activities performed during concrete curing, and formwork is cleaner and more durable. In case of bad weather, movable coverings (figure 1.10) permit continued production, as the only outdoor operations are associated with launching and occur under the deck itself.

A high level of industrialization does not imply prohibitive investments, as most of the equipment is standard (tower crane, gantry crane, movable coverings, general layout) and only the launchers, the launching nose and part of the formwork are specific for the construction, though often reusable.

Finally, incremental launching requires the presence of temporary supports between the formwork and the abutment. These divide the length of this temporary bay into shorter spans compatible with the curing and prestressing degree reached by the deck segments and reduce the flexural rotation of the back-end section of the deck against which the next segment is match-cast. In addition, this deck length permits cantilevering the launching nose on the first span without temporary supports, since, in this phase, the deck acts as a counterweight.

STEEL CAGE PREASSEMBLY

Steel cage preassembly outside the formwork presents many advantages:

- cage assembly is a slow and delicate operation due to the presence of sheaths, anchorages, and minor formworks. Preassembly permits extracting these activities from the critical path and confining them to a specific area and organization;
- by separating the casting yard and a preassembly area, it is possible to reach a better rotation of labor, especially during concrete curing stages;

- after segment extraction, formwork preparation (cleaning, maintenance, oiling, positioning of sliding devices) can be carried out without obstacles, and the steel cage transfer does not compromise results;
- usually, the preassembled cage includes the back-joint formwork, complicated by the presence of tendon anchorages, and at the end of transfer the segment is ready to be cast.

The preassembly yard consists of a scaffolding template which reproduces the external shape of the deck. Its length is generally equal to the longest segment (figure 3.3), but it can reach two segments if prefabricated prestressing tendons are longer than the segment. It can be contiguous to the formwork or separated from it (figure 3.4).

Figure 3.3 - The preassembled cage of the first casting phase is hooked to the transfer lattice. In the background are the internal formworks, ready to be transferred into the formwork by the same portal crane (author).

The contiguous yard does not require lifting equipment and makes possible trailing the cage into the formwork during launching by means of rollers or gantries. A separated yard requires the use of a specific transfer portal crane, which even can be used as a roof and makes the two activities more independent. Cage assembly can occur by transferring preassembled webs and slab panels into the formwork. In this

Figure 3.4 - Cage preassembly of the whole deck segment. Close up of the precasting area of the anchor blocks; in the background is the main formwork (author).

Figure 3.5 - Preassembled cages of the webs (author).

case (figure 3.5), the template is very simple and handling can be assisted by the tower crane only.

Assembly and fastening must be carried out with high precision to avoid excessive deformations during transfer, which could compromise the correct insertion of the cage into the formwork. Assembly times are relatively low, and a tower crane with a tidy stocking yard permits labor costs lower than twenty to thirty hours per ton of reinforcing steel.

CASTING PHASES

The "construction cycle" of a segment is the whole sequence of operations between the launching of two subsequent segments. Its duration varies generally from seven to ten calendar days, though it is generally convenient to reach this lower limit so that the concrete can cure during the weekend.

The sequence of operations depends on the casting procedure, on the eventual cage preassembly, and on the time required for concrete to reach the strength necessary to stress tendons and absorb launching bending moments. With a *B35* concrete, natural hardening is rarely compatible with a weekend duration, and appropriate devices should be used, such as precast anchor blocks (figure 3.6), local concrete thermal cycles with electric wires embedded around anchorages, or steam curing of the whole segment.

Figure 3.6 - Precast anchor blocks permit anticipation of tensioning by distributing anchorage stresses. Notice the interior transfer truss, necessary to maintain a correct geometry of the steel cage before the internal formwork insertion (author).

Therefore, it is often convenient to increase the concrete design strength, achieving the necessary mechanical strength in the scheduled time, and at the same time lightening the deck. In turn, that reduces the influence of launching stresses, unexpected events, and time-dependent phenomena on the structural dimensioning of the deck.

Whatever the chosen solution may be, the deck subdivision into segments, sub-segments, and casting phases always involves the presence of construction joints between the elements. If the adoption of vertical joints is unavoidable, horizontal joints are very frequent as well, as they permit restricting to reasonable limits the quantity of concrete processed daily, carrying out subsequent operations without excessive interference, dismantling the web and diaphragm internal formwork easily, and removal of rubble. Moreover, the presence of a longitudinal joint in the cross section upper node permits support of the top slab internal formwork from the webs (figure 3.2), and avoids the horizontal cracks which often characterize monolithic casting due to the settlement of the fresh concrete of the webs.

The two types of joint have different characteristics, and should be treated separately:

- in vertical joints, the back end of the segment is cast against the formwork and its surface is smooth and weakened by oil. In the subsequent segment, shrinkage and imperfect adhesion tend to produce horizontal cracks orthogonal to the joint surface, which can be avoided by bushhammering or chipping the contact surface, so that roughness acts as distributed shear keys, and by inserting a vertical wire mesh near the joint;
- in horizontal joints, the surface is raked in order to eliminate upper concrete layer weakness, due to a higher W/C ratio induced by bleeding. In the first curing of the second-phase concrete, shrinkage cannot produce vertical cracks, and a careful joint cleaning is sufficient to obtain a good continuity.

In both types of joint, the use of specific shear keys is then unnecessary, since with superficial irregularities of about $10\ mm$, the relative movements necessary to mobilize distributed shear keys are lower than the ones prevented by concentrated keys, and the steel bars crossing the joint increase considerably the shear strength of the interface.

The difference in temperature between the cured segment and the one in the setting phase should be reduced until the concrete has reached the strength necessary to activate the reinforcement and the distributed shear keys, since in these first phases the insufficient adherence does not limit crack width. Therefore, in particularly severe climates it can be necessary to maintain the temperature in the cured segment or to heat it to 15 to $25°C$ as far as 1 to $2\ m$ from the joint, and at the same time to cool the setting concrete of the match-cast new segment or to adopt cements with low heat of hydration. Clearly, the need for these measures depends mainly on the heat of hydration of cement, and will be the highest in the case of particular materials such as High Performance Concrete (HPC hereinafter).

Monolithic Casting

In principle, monolithic casting is adopted for ribbed slabs only (figure 3.7).

In fact, the operations involved in monolithic casting of box girders are rather complex, having to satisfy requirements of cage assembly, concrete pouring, vibration, and internal formwork dismantling, which are often in conflict with each other. Therefore, for long segments this scheme is so rigid as to limit its adoption almost only to long span railway bridges (in figure 3.4, a half-span segment *24 m* long for the French high speed railway "TGV Atlantique") since the higher depth of their cross sections makes the access to the box cell easy.

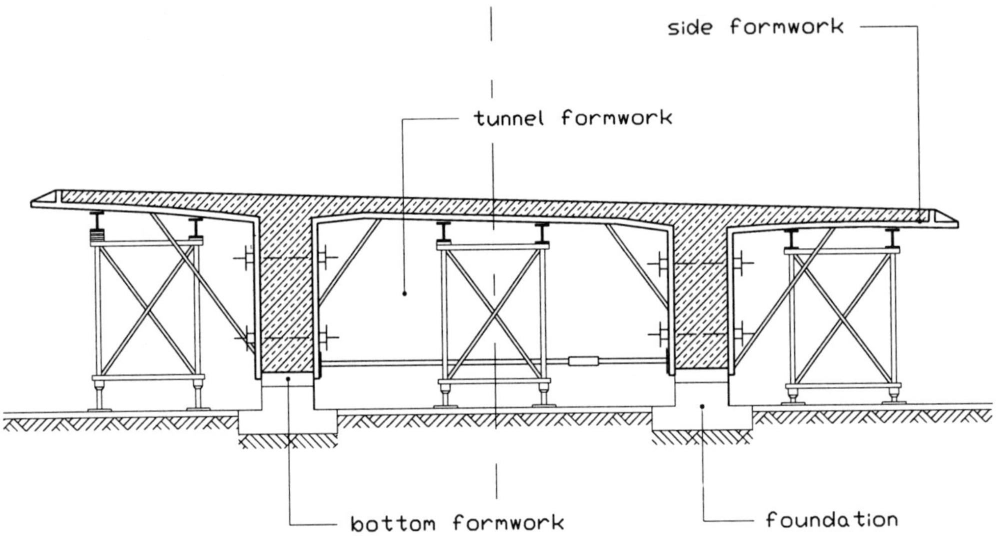

Figure 3.7 - Ribbed slabs are easily built by monolithic casting.

Even the construction schedule is rather rigid (figure 3.8), because of the interference that occurs at the only working location and the dead times involved in concrete curing. Therefore, the steel cage is almost always preassembled, which increases the length of the yard and requires expensive handling equipment.

Since the use of internal left-in-place formworks is economically acceptable only in short decks, the internal formwork is almost always self-extracting and adjustable by means of hydraulic pistons (figure 3.9). It advances with the deck during launching, and is extracted backward only once the preassembled steel cage of the next segment has been transferred into the formwork, to place it in the new casting position. To avoid geometric imprecision which could jeopardize thc rapidity of the operation, the steel cage is assembled on an internal template which stiffens it during transfer (figure 3.6).

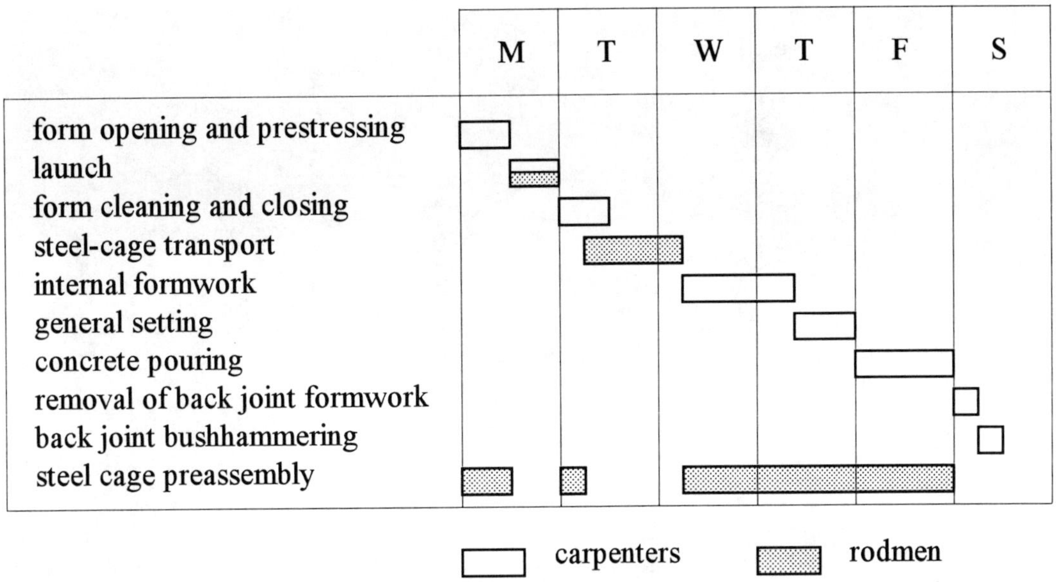

Figure 3.8 - Construction schedule of the French SNCF-TGV Aronde Bridge (figure 3.4), built by monolithic casting in a single formwork. Cage preassembly is necessary to keep labor busy.

The presence of wide formworks and large handling equipment involves high investments: these being amortizable only in long decks, and then only in particular cases; a competitive alternative can be found by subdividing the deck into relatively shorter segments to be built by monolithic casting.

Independently from the span, and sometimes even from the deck length, a more marked deck segmentation than the conventional one (half of the typical span) can be convenient in order to reduce both the cost of the yard and the total investment. The segment length becomes *3* to *8 m*, with peaks over *10 m* only in a case of long decks permitting an adequate amortization of the investment involved in a large hydraulic formwork.

Even with the same general principles, it is possible to single out three different construction schemes which adopt short monolithic segments: the construction in a fixed formwork with the general launch at the dismantling of each segment, the construction in a mobile formwork with launch only after completion of a certain number of segments, and the construction in a fixed formwork with segmental launch independent from the general launch of the deck.

In the first scheme, which has been used in the construction of the Roquebillère Viaduct[186], the formwork is a casting cell similar to the one usually adopted in match-cast segmental prefabrication, and is placed along the deck layout at a certain distance from the abutment. The construction cycle provides for the addition of a new segment match-cast against the back end of the deck in perfect analogy with the case of longer segments, and a general launch.

Figure 3.9 - Back end of the self-extracting internal formwork. The deck diaphragms make this type of formwork rather complex (author).

At its extraction from the formwork, the segment, not prestressed, slides along a curing rail which receives progressively all the segments of a span or of a substantial part of it (figure 3.10). The presence of a long curing rail is indispensable, as without it the delay in clearing formwork produces such long dead times as to make a high casting industrialization useless. Once the planned number of segments is cast, launching prestressing is applied to the whole deck section supported on the rail, which can then proceed toward the first span. As tendon stressing requires a certain curing of concrete, this solution needs a long distance between the fixed formwork and the launching abutment, or, if this is impossible, a temporary falsework which supports the deck in the bank span, and a temporary pier which reduces the flexural stresses in the non-prestressed portion of the deck.

Advantages reside in a small, possibly automated formwork, in reduced foundations, in repetitive operations, warranty of quality and rapidity, and in a better labor rotation. Inconveniences reside in the length of the curing rail, the need for precise regulations and controls (errors accumulate with the number of joints), the multiplication of launch operations which immobilize labor, and the delicateness of the back thrust of sections not completely cured and prestressed, which are not very ductile.

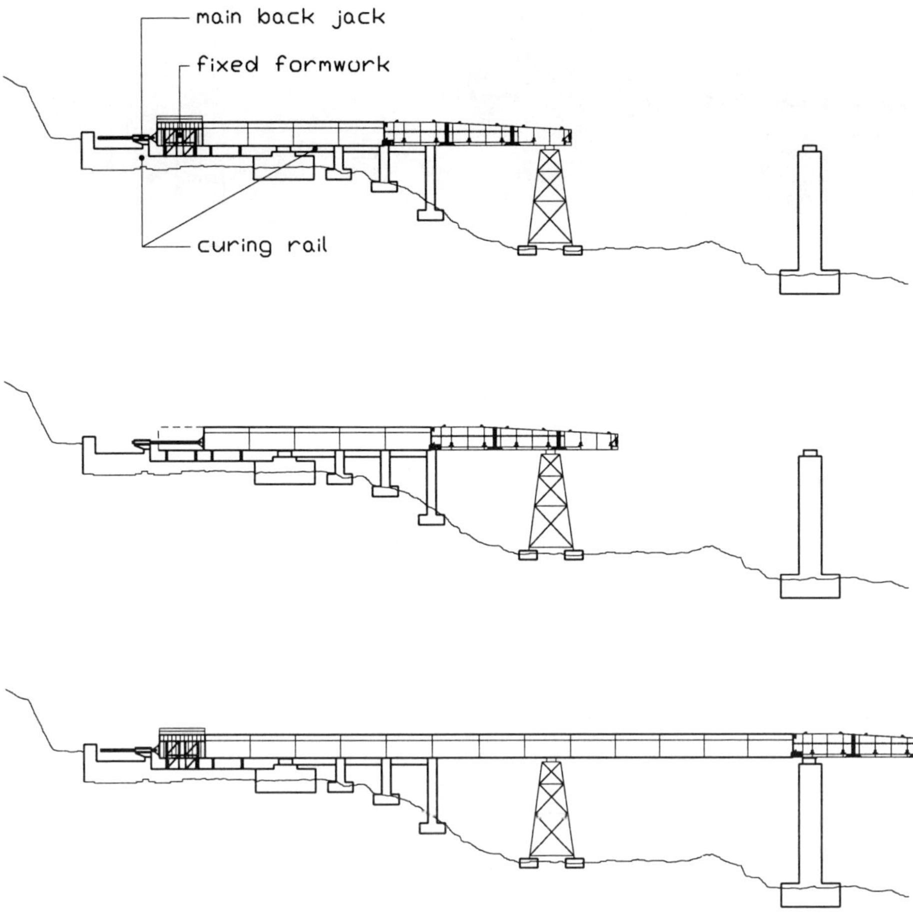

Figure 3.10 - Fixed casting cell and general launch by back thrust.

Using this scheme, in the railway viaduct at Km. 214 of the "TGV Atlantique" in France, the *8.1 m* long segments, one sixth of the span, have been built in two days, reaching a productivity similar to the one of a conventional formwork (half span a week) with lower investments and a better use of resources.

On the contrary, when the yard dimensions do not allow for an adequate rail, it is convenient to simplify the organization of production and to accept the repercussions on the construction duration. For instance, the *9.6 m*-long segments of Shepherds House Bridge[29] have required about two weeks each.

A second scheme, adopted in the launched section of the Poncin Viaduct[87], avoids the frequency of launch operations over relatively short distances, which represents the main inconvenience of the first scheme. In this scheme, the curing rail supports a deck section as long as the whole span, while the short formwork, *5* to *7 m*, shifts along the rail to cast the subsequent segments. Once the deck section is finished, prestressing is introduced in one operation just before launching the deck as a whole (figure 3.11).

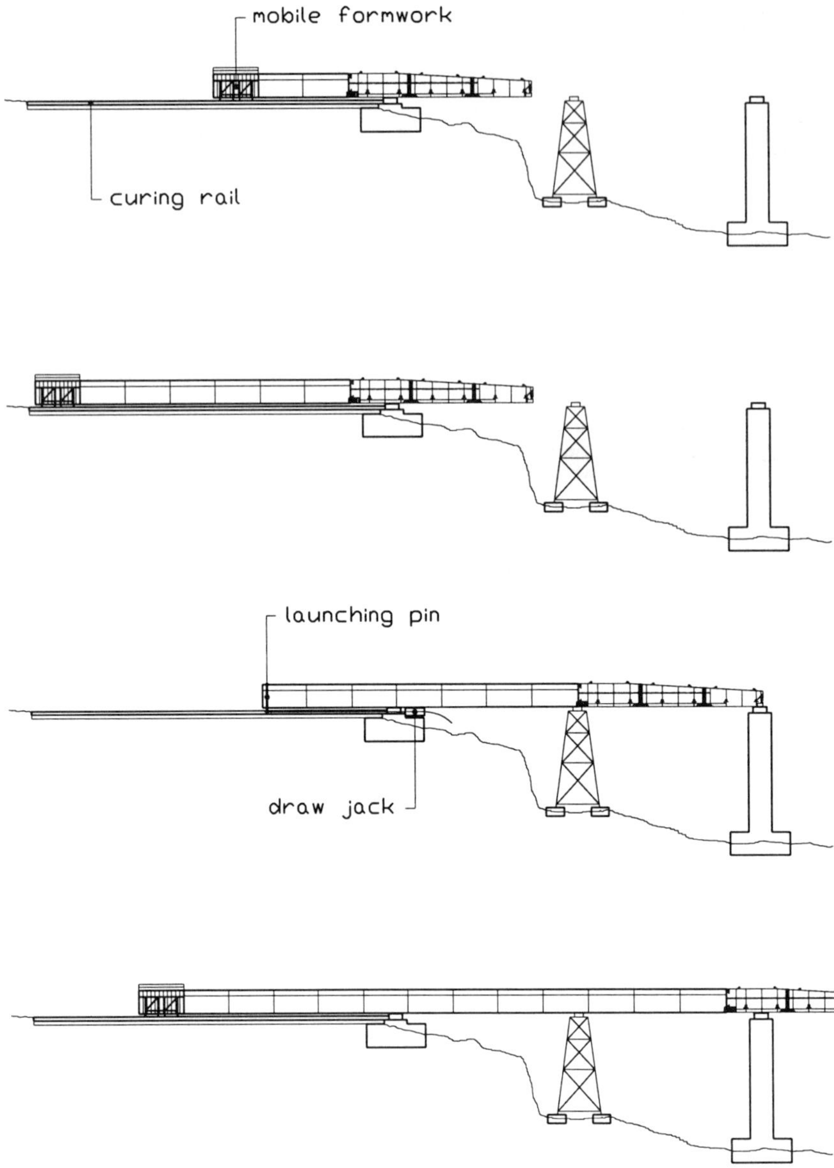

Figure 3.11 - Shifting formwork and conventional launch. The whole span is prestressed in only one operation. This reduces the cost of tendon coupling and makes the use of external launch prestressing easier.

Thus it is possible, with all the advantages of the short formwork, to reduce drastically the number of launch operations and to move the deck by strand drawing instead of by back thrust, without risk of cracks due to friction, the whole deck being prestressed. The disadvantages are a more complex formwork than the one of the previous scheme (it has to shift too) and the higher cost of foundations.

Though casting can occur in two phases (in order to dismantle the web internal formworks by hoisting and to move inside the box cell only the top slab formwork),

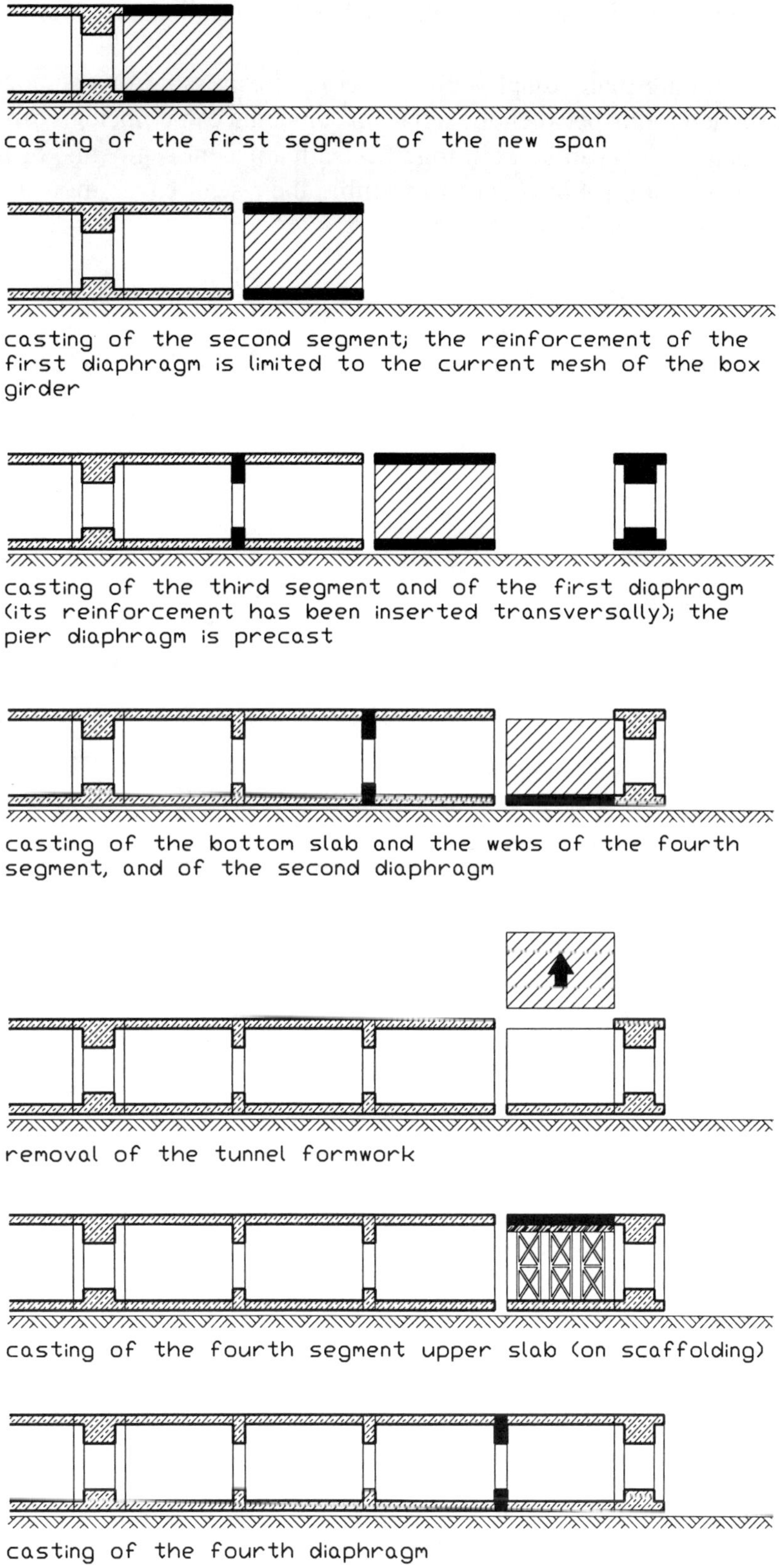

Figure 3.12 - Shifting formwork and conventional launch. Construction sequence for the Amiens Viaduct.

in long decks, whose construction can be more industrialized, the monolithic casting is often convenient.

Where there are many diaphragms (external prestressing or antagonist tendons), the construction joints between the adjacent segments can coincide with diaphragms. Then, the diaphragms can be built together with adjacent segments, or together with the previously cast ones in order to assemble their steel cages more freely through the solutions of continuity in the webs.

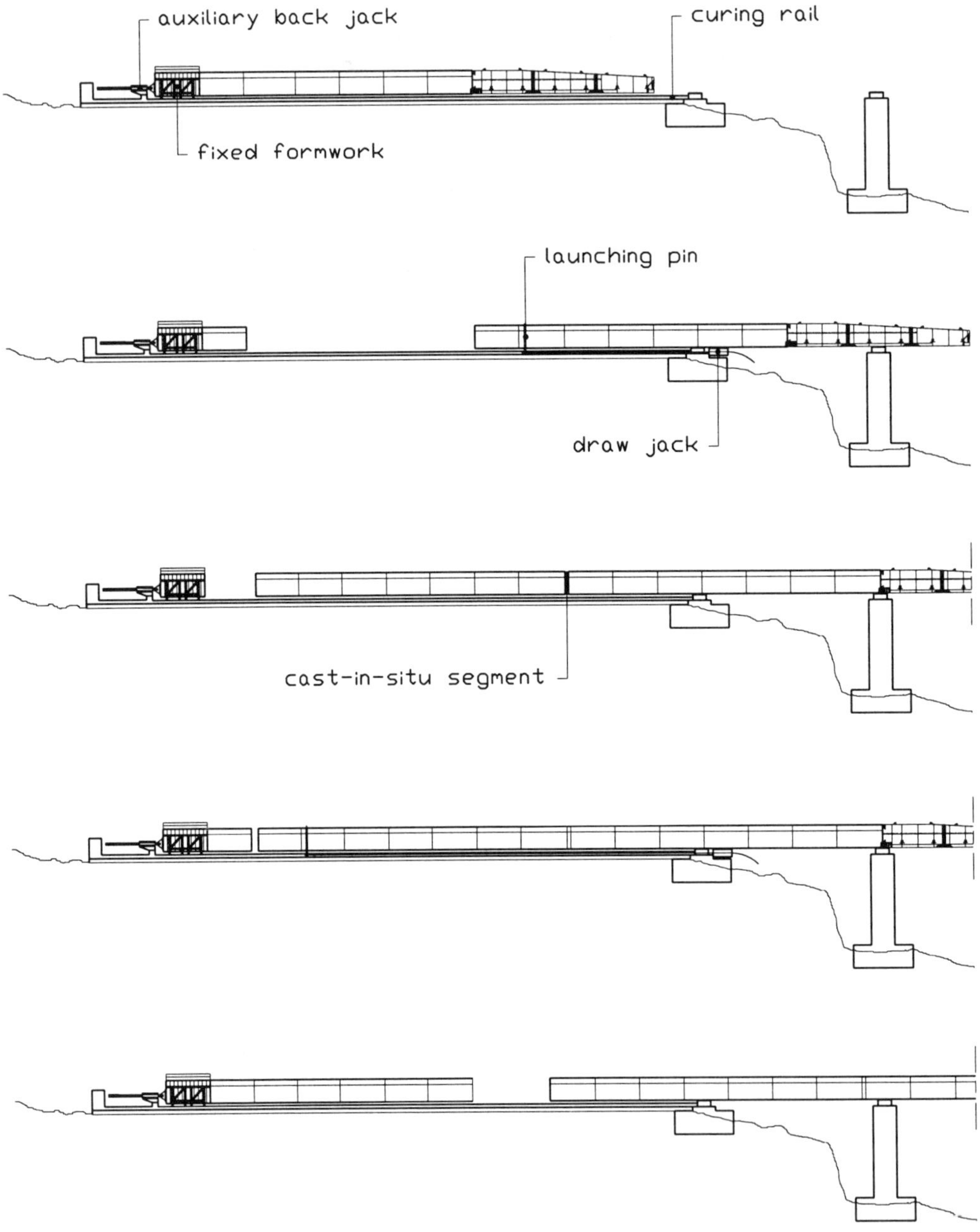

Figure 3.13 - Fixed casting cell, segment launch by back thrust, and conventional general launch. When the curing rail is longer than the launch steps, deck closure, prestressing, and launch do not prevent the construction of new segments.

Even in this case, construction proceeds in opposite sense than launching, and as the last segment is subject to the local stresses of tendon anchorages it is convenient to adopt precast anchor blocks or to anticipate its construction. This solution was adopted for the Amiens Viaduct[83] (figure 3.12).

On the contrary, by proceeding in the opposite direction, launchward, one can rely on cured anchorages, and the last segment built is subject to a distributed and homogeneous compression.

A third construction scheme combines the advantages of the two previous ones and avoids their disadvantages (figure 3.13). The formwork is fixed and placed at the back end of the curing rail, and all the segments composing the span are built by match-casting and launched progressively along the rail by back thrust as in the first scheme. Once the last segment of the span has reached the necessary hardness, the whole span is launched by back thrust to the back end of the deck, to which it is connected by a closure joint, and after the introduction of axial prestressing the general launch can occur. If the distance between the fixed formwork and the back end of the deck is longer than the span, during the operations of closure, prestressing, and launch of the new span it is possible to proceed with the construction of further segments. Therefore, these additional operations do not lengthen the construction schedule.

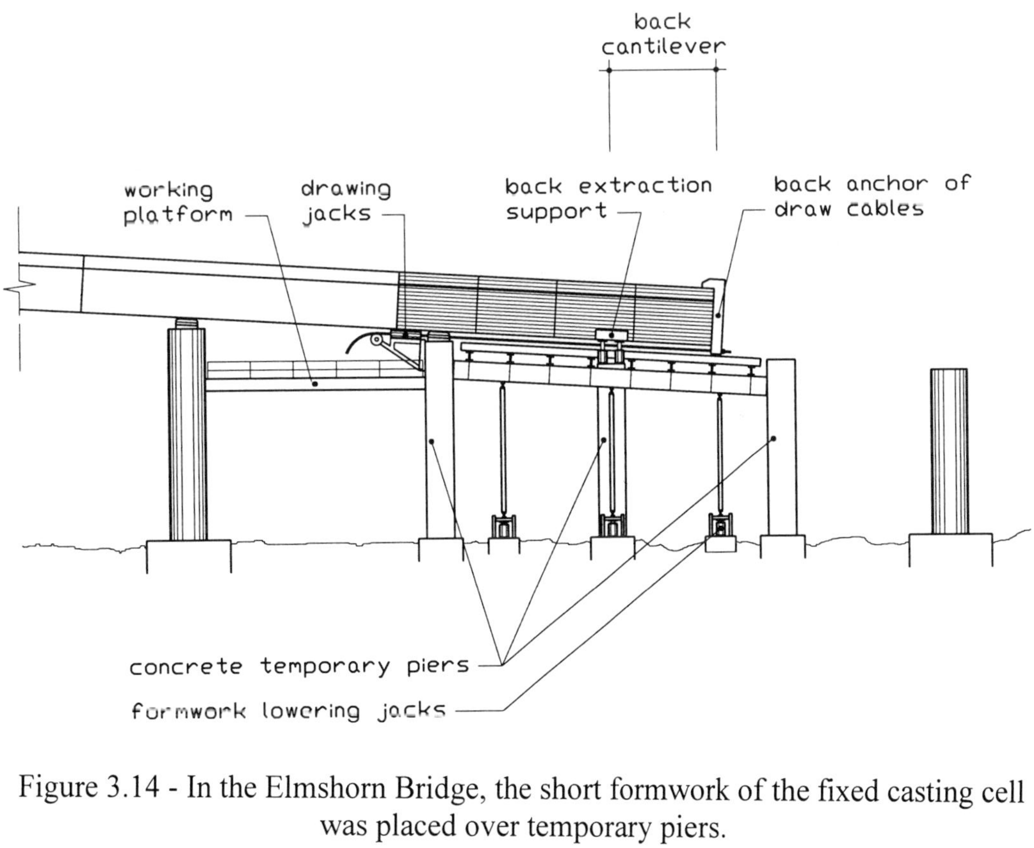

Figure 3.14 - In the Elmshorn Bridge, the short formwork of the fixed casting cell was placed over temporary piers.

With this scheme, the formwork is fixed and potentially automated, foundations for a sliding formwork are avoided, the general launch operations are still concentrated while the single extraction launches are simple and local, and the closure joint against the back end of the deck permits correction of geometric imperfections and avoidance of their accumulation. The disadvantages of the scheme include the long curing rail and the additional closure joint casting (not critical for the activity of the prefabrication cell, however).

From a general point of view, compared with the use of longer segments, all these schemes introduce additional operations, and as the deck length increases this risks compromising the savings in equipment with higher labor and reinforcement costs. On the contrary, a common advantage is the possibility of introducing launch prestressing independent from the production of deck segments, and to reduce the cost of tendon coupling, as prestressing affects the whole span.

Therefore, with internal prestressing, these schemes are applied to relatively short bridges (150 to $200\,m$) in which the investment required by a 20 to $30\,m$-long formwork cannot make incremental launching competitive compared with other construction techniques. The techniques also apply to bridges with a large yard behind an abutment, or in case of yielding or no embankments (figure 3.14), since the construction cost of a small formwork supported on piles or temporary piers is reasonable. On the other hand, in a case of external prestressing and particularly in a case of antagonist launching tendons, construction joints between segments can coincide with deviation diaphragms, and the use of short formworks can be convenient even in very long decks.

Finally, in spite of the lower costs and the higher simplicity, construction duration is substantially the same as for ordinary formworks. A half-a-span-long formwork (that is, about twenty meters) with a weekly cycle permits a daily progress of about $4\,m$. A $4\,m$-long fixed formwork soon reaches a daily cycle (except for the more complex pier segments) and an $8\,m$-long formwork, with a two-day cycle and preassembled steel cages, permits better control of unexpected events.

However, the difficult dismantling of internal formwork generally limits the adoption of these construction schemes to rather deep cross sections. In most cases, therefore, it is convenient to introduce a horizontal construction joint to divide the segment vertically into two casting phases, using a single formwork or two adjacent ones.

Two-Phase Casting in a Single Formwork

In incremental launching, the most popular casting method is the two-phase scheme inside the same formwork (figure 3.15), with a horizontal joint at the level of the top slab, or of the bottom one, or at the midpoint of the webs.

The external and the bottom formworks do not present particular problems: their design can aim at the highest operative simplicity by means of sophisticated devices, or at the reuse of modular equipment easier to amortize.

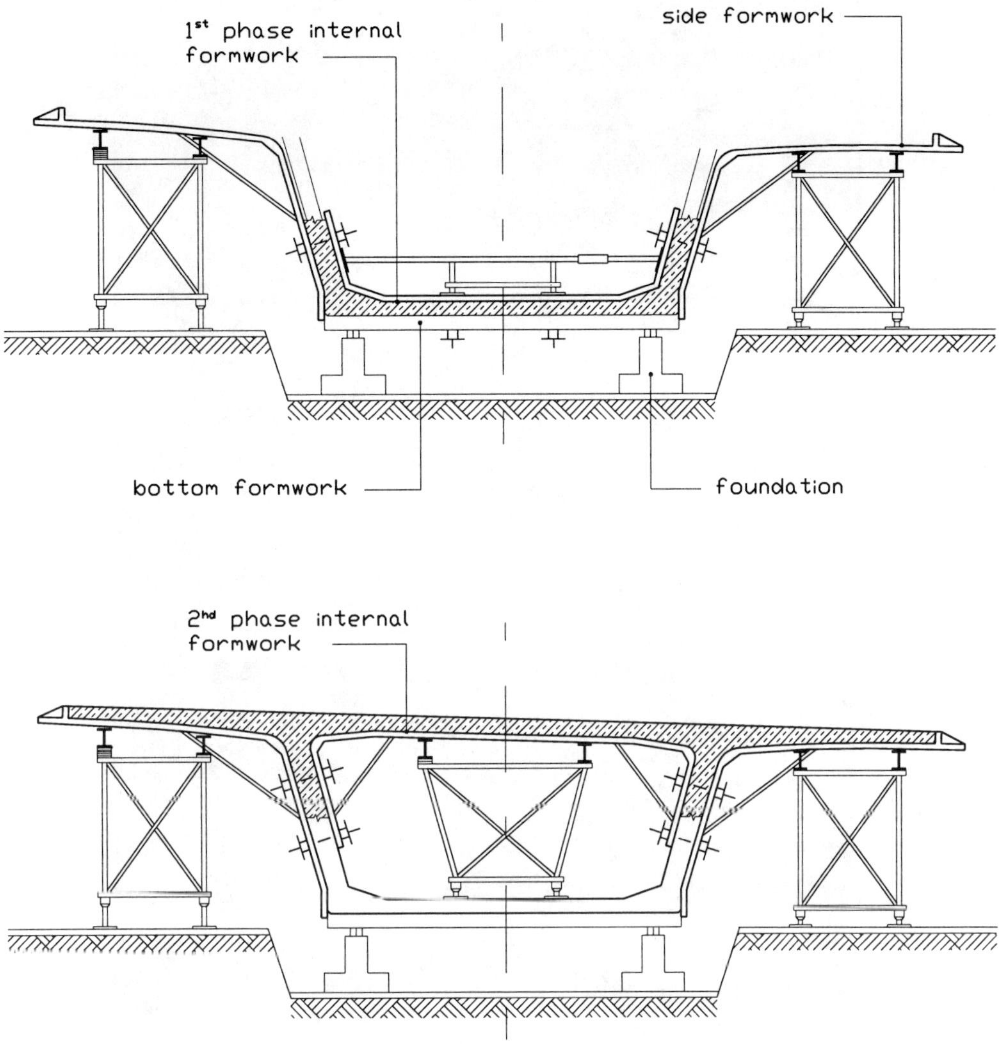

Figure 3.15 - Two-phase casting in a single formwork.

The most expensive type of external formwork consists of rigid steel frames hinged at the base (figure 3.16). Being designed for a specific cross section, it is hardly reusable, and its adoption is justified only in such long bridges as to make the cost of its whole amortization lower than the labor cost that it avoids. Less expensive formworks consist of fixed modular towers which can be dismantled using their manual regulation screws (figure 3.17), or by towers supported on transverse rails and dismantled by side translation (figure 3.18). The simplicity of these schemes and the easy amortization of the formwork are balanced by higher labor costs, which limit their use to small decks. An intermediate type, less expensive and more adaptable than the first since it is reusable for different cross sections, involves the use of modular towers installed on a movable platform supported by hydraulic jacks (figure 3.19). The movable platform facilitates the

Figure 3.16 - Monolithic steel frame formwork. Close up, the parking support of the top slab compacting roll, necessary when the formwork is opened (author).

dismantling of the whole segment, the towers can be reassembled many times according to new sections, and their regulation screws are used only when it is necessary to adjust the alignment of the formwork.

Unlike external formwork, the shape of internal ones depend on the position of the horizontal joints. In a case of joints placed in the upper nodes of the cross section, the web formworks are limited to side panels (figure 3.20), or include a bottom slab counter-formwork in the case of very sloped webs. In short bridges, the top slab formwork is obtained by means of left-in-place materials, as precast concrete planks and corrugated sheet metal or plastic plates, sometimes reinforced by fiberglass. In longer bridges steel or plywood panels are used, supported on

Figure 3.17 - Modular tower formwork adjusted by screws (author).

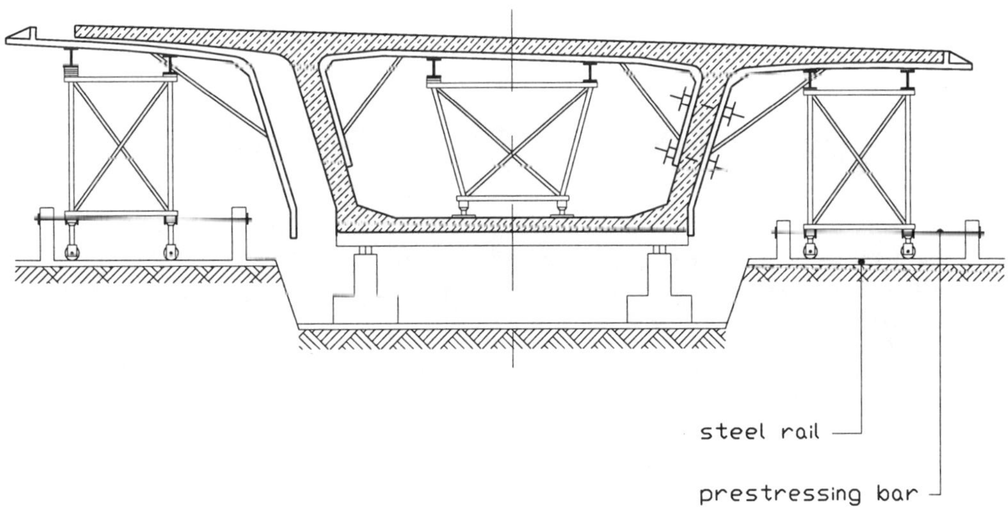

Figure 3.18 - Dismantling by transverse shifting.

modular towers or on self-supporting steel frames sliding along extraction rails anchored to the webs (figure 3.2). In a case of joints placed in the lower nodes of the cross section, the internal formwork is similar to the one for monolithic casting, and therefore rather expensive. In a case of intermediate joints (often required by top slabs with variable thickness in the longitudinal direction in order to dismantle the internal formwork by extraction) the both types of formwork can be used.

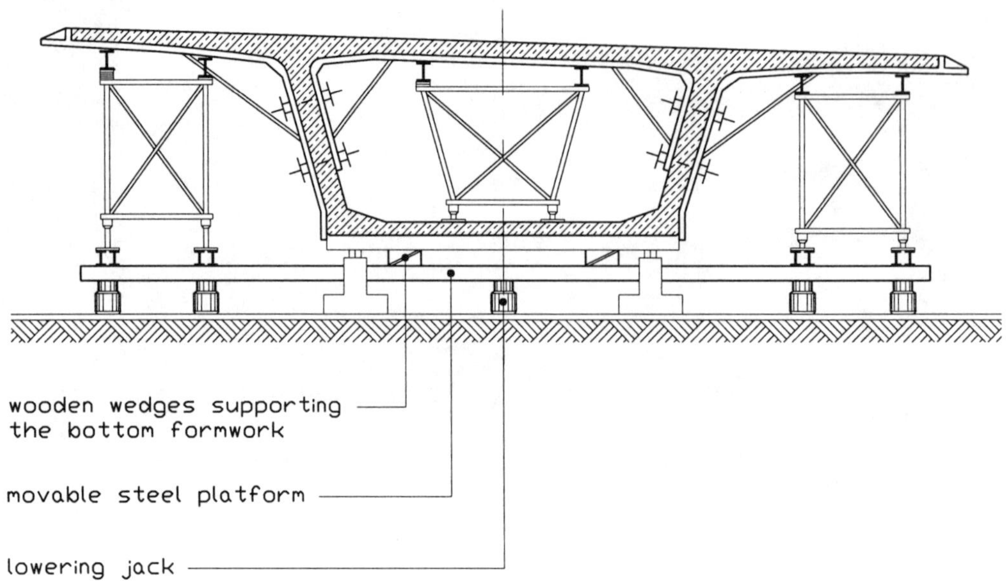

Figure 3.19 - Modular tower formwork placed on a hydraulic platform.

Figure 3.20 - Simple internal formworks for a short deck (author).

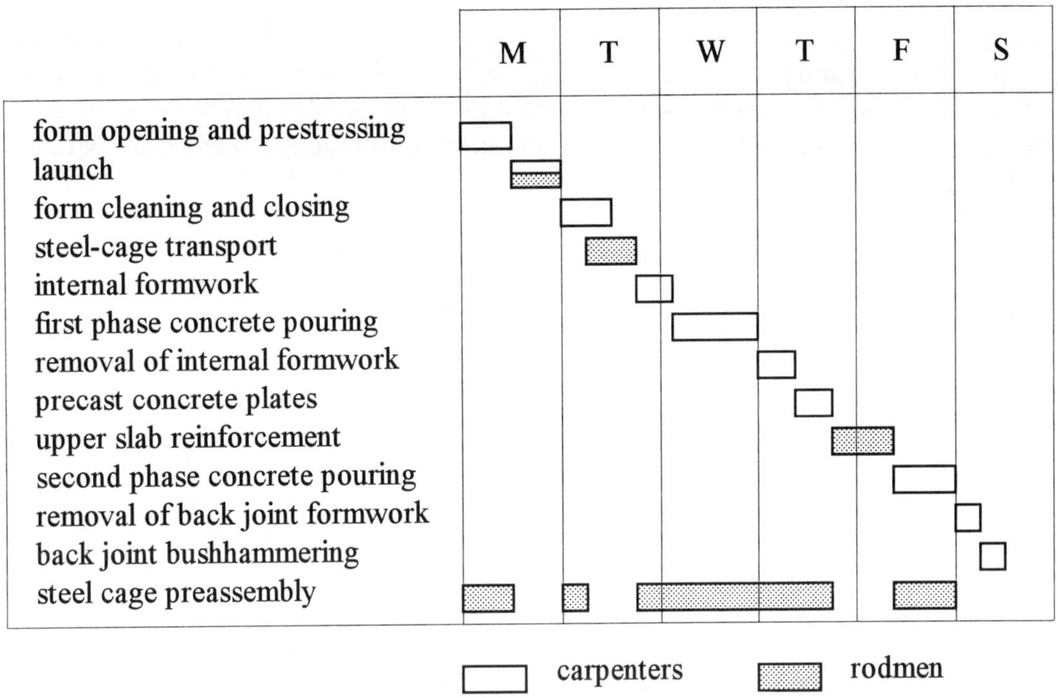

Figure 3.21 - Construction schedule of the Serio River Bridge (figure 1.11), built by two-phase casting in a single formwork.

Figure 3.22 - In the case of parallel bridges launched at the same time, two-phase casting in a single formwork allows an optimum rotation of labor even without steel cage preassembly (author).

Compared with monolithic casting, the scheduling of activities is generally less restrained, and by preassembling the steel cage it is not difficult to achieve the construction schedule of figure 3.21. Finally, two-phase casting in a single formwork is generally adopted when launching two parallel bridges at the same time (figure 3.22) since this scheme produces an optimum rotation of labor even without steel cage preassembly.

Two-Phase Casting in a Double Formwork

Progressive labor specialization has led to the subdivision of workmen operating in the yard into two distinct gangs: the bars workers, often working for a subcontractor, and the carpenters.

In the case of single formworks, the presence of a steel cage preassembly yard prevents the activity of one gang from interfering with or impeding the activity of the second one. But even so there is interference: the adjustment and the completion of the steel cage after its transfer into the formwork, the sheath connection, and so on. In a case of long segments, the activities of the two gangs can be staggered longitudinally (for instance starting the concrete pouring on one side while cage assembly is still going on at the opposite side of the segment) but this is generally a remedy for inadequate scheduling rather than an efficient organization. In a case involving short segments, these solutions generally lead to chaotic situations which are in conflict with the cadenced cycles which characterize incremental launching. On the contrary, a very long external formwork permits work in different points, and the internal formwork may be much shorter (figure 3.23), but costs are generally high.

When the yard is long enough, it can be convenient to divide the production into two distinct formworks, in which different elements of the segment are built in operational sequences staggered both in time and in space. Many are the advantages of this solution: by operating in two distinct areas, steel cage assembly in one formwork does not jeopardize the activity of carpenters in the second formwork and does not require preassembly and hoisting equipment, the time available for each activity increases, the critical situations reduce, the average curing at prestressing increases, and the concrete surface finish improves, since dismantling may be delayed. The only disadvantage is the higher cost of the formwork and of the eventual protective structures, whose length is double.

In the less expensive configuration, with horizontal joints at the top slab level and transfer of the segment from the back formwork into the front one on single bearings, the yard is divided into the two zones of figure 3.24: a back zone for the construction of the bottom slab and the webs, shorter than the segment, and a front zone for the construction of the top slab and the remaining small segment of the first-phase casting.

The bottom formwork of the back zone is a concrete slab or a steel panel supported on foundation beams by means of wooden wedges which permit its adjustment in case of settlement. The web external formwork is composed of panels hinged at the base, and the internal one is quite conventional.

Figure 3.23 - The Charix Bridge was built with an external formwork as long as the whole span, and a shorter shifting internal formwork (reproduced with permission, SCETAROUTE - Berenguier).

The first-phase concrete is not match-cast against the previous segment, the presence of a solution of continuity and the adoption of a specific working cycle being necessary:

- during the curing of the first phase casting, the web internal formworks are dismantled, and the rails for the top slab internal formwork are bolted to the concrete webs;
- once the necessary curing has been reached, the back segment is hoisted from the fixed bottom formwork in order to introduce neo-flon plates between the deck itself and the extraction bearings. The solution of continuity in the bottom slab and in the webs avoids the concrete to stress;
- the back segment is connected to the front one by means of two lower tendons of the launching prestressing (the low tendon stressing, required only to draw the segment during extraction, is balanced by steel struts located into the solution of continuity) and it is drawn forward by the deck launch, sliding above the fixed bearings;
- once the extraction is completed and the drawing tendons released, the segment is lowered on continuous neo-flon plates arranged along the extraction rails of the front formwork, and in this way its soffit aligns with the front section of the deck;

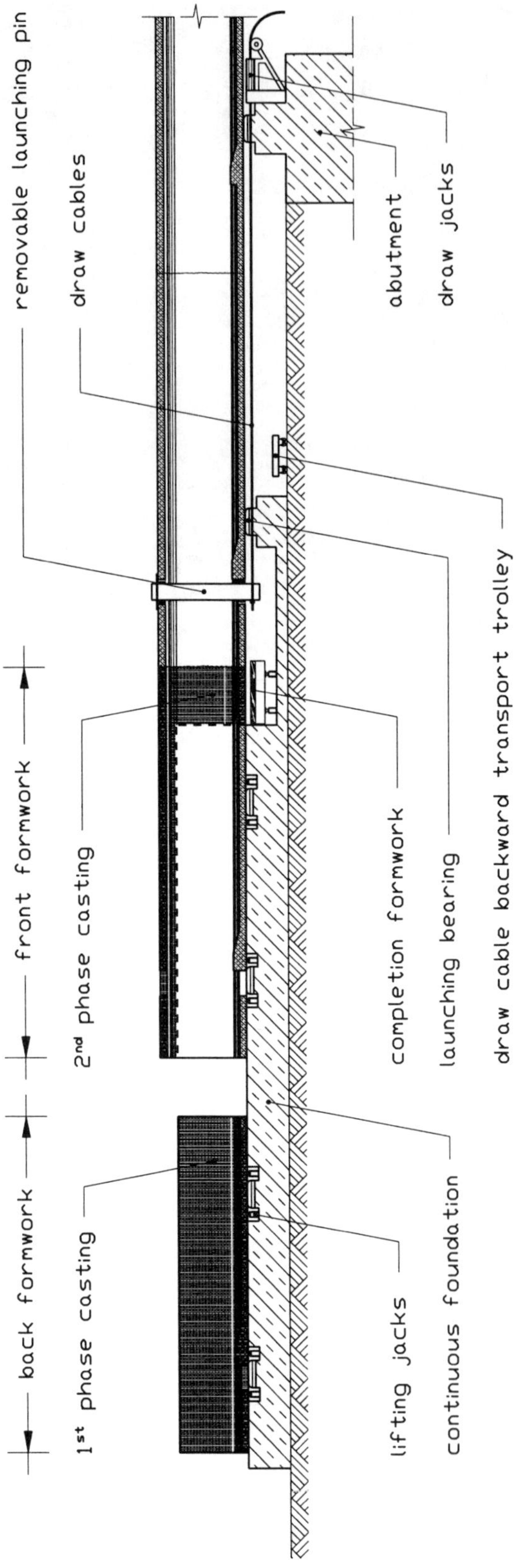

Figure 3.24 - Two-phase casting in double formwork, with transfer of the back segment on individual fixed bearings.

– the top slab internal formwork, which is still inside the previous segment, is drawn backward and placed above the first phase casting; at the same time, the formworks for the cantilevers and the solution of continuity are set up.

This organizational scheme allows construction of a *20* to *30 m*-long segment per week, by operating five working days with the time schedule of figure 3.25.

The segment extraction on single bearings has been progressively replaced by the use of continuous extraction rails, which, in spite of the higher cost of the rails, the neo-flon plates, and the mobile bottom formwork, do not require hoisting nor the solution of continuity in the first phase casting, which can be match-cast.

In this case, the prestressing tendons contained in the bottom slab of the first-phase back-casting are eccentric with respect to the centroidal axis of this partial U-section. Since the match-casting of the next segment covers the anchorages placed at the construction joint, and the partial tensioning requires expensive blisters or tendon couplers, these tendons are completely tensioned before launch, and this

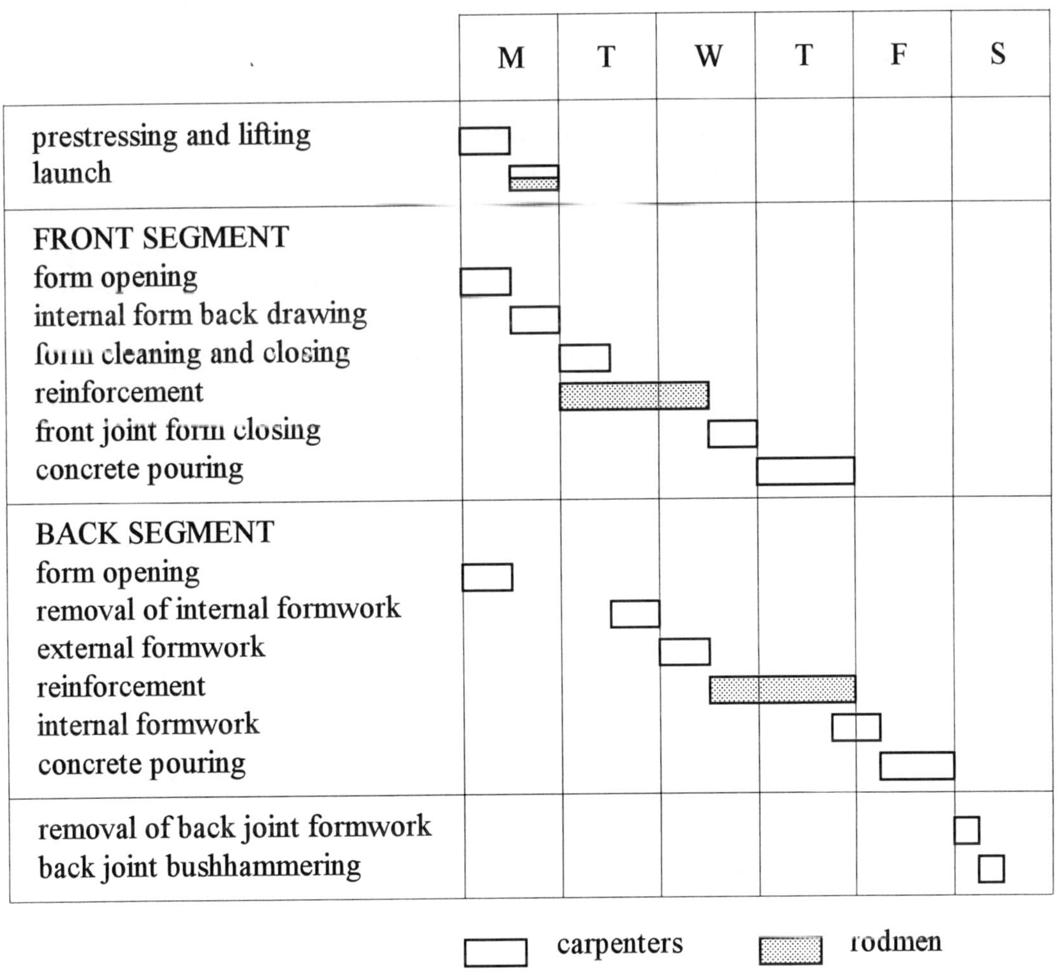

Figure 3.25 - Construction schedule for two-phase casting in double formwork.

produces a negative bending moment which tends to crack the horizontal joint surface vertically. After one week, the stressing of the tendons contained in the top slab centers the prestressing resulting force on the final centroidal axis, and the low residual stresses at the joint are eliminated by creep, very high at short curing. As an alternative, it is possible to prevent the problem by placing the horizontal joints in the lower nodes of the cross section, with higher costs of the internal formwork.

Assembly of Precast Segments

The very first applications of incremental launching were for structures composed of precast segments arranged along concrete beams and connected by means of closure joints. In effect, the assembly of precast segments presents many advantages:

- the possibility of making most of the deck construction independent from the construction of the piers, by stocking segments elsewhere;
- the use of minimal labor and inexpensive formworks in a series of highly repetitive operations with the best control of manpower and materials, especially if in a prefabrication plant instead of in the yard, and with the best amortization of investments;
- the speed of assembly and launching of the deck;
- the higher dimensional stability of the deck deriving from the higher average age of precast segments compared with the cast-in-situ ones.

Although there are no strict dimensional limits for precast segments, hoisting and handling difficulties generally limit their length to between *2* and *5 m*. Segments are rarely coupled directly by means of dry joints or epoxy because of many problems arising from this solution:

- the difficulties deriving from the strict geometric tolerances during both segment construction and their reassembly in the yard, which impose the construction by match-casting and therefore the presence of long support rails in the prefabrication plant;
- the higher consumption of longitudinal prestressing, up to *20* to *30%*, because of the necessity to avoid joint decompression with an adequate safety factor even under infrequent load conditions, such as combined thermal gradients and differential settlement of bearings during launching, or the most unfavorable combinations of live loads in service;
- the problems that may arise in the ultimate-state structural behavior or in case of rapid variations of thermal gradients.

To avoid these problems, segments can be precast with reinforcement emerging from the joint surfaces and arranged in the yard along a pair of continuous launching rails at a distance of *0.20* to *0.60 m*, in order to splice the reinforcement and to fill these solutions of continuity with high strength concrete (figure 1.2). Yard operations are limited to the placement of segments along the rails, joint sealing, and the introduction of prestressing in the deck section supported on the rails. The construction speed can double compared with casting in-situ, though with additional labor costs because of these additional activities. Shrinkage cracks in the closure joints can be avoided by means of modest local thermal cycles, or, as in the Rio Caroni Bridge[20], by casting the joint segments in the second half of the night in order to rely, during their setting, on the thermal expansion of precast segments produced by the daytime temperature, and by limiting coupling to few segments each time[127].

SEGMENT EXTRACTION FROM THE FORMWORK

Although the cylindrical shape of the deck can remind continuous construction processes such as extrusion or slip forms, the deck construction is cadenced, and the segment casting is followed by its curing inside the formwork. At dismantling, therefore, the deck segment must be supported by bearings which permit its extraction.

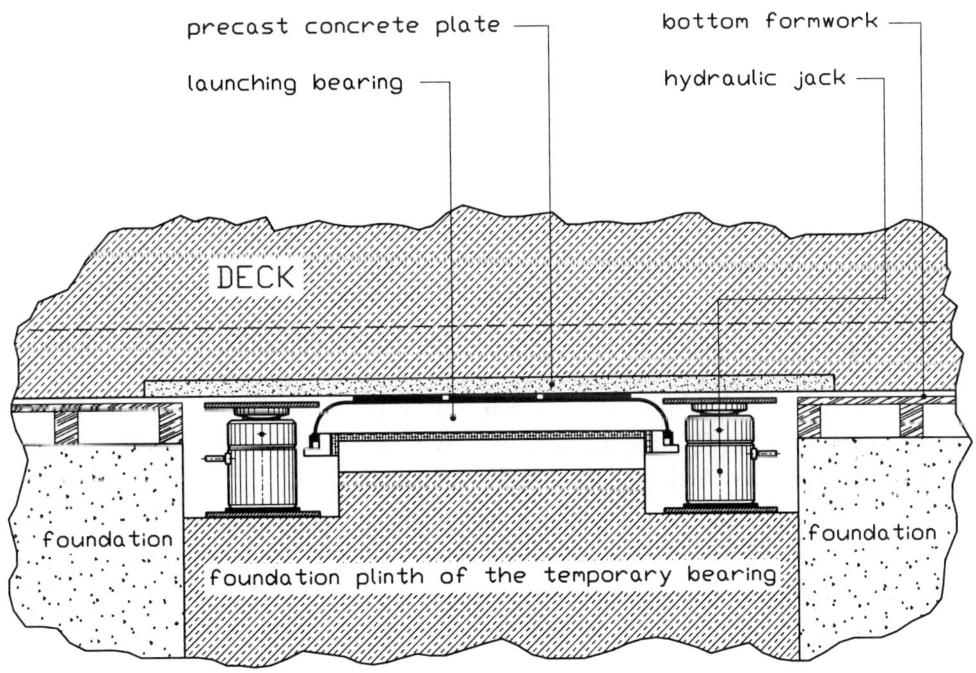

Figure 3.26 - In the case of segment extraction on fixed bearings, precast concrete plates protect the launching saddles during the segment casting.

The problems to solve are essentially two: to reduce the friction during extraction to prevent the deck from dragging the formwork, and to reduce longitudinal and transverse flexural stresses in the young concrete.

In order to reduce friction it is indispensable to separate the formwork from the concrete surface, and this can be accomplished in only two ways: by hoisting the segment or by lowering the formwork.

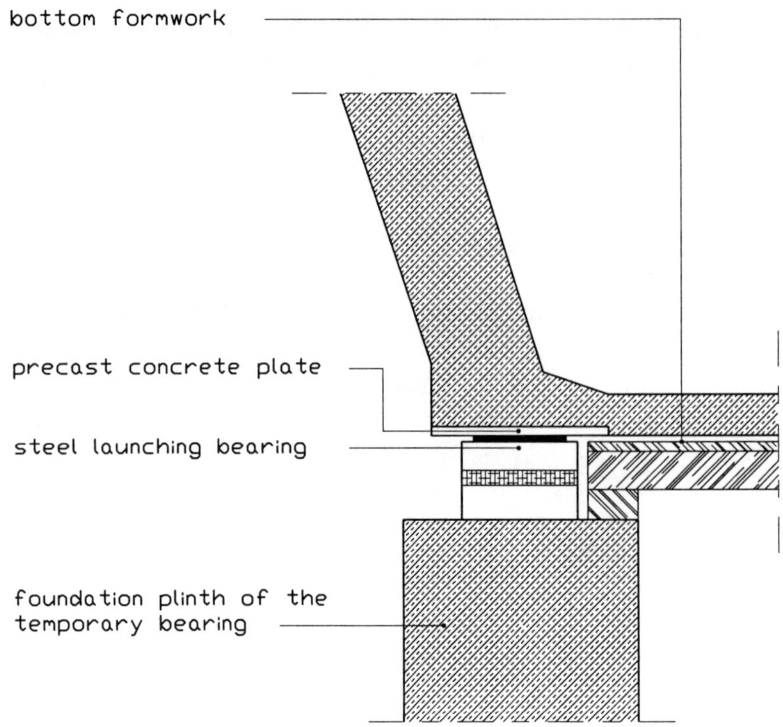

Figure 3.27 - Cross section of the extraction bearing.

The hoisting of the segment permits insertion of bearings between its soffit and the bottom formwork surface. These bearings can be mobile or fixed during the segment extraction depending on whether they are integral to the deck or to the formwork. In case of mobile bearings, the deck is hoisted by means of four hydraulic jacks placed at the corners of the formwork. The two front jacks support launch saddles and remain fixed, while the two back ones permit introduction of the neo-flon plates which constitute the mobile back support and slide during launching along the formwork. In the case of fixed bearings, the segment is always dismantled by hoisting, and neo-flon plates are introduced between the segment itself and special launching saddles protected during casting by precast concrete plates (figures 3.26 and 3.27). Their position is chosen so that the deck back cantilever is statically compatible with the mechanical characteristics of its cross section, which is only partially cured and prestressed. Figure 3.14 illustrates the layout used for the Elmshorn Bridge, and figure 3.28 shows a detail of the back hydraulic support.

In both these support schemes, the bottom formwork can be obtained by means of a simple concrete slab covered by a plastic sheet, no vertical movement being required. The high economy of equipment and labor, however, can be compromised by the settlement of the slab and by the stresses caused by hoisting in the young concrete.

Figure 3.28 - Hydraulic extraction support of the Elmshorn Bridge (figure 3.14) (reprinted with permission, Dywidag Systems International[229]).

Segment hoisting with extraction on single bearings has been progressively replaced by the lowering of the external formwork (figure 3.29) and by the use of a continuous fixed support which avoids vertical deck movements. Hoisting is now adopted practically only in case of aerial formworks, as it permits concentration of support reactions and therefore savings in temporary piers (figure 3.14).

In the shortest decks, which suffer fewer problems due to friction, upon formwork dismantling the deck segment is supported on two continuous extraction beams above which steel plates are aligned on a layer of grease, hydraulic oil, soap or other anti-friction materials before the steel cage is installed.

In longer decks, on the other hand, extraction rails are covered by stainless steel plates above which neo-flon plates are aligned in order to obtain a continuous steel-Teflon contact. Their upper face is protected by sheet steel or greased paper which prevents their sticking to the fresh concrete, match-cast above them.

In both cases, the lowering of the bottom formwork central strip places the segment on a low-friction longitudinally continuous support. The deck launch extracts the segment from the formwork and trails it on the first curing supports, and axial prestressing, introduced before launching, absorbs the low tensile stresses produced by friction and by the first bending moments.

Figure 3.29 - Launch of the third segment of the cable-stayed Milan-Palizzi Bridge. The central strip of the bottom formwork is lowered (author).

Neo-flon plates advance with the deck. Once reaching the front of the formwork, they fall on the ground, and are ready to be cleaned and arranged again along the extraction rails for the construction of the next segment.

The positioning of the extraction rails must be extremely accurate, as the deck surface, which during launch comes in contact with the launching bearings placed on the piers, is match-cast directly against their upper surface. An excessive discordance between the actual alignment and the theoretical layout of the launching directrixes, along which the launching bearings are positioned, produces secondary stresses in the continuous beam during launch. Local defects produce stress accumulation, while larger defects have consequences similar to the differential settlement of bearings. Therefore, extraction rails must be very stiff, supported on foundations as stiff as possible, and provided with regulation devices which permit setting their position with differences from the theoretical profile not larger than 1 to $2\,mm$. It is necessary to frame the yard in a wide closed traverse, to use continuous foundations instead of spread footings, and, in case of settling soils, to use piles and a light prestressing of the foundation which avoids the deformations produced by concrete cracking.

Compared with the techniques which involve the hoisting of a segment, this type of formwork is more complex and expensive. On the other hand it allows more precise vertical adjustments, the possibility of compensating for eventual settlement of the platform, and better handling of concrete.

TEMPORARY SUPPORTS IN THE CURING YARD

Before being extracted from the formwork, the segment almost always receives just a portion of the launching prestressing, since curing is only partial. Therefore, the formwork yard is generally separated from the launching abutment in order to let the concrete cure enough to resist the whole axial prestressing and the stresses deriving from its launch on the bank span.

In the area between the formwork and the abutment, the deck must be supported by temporary bearings which reduce the spans to values consistent with the grade of curing and prestressing progressively reached by the deck segments. These bearings generally consist of launching saddles located on small reinforced concrete pillars, and their number must be as small as possible because of both their cost and the labor they immobilize during launching. Particular care must be taken in positioning the first support ahead of the formwork, which is subject to several restraints:

- the necessity that at the end of each launch the back end section of the deck (that is, the construction joint against which the next segment will be match-cast) be orthogonal to the theoretical deck axis, without tangential discontinuities which would generate sharp points and therefore stresses in the continuous beam during launch. This requires that the first support be placed at such a distance from the formwork as to cancel the flexural rotation produced by dead load;

Figure 3.30 - The optimum distance between the formwork and the first curing support depends on several factors (author).

- the necessity that the back end of the deck remain firmly supported on the formwork even in the presence of thermal gradients and vertical misplacement of temporary bearings (hoisting creates a vertical discontinuity in the deck soffit which makes launch difficult) requires, in turn, a certain minimum distance from the first support;
- its role as transverse guide during launching suggests its positioning as close as possible to the formwork head, in order to use its foundations as transverse anchorage;
- an excessive distance makes it difficult to extract the last segment of the deck from the formwork because of the local stress concentration at its back head due to the support reaction supplied by the formwork, which often breaks the soffit at its exit (figure 3.30). This can be avoided by rounding off the extraction rails in order to make the reduction of the support reaction progressive, or by supporting the deck by means of hydraulic jacks which slide over Teflon plates. The same problem arises when the deck back end leaves the curing yard launching saddles (figure 3.31).

Figure 3.31 - When the back end of the deck leaves the curing yard launching saddles, the concentration of the support reaction tends to break the deck soffit (author).

Even the distance between the last curing support and the launching abutment must be studied carefully, having to guarantee at the same time an adequate damping of the bending moment produced by the continuous beam bank span and a positive reaction on both these supports. Otherwise, they being unilateral constraints, the deck would raise and the static scheme would change.

Once these two supports have been positioned, the spacing of the eventual intermediate ones is defined according to the development of mechanical strength of concrete, the grade of prestressing progressively applied, and the available space. When their number is excessive, the introduction of axial prestressing may be accelerated by adopting shorter and less overlapped tendons, or a temporary pier may be introduced in the bank span in order to bring the formwork nearer to the abutment without jeopardizing a correct damping of the flexural stresses due to dead load.

Finally, the spacing of the curing supports could turn out excessive in the static schemes that occur during the final phases of the deck launch, when the back end of the deck cantilevers out for as much as the whole distance between supports.

In this case, rather than reducing these temporary spans, it is convenient to introduce additional intermediate supports to be used only in the last launch phases, disengaging them during the previous stages by not inserting neo-flon plates (taking care that the elastic deflection of the deck does not cause contact).

As an alternative, a steel rear nose may be used, most often in case of launching devices operating by friction (figure 4.16).

Chapter 4

LAUNCHING TECHNIQUES

The handling of a prestressed concrete deck involves enormous forces and requires the guide and control of big volumes.

The force necessary to launch the deck is linearly proportional to its weight, as both friction and the longitudinal force produced by the slope of the launching surface are a function of it. In the longest bridges, and especially in railway spans, this force can exceed $10\ MN$, but in most cases it is only a few MN. This being nearly the same strength as prestressing tendons, in the first applications of incremental launching it was natural to use the prestressing materials and equipment already available in the yard. From here came the development of drawing devices composed of one or more jacks anchored to a foundation and acting on sheaves of strands or bars anchored to the deck.

With time, the inconveniences of the drawing systems stimulated the development of new thrust devices, some applicable to relatively modest loads, like back jacks which self-clamp to formwork extraction rails, some suitable for higher loads which engage the thrust force by friction, exploiting the support reaction of the deck above them.

The equipment currently used presents mechanical characteristics, power, and cost so variable as to define substantially its field of application. The less expensive types apply the tow or thrust force directly by means of strands or bars or by self-clamping and are adopted for the shortest decks. The most expensive types transfer the horizontal force by friction and are generally reserved to the movement of large masses in short times.

Independent from the transmission modality of the launching force, every launching equipment requires the presence of an anchor element restrained to the ground. The foundation of the abutment is generally suitable for this use, since it is usually dimensioned for the absorption of most of the longitudinal service loads. When the launching force exceeds the capacity of the abutment (a condition not infrequent in the final launching phases of highway bridges), oversizing can be avoided by joining the yard foundations with the abutment in order to create a large concrete bed which cooperates by friction. As an alternative, specific foundation blocks which couple to launching jacks directly and avoid the application of horizontal forces to the abutment have been tested in some particular applications.

LAUNCH OF SHORT BRIDGES

The launch of short decks is always characterized by the use of inexpensive equipment because of the low forces involved, the impossibility of amortizing high investments, and the necessity of assembling and demobilizing the yard in short times and with low costs.

Drawing by means of prestressing bars or strands is the most common solution both in incremental and monolithic launching of complete decks, although it presents some disadvantages, especially when launching along sloped surfaces.

In fact, the ordinary longitudinal slopes of highway bridges fall in the range of variation of the steel-Teflon friction coefficient, 1 to 4%. Therefore, when the deck launch occurs with new or well-greased neo-flon plates, it can be necessary to brake the deck or to lock it during the construction of segments to prevent the component of its dead load along the launching surface from mobilizing the deck.

As an alternative to towing, thrust devices are constituted by hydraulic pistons which act directly on the deck back end by self-clamping to the formwork extraction rails, or more simply, by blocking in racks housed between twin extraction I beams.

None of these devices permit pushing the deck back in case of excessive advance, nor its braking in launching downgrade, and therefore they must be used with a certain prudence.

Tow Systems

Coupled prestressing bars and hydraulic jacks anchored to the abutment (figure 4.1), or to the deck (figure 4.2), have been used many times in order to reach launch forces of about 0.8 to $1\ MN$.

The short stroke of the conventional prestressing jacks, usually about $0.20\ m$, suggests coupling the jacks so that they alternate (figure 4.3): during the stroke of the first jack the second one returns to idle to be ready on its turn to pull the deck, avoiding relieving the bars.

When the launching surface is almost horizontal, single jacks can be used by relieving the bars at the end of each stroke, and therefore loosing in each launching cycle the stroke necessary for their new stressing. The bar shortening is obtained by replacing their nuts on the steel anchorages clamped to the deck. In this case, the jacks are sized specifically for deck launching, with a longer stroke of about $1\ m$, (figure 4.4).

In the case of launching up-grade, the deck must be locked during each return stroke of the jacks, and the modest weight of short decks allows utilization of the draw bars themselves for this purpose.

While the use of coupled jacks avoids any problem, in the case of single jacks placed in front of the abutment, the bars are divided into coupled segments slightly shorter than the jack stroke, and specific anchorages are embedded in the abutment. Once the jack limit stop has been reached, the bars are locked in these anchorages, the jacks are released, their rods are reentered, and once the superfluous bar elements are eliminated, the new heads are connected again to the jacks.

Tow bars are placed under or next to the deck and can be anchored to it in several ways. In virtue of the low forces, two steel plates placed symmetrically on the two deck sides and connected to each other by strands passing through the bottom slab are generally sufficient.

In case of higher launching forces, bars are substituted with sheaves of strands. In principle, by using strands the tow force is limited only by the capacity of the jacks,

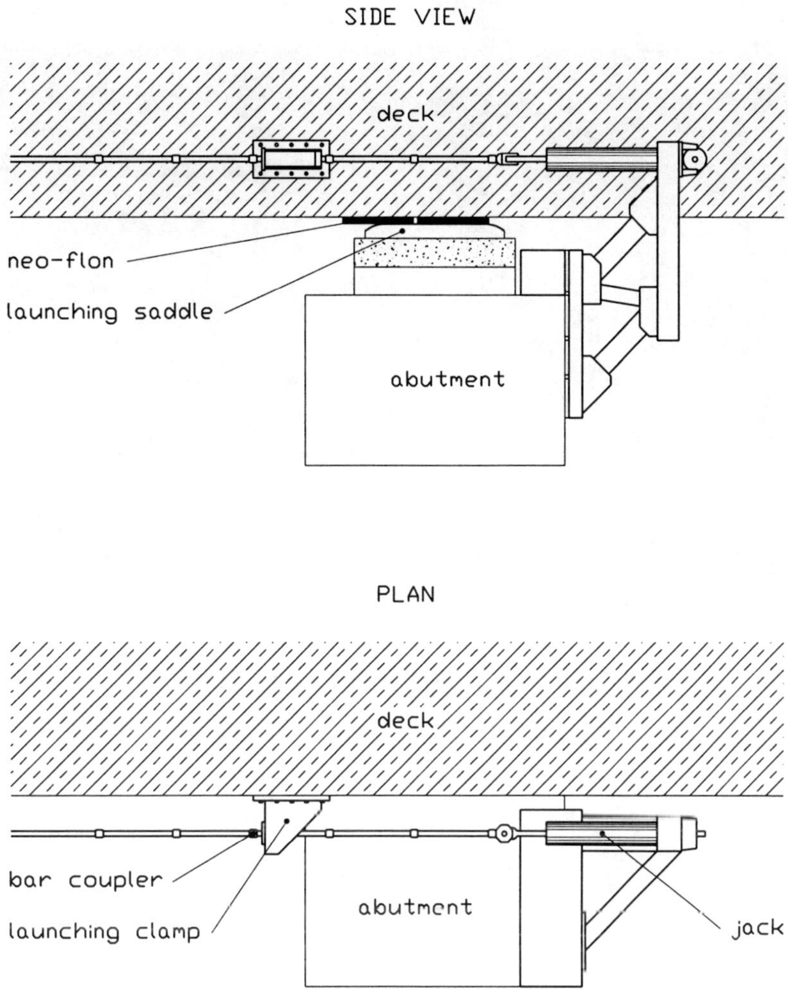

Figure 4.1 - Movable high strength bars are pulled by jacks anchored to the abutment.

the addition of further strands being always possible. However, it is generally convenient not to exceed the force of *2.5 MN* in each tendon in order to use ordinary *19T15* prestressing jacks with adequate safety factors, and, if needed, it is better to increase the number of tendons.

There are many possible schemes for applying thrust force to the deck, and physical principles for load transfer. The use of friction between the deck and the draw tendon anchor plates (mobilized by hydraulic local compression (figure 4.5), by prestressing bars (figure 4.6), or by hoisting the deck until obtaining an adequate vertical reaction) is limited to the shortest decks, which require low draw forces. Higher forces require mechanical anchoring: transoms anchored to the deck back end (figure 4.7), or through pins (figure 4.8).

Figure 4.2 - Fixed high strength bars anchored to the abutment are pulled by jacks anchored to the back end of the deck (author).

Load transfer devices induce high concentrated stresses in areas already loaded by prestressing, and therefore require careful design and proper local reinforcement. Moreover, at the end of each construction cycle they must be relocated to the subsequent segment, and draw tendons must be brought back to their initial length. Their self-weight makes this operation difficult, and in case of long decks it can justify the adoption of electrical systems with service trolleys (figure 4.9).

The problem of deck locking in up-grade launch can be solved, as with bars, by locking strands during the return stroke of the jacks. This can be obtained easily by means of ordinary mono-strand anchorages. As an alternative, it is possible to use two oversized draw cables and to take advantage of the difference between static and kinetic friction, by relieving the first tendon while keeping the second one stressed, helped during this operation by the static friction backward.

In the longest bridges it is preferable to avoid anchoring the deck by means of the draw tendons and to adopt special friction devices. These guarantee a higher safety when it is necessary to substitute some strands, and are anyway indispensable during the segment construction in the case of towing with the back transom.

The first friction anchor devices involved the deck hoisting at the abutment by means of jacks with knurled top plates able to mobilize a friction coefficient of about *0.60* against the concrete deck. This technique has a variable effectiveness, braking force being proportional not only to the friction coefficient, but also to the continuous beam support reaction at the abutment. This remains practically constant during most of launching but reduces considerably just in the final phases, when the

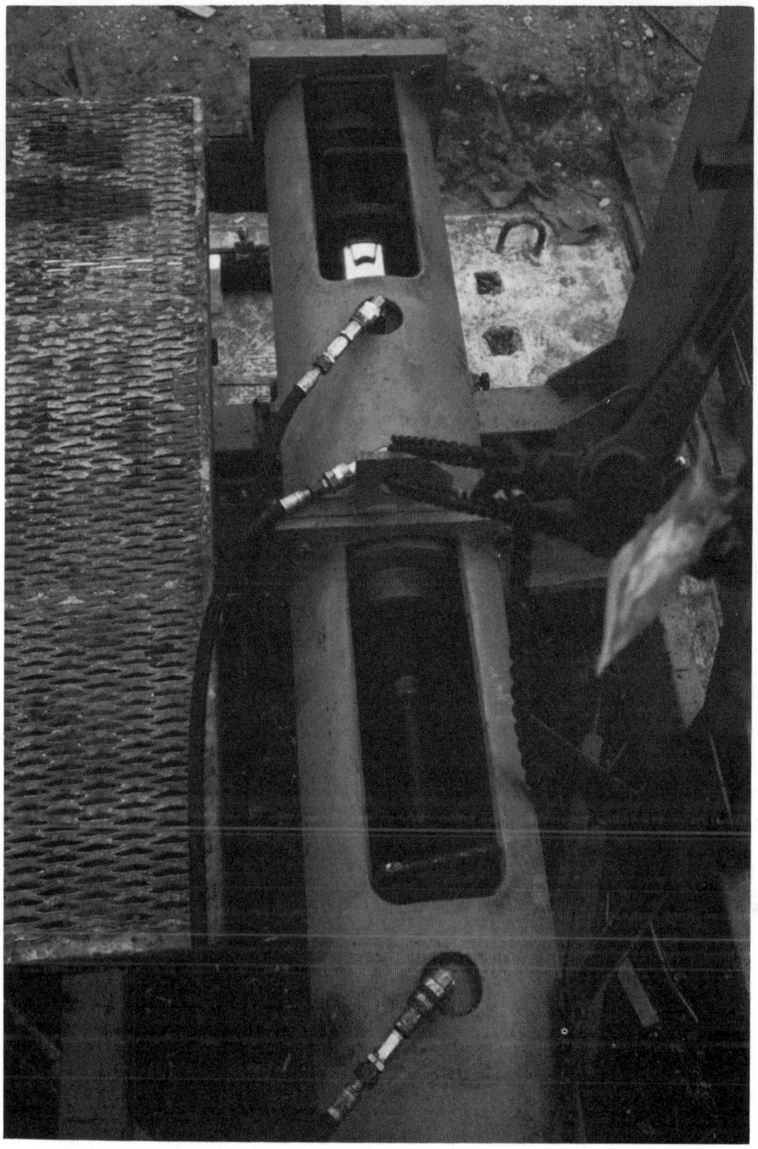

Figure 4.3 - The use of coupled jacks avoids bar relieving: when one jack is reentering, the second one is pulling (author).

weight in movement is the highest and, with it, its sliding vectorial component along the launch direction – that is, the force to balance.

Therefore, friction anchoring has been improved by using horizontal jacks acting on both the external sides of the bottom slab, so that the transferred force does not depend on external constraint reactions.

In case of steep slopes, it can be convenient to launch downgrade maintaining the deck braked, as the braking force is lower than a tow force required by uphill launch, friction working in favor of equilibrium. In reality, however, the possibility

Figure 4.4 - Long stroke bar jack (reproduced with permission, Dywidag Systems International).

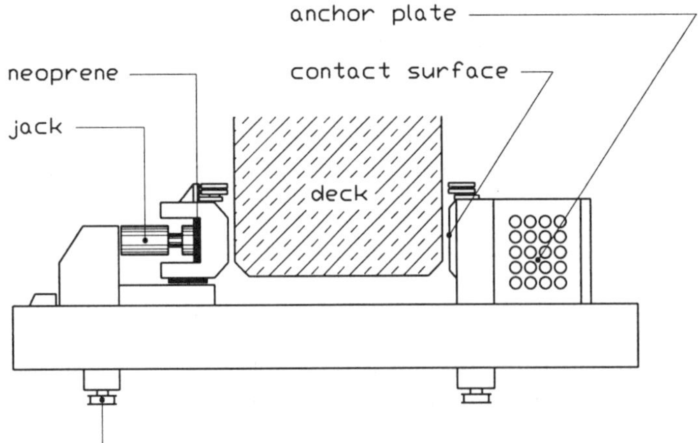

Figure 4.5 - Friction transfer of the thrust force by means of local hydraulic compression.

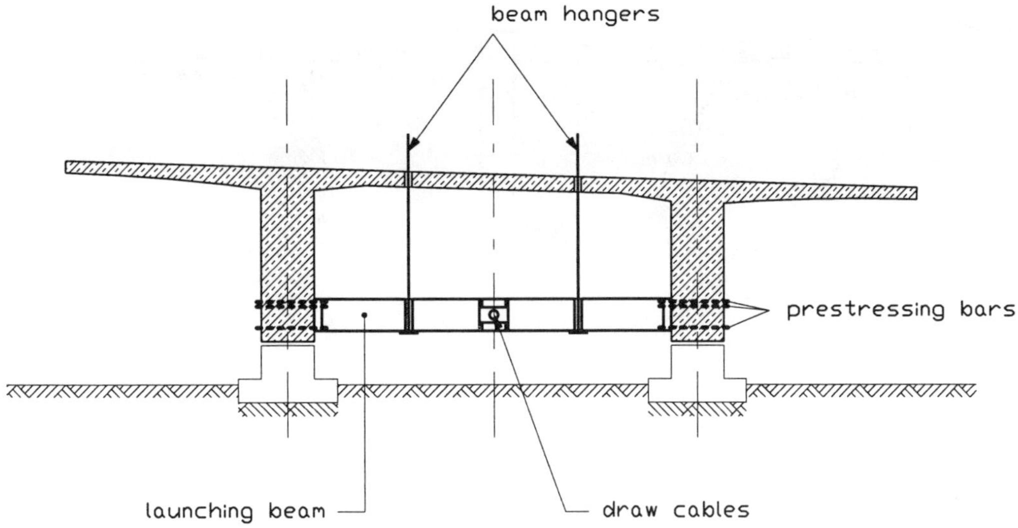

Figure 4.6 - Friction transfer of the thrust force by means of high strength bars.

Figure 4.7 - Back anchor of the launch strand jacks (author).

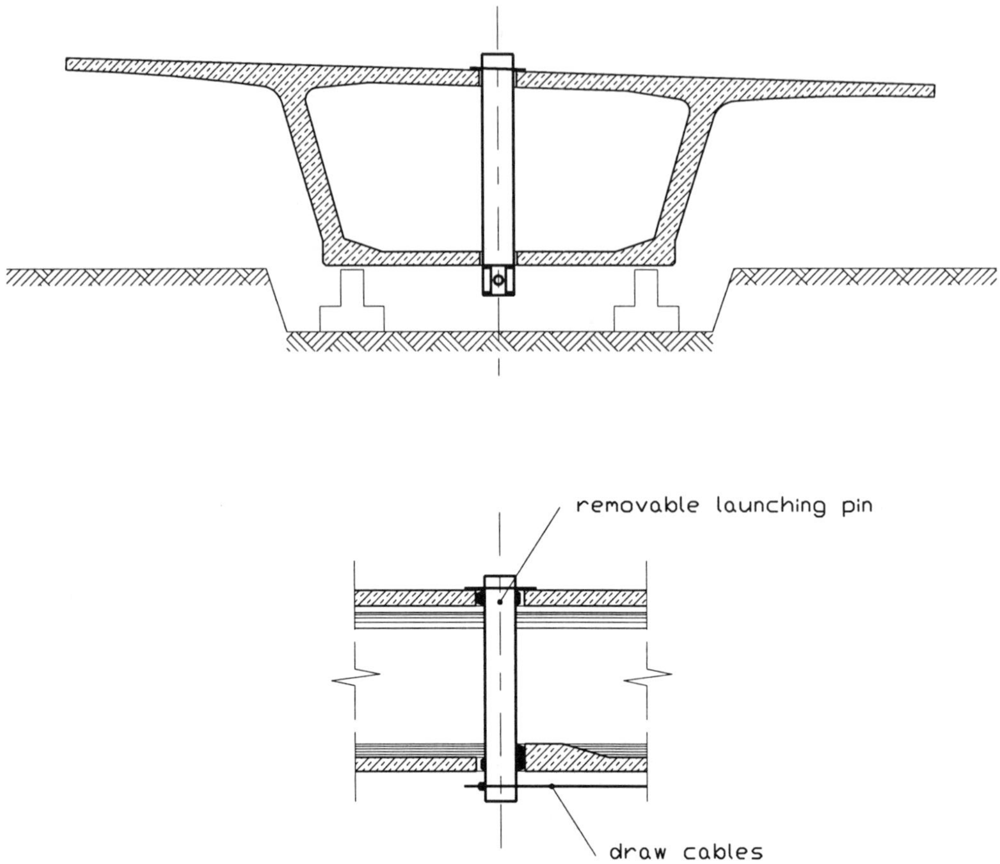

Figure 4.8 - Launching pin.

of pulling the deck uphill should be taken into account when sizing braking tendons, since it facilitates both the release of the anchorages the deck is hooked to during the jack re-enter or segment construction and the correction of an accidental excessive advance.

In case of decks with curved plan layout, draw systems present many limits, as the tow force is applied along the chords subtended by the jacks and the tow tendon anchorages and not along the local tangent at the deck layout. This produces varying angular discontinuities in the strands both at their entry in the jacks and in the opposite anchorages, which must be reduced by deviation guides (figure 4.10). Since, in this case, the strands cannot slide inside deviators (friction would be excessive), it is necessary to place launching jacks on the back movable anchorage, and tendon deviators must be progressively disassembled in order to permit the jack passing. Even so, the tow force applied along the subsequent chords tends to produce drift phenomena in the deck, which require the presence of many transverse guides in the formwork area. In any case, the bars or the strands cannot be placed

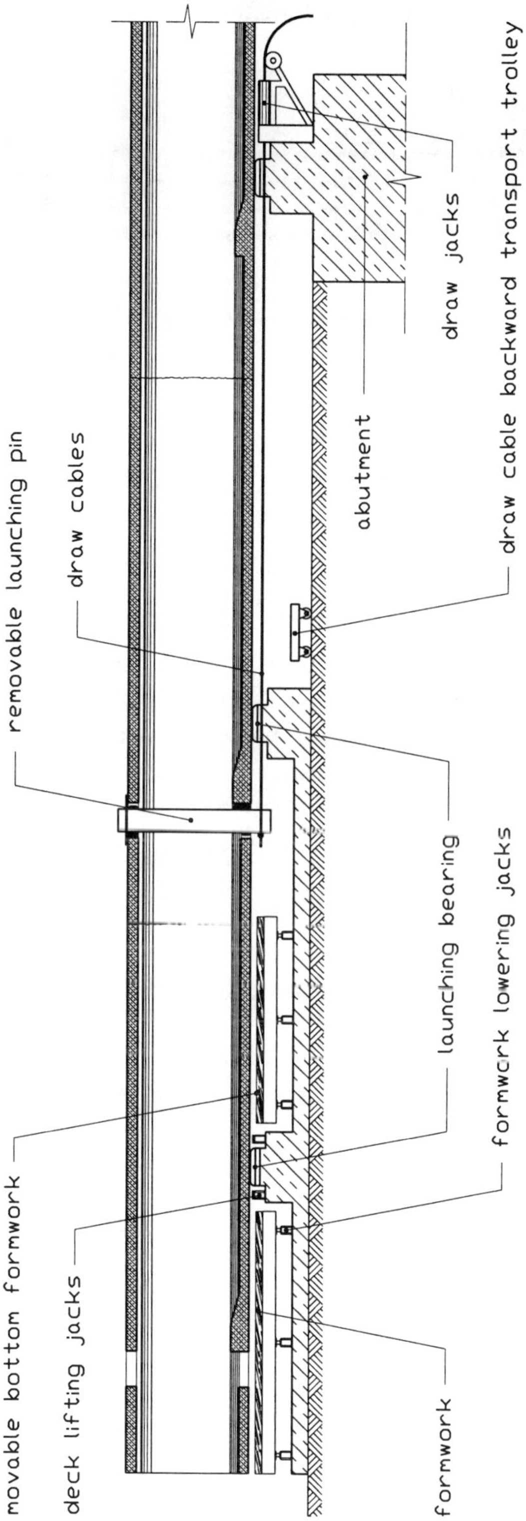

Figure 4.9 - Strand relocation trolley. It shall be noticed that fixed extraction bearings can be used also with a movable bottom formwork.

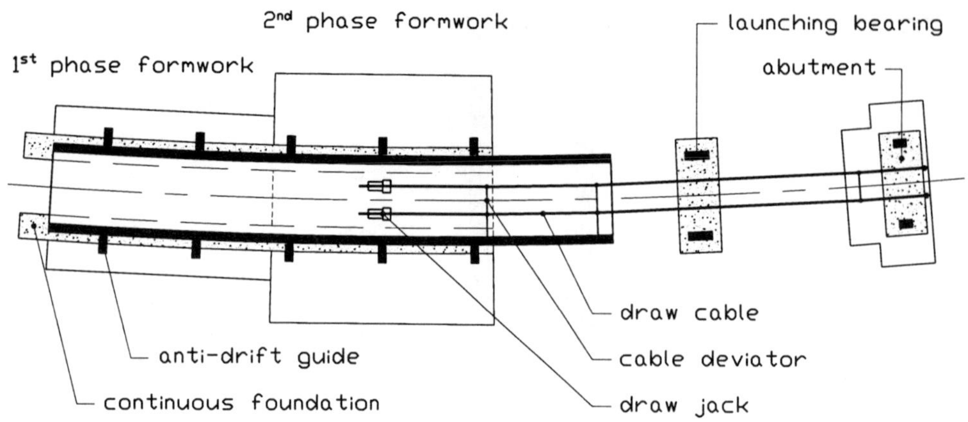

Figure 4.10 - Launch cable deviators and guiding devices are necessary in the case of decks with a curved plan layout.

along the deck (this layout is the less expensive because of the good accessibility) as they touch the deck along the external side of the curve.

To conclude, in spite of the attentions required by particular cases (steep grade, curved alignment) and a certain operative slowness, the tow systems remain the less expensive and most efficient solution in case of short decks built both by incremental launch and by monolithic launch (symmetrical approach and rotation), since they require simple equipment and are easy to assemble and transfer from one bank to the other.

Self-Clamping Back Systems

An alternative to draw systems, for short decks, is the use of back pistons acting directly on the deck back end to transfer the thrust force.

In this case, it is necessary to anchor these devices at many locations along the formwork, and the presence of the longitudinal extraction rails allows resisting the thrust reaction by means of friction mobilized by hydraulic compression (figure 4.11), or directly by means of a continuous mechanical contrast (figure 4.12). The utilization of these devices is generally limited to short bridges, as the high thrust forces required for long decks create problems which aggravate the intrinsic defects of this type of equipment:

- the deck backward movement is impossible, as in the tow systems. Therefore, for instance, the recovery of the elastic deflection of a pier caused by the seizing of a neo-flon plate is an extremely delicate operation, and the wrong insertion of a plate requires the hoisting of the deck. Both these operations increase the duration of the launch and immobilize labor on the piers;
- the presence of back pistons slows down the formwork set up and interferes with the distribution of the joint reinforcement and the prestressing tendons, as the

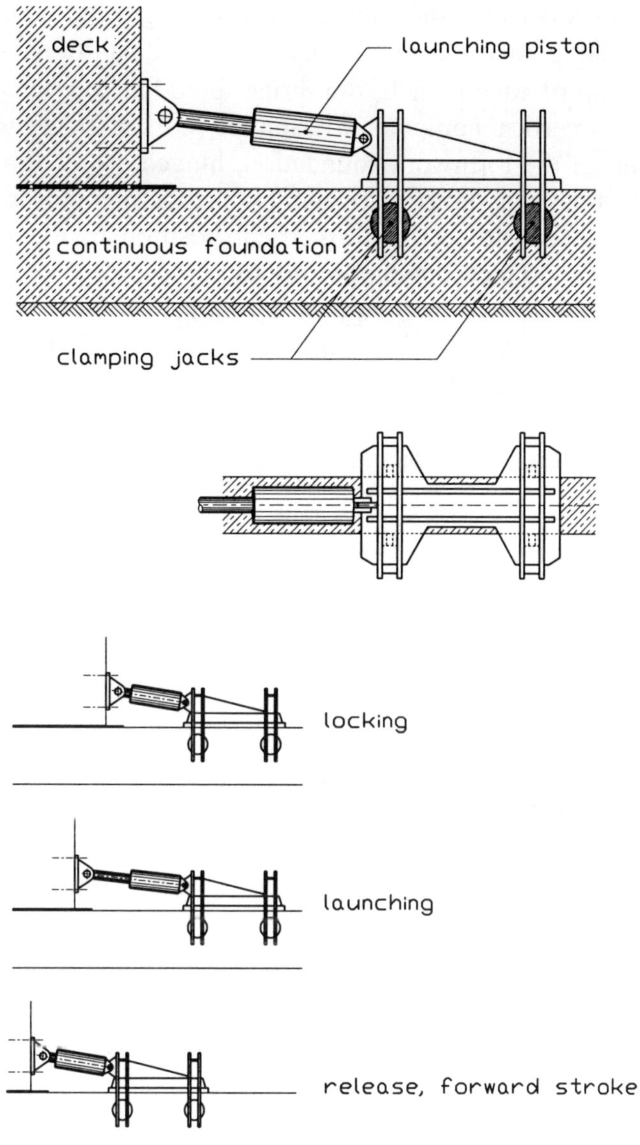

Figure 4.11 - Self-clamping back piston and operational sequence.

necessity to limit the local compression in young concrete requires large and complex contact plates;

- because of the piston slope, the thrust force has a vertical and a horizontal component, and the flexural stresses it produces after only a few days of curing and with partial prestressing dictates launching the whole cross section; that is, the use of double formwork is discouraged;
- the formwork front head is generally placed near the abutment, in order to transfer the thrust force directly to its foundation. This causes large rotations of

the joint sections between the subsequent deck segments which often require the insertion of a temporary pier in the bank span, with additional costs;

- at the beginning of each launch, the thrust force has to overcome the high static friction, and its vertical component is applied to the extraction rails very far from the abutment. In the formwork foundation, hinged at the abutment, the moment produced by this long lever arm and by the segment dead load can produce rotations which jeopardize the altimetrical position of the whole casting yard, unless expensive indirect foundations are used.

On the whole, the investment required by these schemes, certainly functional in short or medium bridges, increases quickly with the deck length to soon become uneconomical. In order to reduce this gap, in long bridges the back thrust can be combined with strand drawing sized for kinetic friction, so that the back pistons cooperate only in overcoming the static friction at the launch beginning.

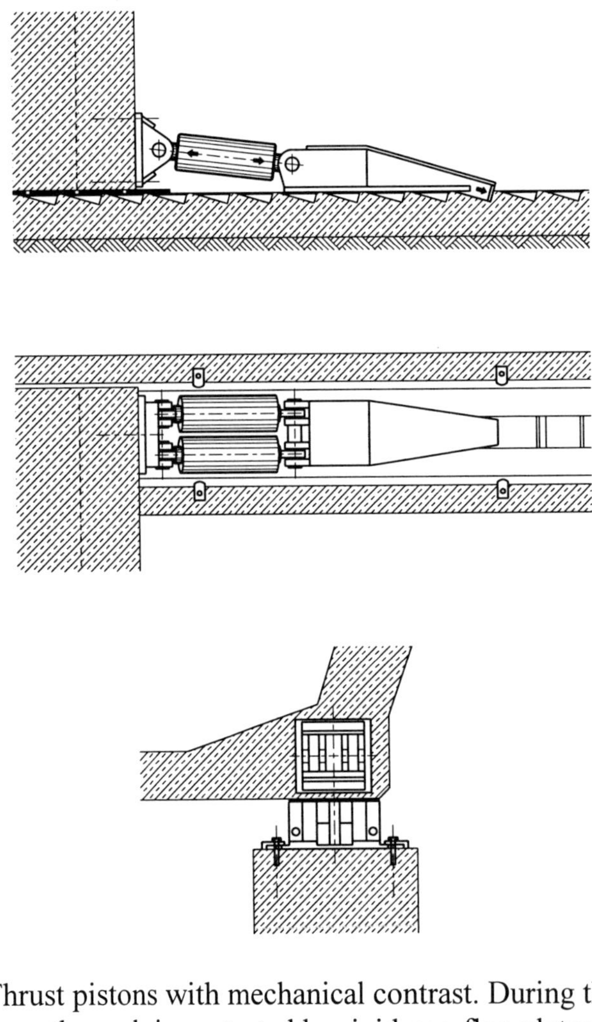

Figure 4.12 - Thrust pistons with mechanical contrast. During the segment casting, the rack is protected by rigid neo-flon plates.

FRICTION LAUNCH OF LONG BRIDGES

In the longest bridges, thrust force is transferred by friction using one or more pairs of "launchers" placed under the webs of the deck.

In the simplest version, a launcher is composed of a vertical hydraulic jack pushed along a lubricated surface by a horizontal piston acting against a foundation block which supports the deck in the pauses between two subsequent launches. In the most advanced version, a launcher is a monolithic device composed of a sledge containing the vertical jacks which slides between guides along a steel-Teflon contact base under the thrust of one or two pistons pivoted to a back steel block.

In both cases, launch develops according to the cycle of figure 4.13:

- using the vertical jacks, the deck is hoisted a few millimeters from a temporary support placed near the launcher, and the support reaction of the continuous beam is transferred from the temporary support to the movable sledge;
- using the thrust pistons, the sledge advances by sliding on a low friction surface. The transfer of the thrust force to the deck occurs by friction, by means of the deck support reaction on the sledge and the high friction coefficient between the concrete surface of the deck and knurled steel plates placed above the vertical jacks;
- once the piston limit stop is reached, after a stroke which varies from $0.25\ m$ to $1.00\ m$, the deck is lowered onto the temporary support and the sledge, unloaded, returns idle to the original position to start this cycle again.

The high loads suggest placement of launchers directly on the foundation block of the abutment, which therefore can be finished only when the deck has reached its final position and the launchers have been removed.

For the deck to be trailed by the launchers it is indispensable that the ratio of the thrust force, F, necessary to move the deck to the vertical hoisting force, R_V, be lower than the friction coefficient between the knurled plates of the launchers and the deck surface, which is usually about 0.60. In the opposite case, the sledges would slip under the deck damaging its surface, but without moving it:

$$\frac{F}{R_V} \leq 0.60 \qquad (4.1)$$

As the vertical force, R_V, is the continuous beam support reaction at the launchers, it depends on the static scheme of the deck – that is, on the position of the launchers with respect to the adjacent supports. Therefore, in very long bridges, the high thrust force requires a considerable distance between the abutment which supports the launchers and the last front support of the curing area. This restrains the distribution of most of the temporary supports of the curing area, and therefore the sequence of introduction of the launch prestressing.

The calculation of R_V can be carried out by considering aligned supports, as the launchers hoist the deck only a few millimeters. If insufficient, forcing the deck upward can increase this reaction, and the related stresses will be checked for the conditions of partial curing and prestressing. Hoisting jacks are generally oversized with respect to the highest theoretical value of R_V, with a safety factor which permits hoisting the deck even when support misplacements or thermal gradients increase R_V, as otherwise the launch could not proceed.

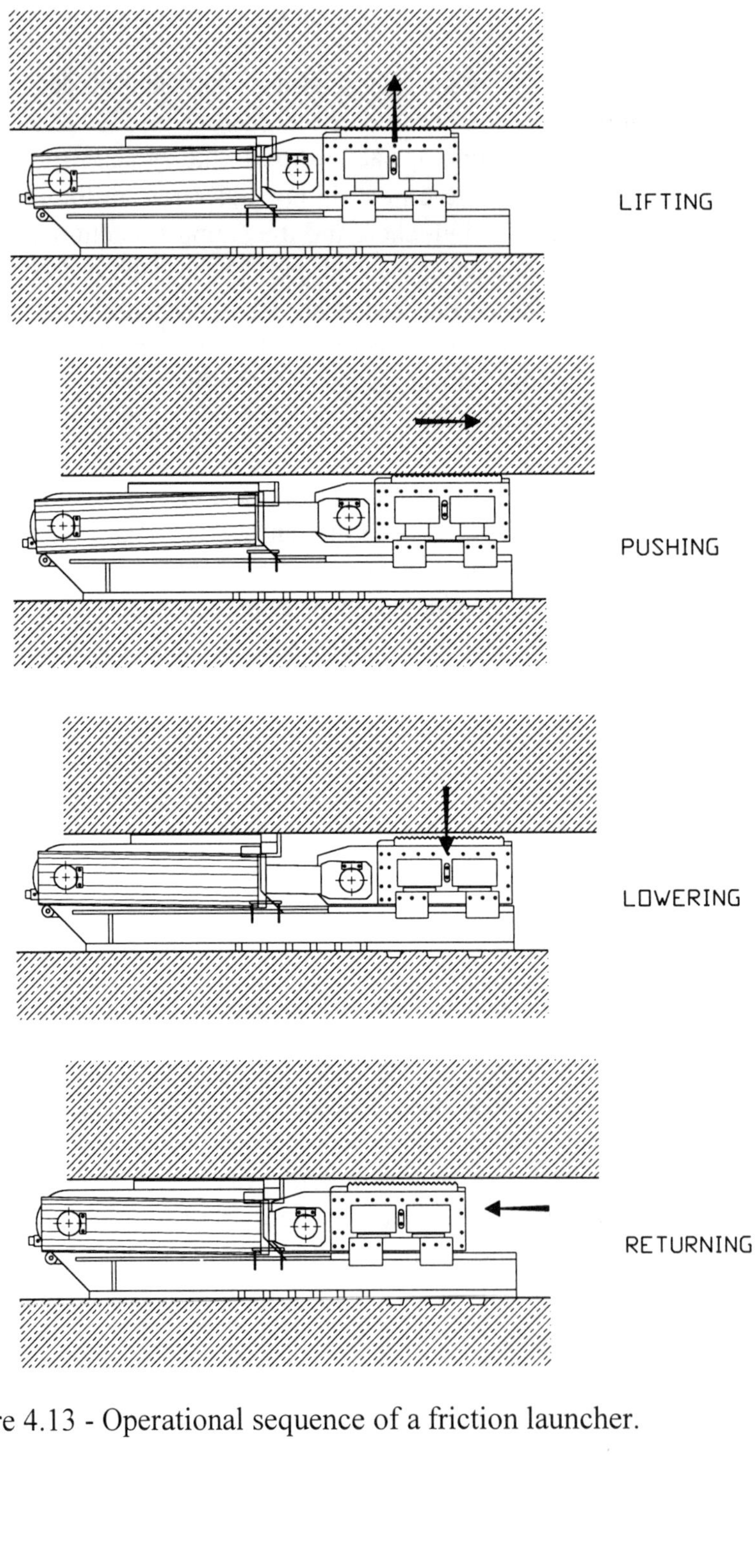

Figure 4.13 - Operational sequence of a friction launcher.

Figure 4.14 - At the beginning of launch, the deck is trailed onto the friction launchers by means of strands (author).

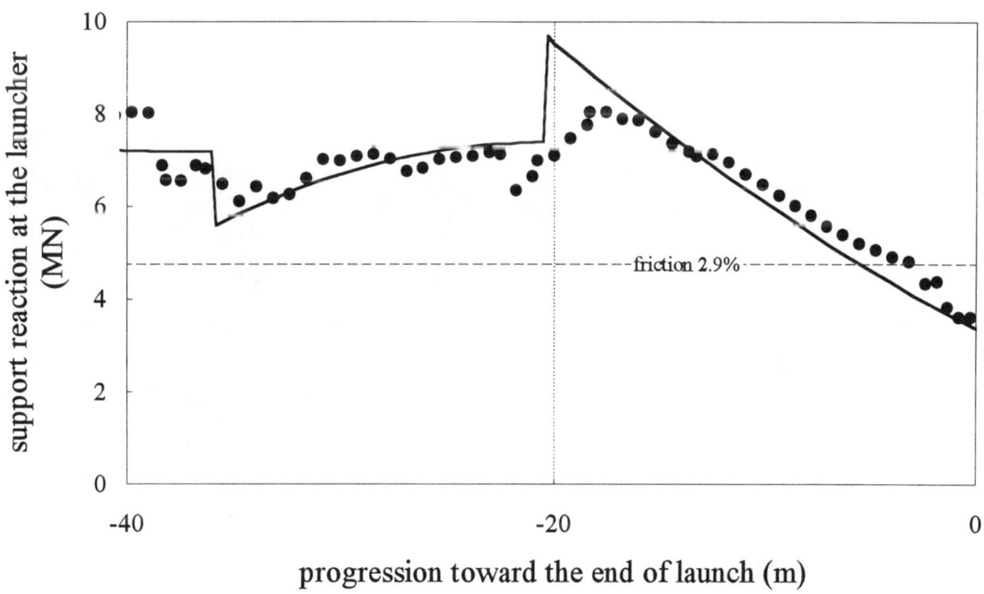

Figure 4.15 - Serio River Bridge. Reduction of the support reaction at the launcher in the final launch stages. Continuous line is theoretical, black dots are recorded data. The reaction needed to move the deck with the actual overall friction coefficient of 2.9% is dotted. Before using the rear-nose (5.5 m before the end of launch), the support reaction was increased by forcing the deck upwards.

In turn, thrust pistons are sized for the highest launch force considering the friction of the formwork, the launching bearings, and the launchers themselves, in addition to the longitudinal slope of the average launching surface.

The use of friction devices involves specific problems in the initial and final phases of launch:

- at the beginning of the launch, the curing area separates the formwork from the launchers, and the need for a perfect connection between the launching nose and the deck is generally solved by match-casting; that is, by assembling the launching nose against the formwork front head. This requires that the launching nose and the front section of the deck be trailed onto the launchers until a support reaction sufficient to continue launch by friction is obtained. For this purpose, sheaves of strands anchored to the launching nose and to the launchers permit use of the latest thrust pistons (figure 4.14);
- in the final phases of the deck launch, R_V decreases until the reaction of the continuous beam end support is reached, generally insufficient to satisfy the highest thrust force demanded by friction (figure 4.15). This gap is solved by means of a second couple of launchers placed on a more advanced support (generally a short internal pier) or of a steel rear-nose integral with the deck and connected to the launchers by sheaves of strands (figure 4.16), resulting again in a tow scheme[171]. In this case, the draw tendons are sized only for a share of the final thrust force, as the continuous beam has a support reaction R_V to exploit by

Figure 4.16 - Serio River Bridge. In the final launch stages, a steel rear-nose permitted builders to apply an additional launch force by means of strands pulled by the friction launcher itself (author).

Figure 4.17 - The front drawing is a competitive alternative to the use of a rear-nose (author).

friction, and when the back end of the deck approaches the abutment the launch is no longer subject to the extraction friction from the formwork and the friction on the temporary bearings of the curing area, which together represent a considerable share of the total friction;

- as an alternative to the rear-nose, front drawing by means of strands and ordinary prestressing jacks permits solving the problem in a less expensive way (figure 4.17), although with some interference with the launching nose.

The utilization of friction launchers (single, figure 4.18; coupled, figure 4.19; or distributed in more couples, figure 4.20; controlled by a numeric-control hydraulic plant, figure 4.21) results in enormous synchronized thrust forces, and therefore an ability to move huge masses, such as the ones of the longest railway decks.

Compared with tow and back thrust systems, the utilization of friction launchers presents many advantages:

- in tow systems, the breaking of any mechanical component leaves the deck free on low friction supports, and this forces adoption of high safety factors which involve oversizing of equipment and operative slowness; that is, higher investment and labor costs. On the contrary, friction launchers guarantee an absolute intrinsic safety, as the worst consequence of mechanical or hydraulic breaking would be the launch stop and the deck descent onto the temporary supports near the launchers, which act as friction constraints. Equipment can be

Figure 4.18 - In the two-cell Serio River Bridge, an axial friction launcher moved a 727 m long bridge (author).

Figure 4.19 - Single cell box girders need two synchronized friction launchers placed under the webs (in this initial launch stage, they are under the launching nose) (author).

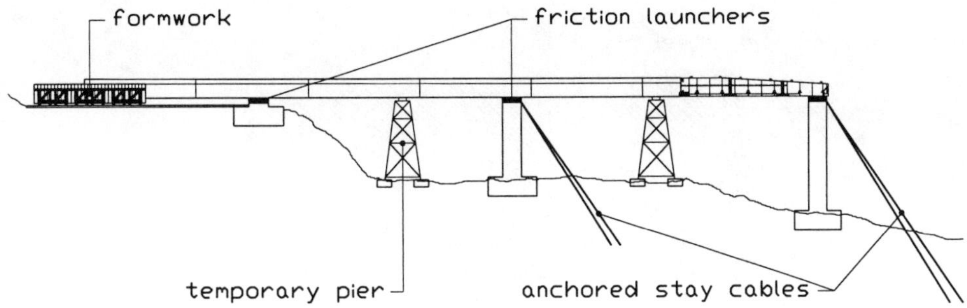

Figure 4.20 - Several pairs of friction launchers distributed along the deck allow both to launch the heaviest bridges with small launchers, and to launch by friction when the formwork is so close to the abutment as to produce an insufficient support reaction.

Figure 4.21 - Numeric-control units allow one to record all the operational parameters (launch and lifting forces, deck raising, progression of launch) as well as to set limit values, and to automate the launch cycle (author).

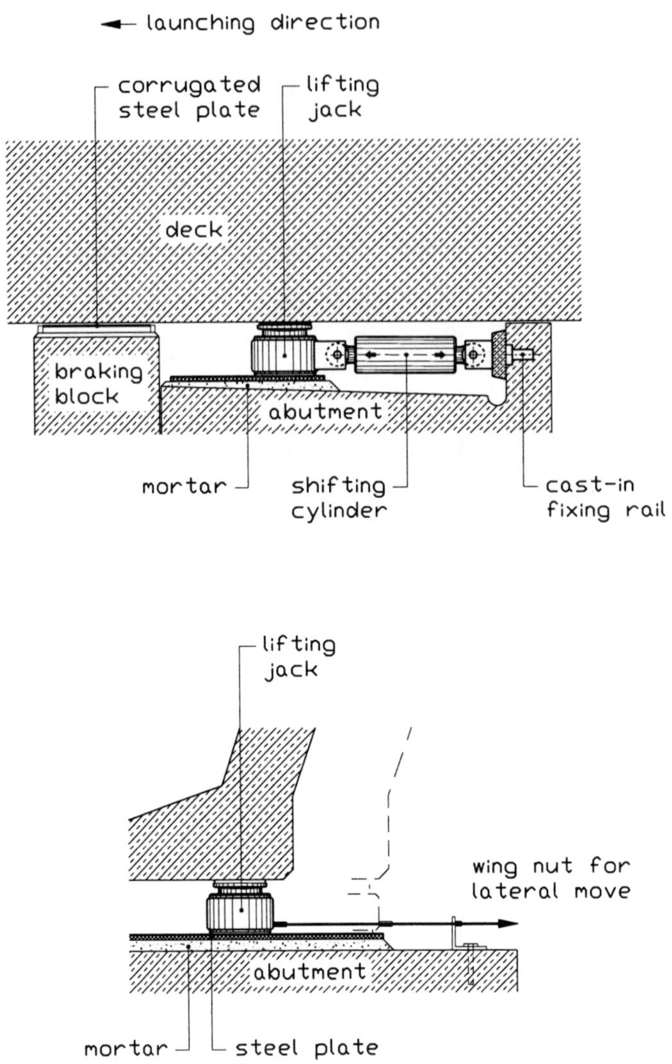

Figure 4.22 - Shiftable launchers allow one to launch variable width decks by friction.

not only sized for the actual launch loads, but can even be overloaded without excessive worry in case of unforeseen events;

- the electronic control of the hydraulic circuit permits synchronization of the action of launchers and recording in real time of the forces produced. By monitoring R_V, it is possible to verify the local accuracy of the deck static scheme. By relating R_V to the deck raising, it is possible to estimate indirectly the elastic modulus of the concrete[171]. The possibility of setting a limit value for the thrust force permits avoidance of pier overloads produced, for instance, by the upside-down insertion of neo-flon plates. This mistake seems ridiculous, but it is rather frequent and potentially very dangerous, as the friction coefficient increases dramatically when passing from a greased steel-Teflon contact to a steel-neoprene one;

- launch cycle automation by means of electronic limit stops facilitates even more the operations, which can be overseen by one technician only, and increases the launch speed, which can easily reach *10 m/h*;
- labor costs are definitely lower, due to the higher launch speed and the possibility of avoiding tow tendon repositioning and deck anchoring during these phases. Moreover, after launching, the formwork is immediately ready for the construction of the next segment;
- the transverse guides of the launcher sledges act, by friction, also as transverse guides of the deck, and avoid the need for further external devices and their related labor costs. As an alternative, the vertical jacks can be left free to shift transversally (figure 4.22), by entrusting external devices with the deck transverse guide as in the tow or back-thrust systems, and it is thus possible to launch by friction even varying-width decks.

In order to transfer the thrust force without stress concentration and deck surface damage, the contact plates should be articulated so as to adapt perfectly to the deck, and this can be obtained by connecting in series the jacks of a sledge and by equipping them with ball-and-socket caps.

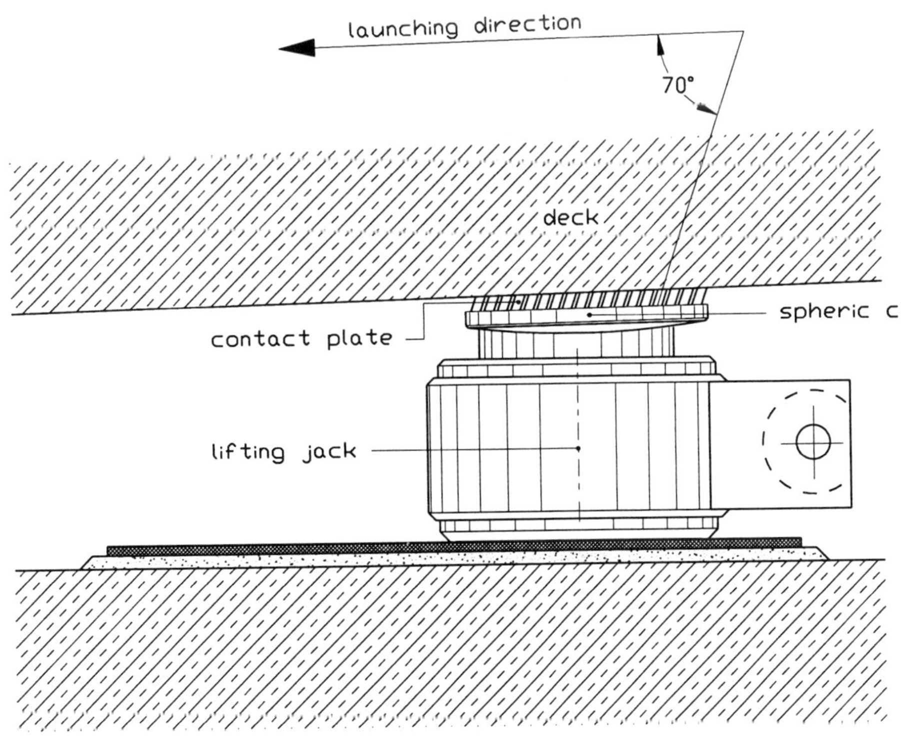

Figure 4.23 - The engraved contact plate should be disposed so as to transfer the thrust force by bending.

The contact plates should be relatively "soft" to compensate for concrete irregularities. Their surface is generally deeply engraved in the opposite direction of the launch (figure 4.23), so that the single small plates so obtained can transfer thrust by bending, while the opposite arrangement could easily break the concrete surface because of local stress concentration.

The original vertical load friction launchers have evolved into different schemes. One of the most interesting ones has been utilized in the Olifant's River Bridge in South Africa. It is composed of horizontal thrust pistons acting on a movable transverse transom placed under the deck, which clamps its bottom slab by means of horizontal jacks which create the surface compression necessary for friction transmission. The operative cycle, similar to the one of vertical load launchers, involves deck clamping, thrust, deck release, and finally transom return to the initial position. Not depending on the continuous beam support reaction, this scheme has the advantage of not restraining the curing area support positioning and not breaking down at the end of the deck launch.

LAUNCHING SADDLES

The technological evolution of incremental launching led to a clear-cut simplification of deck support and guide devices, all based on the original principle of the steel-Teflon contact.

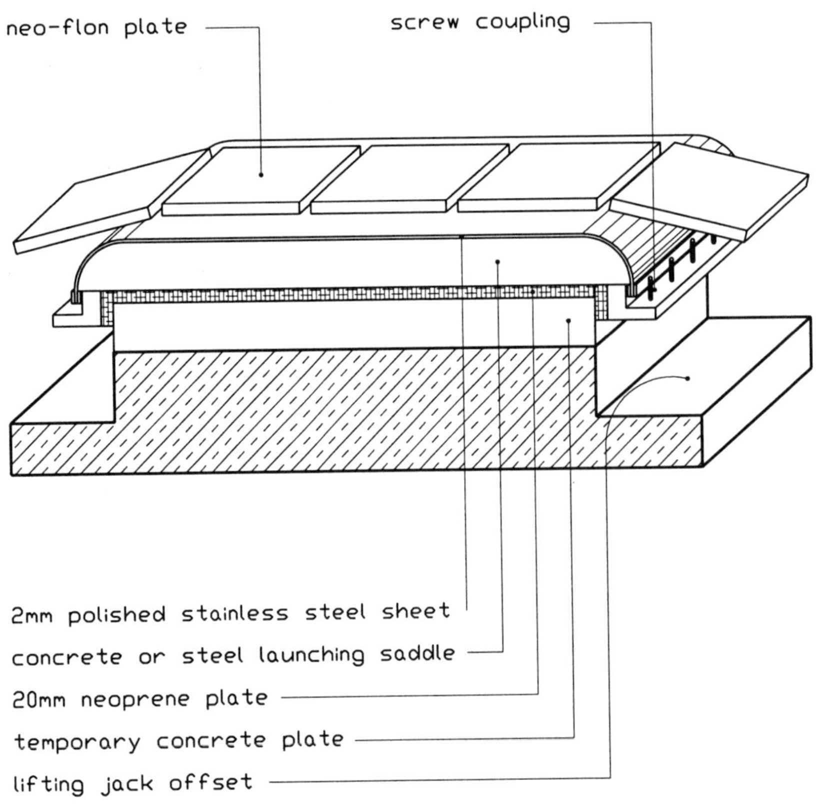

Figure 4.24 - Launching saddle.

In principle, by attaching stainless steel continuous strips to the deck and placing Teflon bearings integral to the piers under these strips, it should be possible to reduce labor costs considerably, since pier workmen are no longer necessary. This solution clashes with the necessity of guaranteeing a perfect link between the steel strips and the deck, and no application of this scheme has been found in literature.

In the very first launched bridges, sliding occurred on bearings integral to the deck and supported on stainless steel plates integral to the piers with the discontinuous sequence of figure 1.5. The length of each launch cycle was limited by the pier cap longitudinal width, and once limit stops were reached the launch had to be interrupted and the deck had to be hoisted, pier by pier, in order to reposition the bearings at the beginning of their stroke. The time and costs involved in these operations soon suggested adoption, instead of mobile bearings, of a specific type of launching bearings composed of neoprene-Teflon (neo-flon) plates inserted between the deck and fixed stainless steel saddles with a convex surface which facilitates plate insertion and automatic expulsion in the launch direction (figure 4.24).

Neo-flon plates are composed of neoprene sheets alternating with steel sheets and vulcanized to a lower Teflon plate. Reinforced neoprene improves the uniformity of the contact pressure and absorbs deck rotations and surface irregularities, while Teflon reduces sliding friction between the deck and launching saddles (figure 4.25).

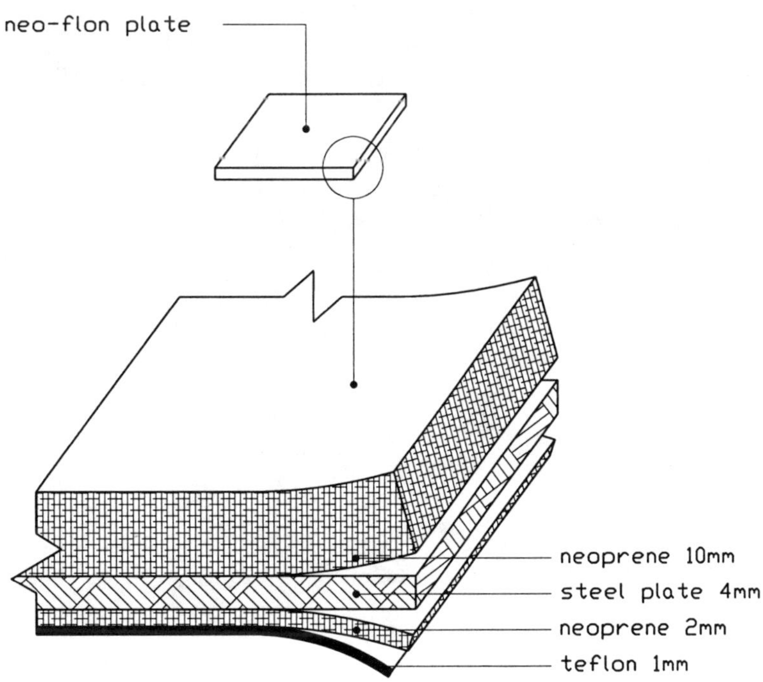

Figure 4.25 - Neo-flon plate.

The possibility of introducing neo-flon plates without hoisting the deck makes launch continuous: it is interrupted only in order to reposition the tow or thrust devices and the total speed is more than double, since in discontinuous launching more than half of the launch duration was required for lifting and lowering the deck. Many other disadvantages are also avoided, such as load eccentricity on the piers produced by longitudinal displacement of the deck support forces and which, compounded with horizontal friction forces, often required pier reinforcing or cable-staying.

Launching saddles are generally reinforced concrete precast plates or steel plates about *50 mm* thick, in both cases covered by a *2 mm* stainless steel sheet attached with screw couplers. They are embedded in a high-strength rheoplastic mortar with small positional tolerances (longitudinal, transverse, vertical, and of slope) by means of adjustable temporary supports. Concrete saddles tend to have a less rigid action than the steel ones, especially if built with a weaker material than the deck in order to assume a role of "fuse" against excessive stress concentrations and, if needed, to break first. Other authors[108] suggest use of high strength concrete (about *60 MPa*).

In some cases (especially with steel saddles, as the concrete ones tend to break) the launching bearings are placed on neoprene plates which improve the stress distribution in case of irregularities, such as rotations or surface defects of the deck.

Figure 4.26 - Launching nose is approaching a pair of hydraulic bearings (author).

Sometimes launching bearings are supported on jacks fed by a common hydraulic circuit which can be adjusted to control the torsional couple produced by the support and to limit or cancel the effects of vertical misplacement of the bearings (figure 4.26).This scheme is often adopted on temporary piers, since the control of the vertical force transferred by the two launching saddles permits limitation of transverse bending in the pier.

Finally, decks with varying curvature or varying width require a transverse shift of the bearings during launch (figure 2.2), and, if needed, the pier cap can be widened by means of temporary side piers (figure 4.27).

Figure 4.27 - In the Schnaittagh Bridge, the Cornu spiral front segment of the deck required side temporary piers which widened the caps transversely. The opposite deck, launched downhill, has a constant curvature plan layout (author).

In this case, hydraulic bearings may be placed in sledges which also contain the service hoisting jacks (figure 4.28).

The possibility of inserting hoisting jacks under the deck webs is indispensable both during launch, in order to substitute damaged saddles or incorrectly inserted neo-flon plates, and at the end of launch, in order to remove the launching saddles and to insert the final bearings. In both cases, the jack stroke is limited to few tens of millimeters.

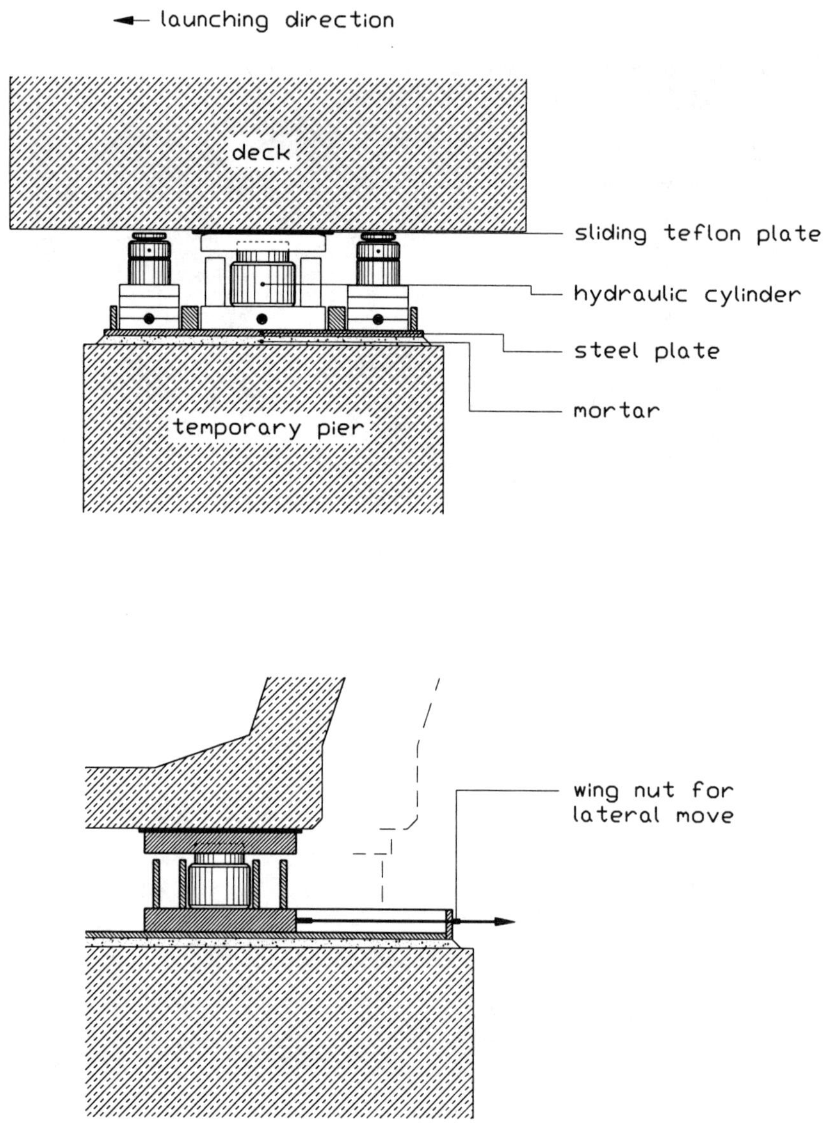

Figure 4.28 - Transversely movable hydraulic launching bearing.

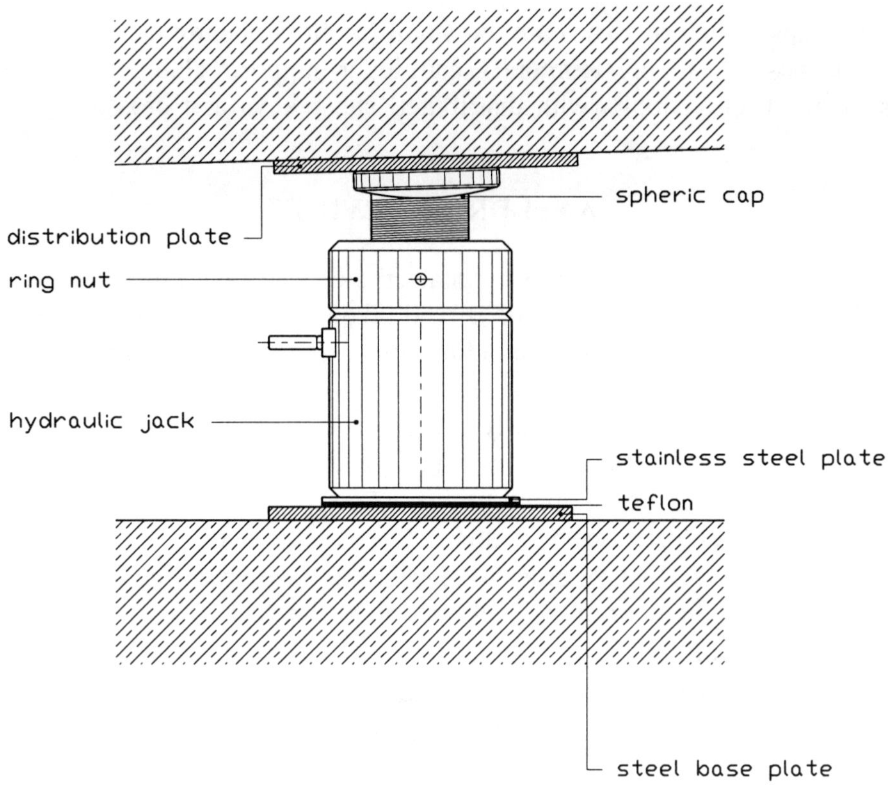

Figure 4.29 - Hoisting equipment placed on a low friction support.

In the temporary piers, on the contrary, once launch is finished and the parabolic prestressing tendons are stressed it is necessary to eliminate the residual vertical force transferred by the deck before beginning pier removal. In this case, the jacks must guarantee longer synchronized vertical strokes.

As in friction launchers, jacks have ball-and-socket caps which permit the distribution plates to adapt correctly to the deck, often irregular and/or sloped (figure 4.29). In addition, jacks should be provided with safety ring nuts which limit the hydraulic support to the phases of lifting and lowering of the deck.

It is convenient to place jacks on stainless steel plates over neo-flon plates, in order not to concentrate on the jacks the horizontal force deriving from the elastic deflection accumulated by the pier because of the launch sliding friction and the longitudinal slope of the launch surface. As the steel-Teflon contact so obtained has generally a lower friction coefficient than the neo-flon plates, often dirty, the pier tends to recover a portion of this elastic deflection during hoisting. It is advisable to take into account these movements when arranging the hoisting equipment on the pier cap, to avoid risks of fall of the external jacks.

In many cases, the use of hoisting batteries composed of many small jacks is more convenient than the use of few strong units, as it permits easier manual jack shifting under the deck and a less-restrained design of the hoisting scheme even in small pier caps. When pier caps are too small to allow placing jacks under the deck webs, it is possible to use temporary cantilever supports clamped to the pier by means of dry friction joints with distributed shear keys and crossing prestressing bars.

ANTI-DRIFT DEVICES

Transverse guides, necessary to maintain the correct plan alignment of the deck during launch, especially in a case of curved layout, are placed beside the launching saddles. They represent a fixed or adjustable horizontal constraint (figure 4.30), which utilizes steel rollers or a steel-Teflon contact similar to the one of the launching saddles, and are generally placed outside the webs, or inside special offsets in the case of cross sections devoid of sharp corners (figure 4.31).

Though many piers are equipped with transverse guides, the correction of deck alignment occurs in most cases by acting in three points only:

- near the formwork head, in order to correct the position of the construction joint with the next segment, or of the back section of the deck;
- on the launch abutment, especially in tow systems with bars or strands;
- on the most advanced pier reached by the deck, in order to correct the launching nose position at its contact with the next pier.

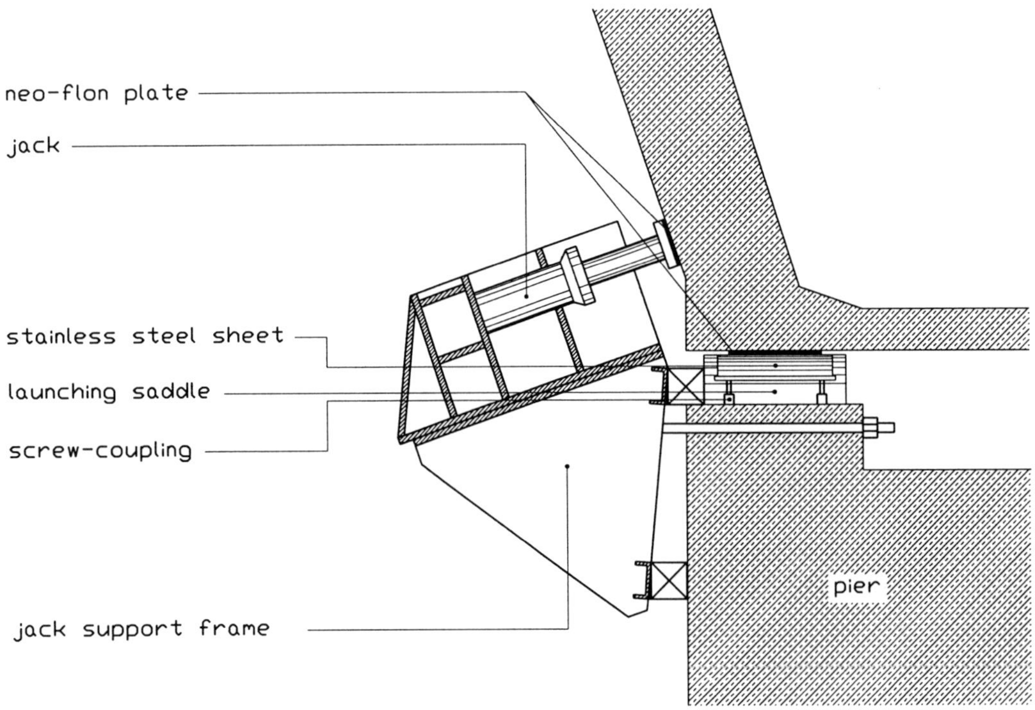

Figure 4.30 - Adjustable transverse guide.

Figure 4.31 - Axial guide (author).

While the launching saddle design load results from the envelope of the continuous beam support reactions and is relatively simple to compute, the transverse guide load is empirical, being based on the analysis of the force necessary to move the deck by acting at the same time at the appropriate points during each launch stage. Moreover, in case of a curved alignment, each guide is subject to a radial force produced by the vectorial component of the thrust force. This system of forces (of elastic nature and therefore proportional to the stiffness of each pier) constitutes an indeterminate system which is solved by considering the continuous beam on elastic supports, and the forces so obtained are added to the ones estimated for the deck alignment correction.

Finally, as the launch operations are delicate and require many work stations distant from each other, these will be connected by radio or telephone lines. As most of the thrust devices produce irreversible movements of the deck, launch operations must be carried out rigorously and in the foreseen order, and it is advisable to equip each pier cap with emergency electric circuit breakers which act directly on the thrust device supply mains, in order to permit each worker to stop launch. These circuit breakers can be integrated with pilot lights which inform pier workers on the current operations (emergency stop, rod reentry, soon to start launch, launch) and by acoustic signals of launch start.

THRUST FORCE

The force necessary to move the deck depends on the average longitudinal slope of the launching surface and on the total friction forces which oppose launch.

While the effect of the longitudinal slope is easy to evaluate and unquestionable, the total friction is influenced by many factors. By assuming the deck is horizontal, on a generic support the minimum tangential force, F_{min}, necessary to produce sliding depends on the orthogonal force, N, transferred by the deck to the launching saddle, and on the friction coefficient, C_f, between the neo-flon plates and the stainless steel surface of the saddle:

$$F_{min} = C_f N \qquad (4.2)$$

Friction can be studied by means of the friction angle:

$$\phi = a\,tan \frac{F_{min}}{N} \qquad (4.3)$$

that is, the angle that the vectorial resultant of the forces N and F_{min} has to exceed to produce sliding. It depends on many factors, since the behavior of Teflon in contact with a polished stainless steel surface is rather variable:

- the average contact pressure: if pressure increases, the friction coefficient reduces, halving from *2%* to *1%* passing from *10 MPa* to *15 MPa*;
- temperature: increase in temperature produces effects similar to increase in load;
- lubrication and surface cleanliness: an imperfect cleaning or lubrication of neo-flon plates and/or of the launching saddles, or their excessive wear, can increase friction substantially.

Beyond the effects of the single contact surfaces, it is then practice to consider a "global" friction coefficient to apply directly to the self-weight of the deck section to be launched to obtain the force to be produced by thrust devices. The development of the overall friction force, F , depends on many factors:

- the force F_f opposed by the formwork during the segment extraction. It is rather high, compared with the segment self-weight, as this support has a low contact pressure and is dirtied by the grout of concrete pouring, which increases both the sliding friction and the neo-flon plate wear. The friction coefficient can be about *10%* in case of a continuous stainless-steel-Teflon contact, and higher in case of ordinary steel or of a lubricated steel-steel contact;
- the force F_i^0 opposed by each launching saddle at the time of breakaway, which soon reduces to the kinetic value. The friction coefficient C_f for the support reaction is actually between *2%* and *3%*, but when sizing launchers and piers it is prudent to use at least a static value of *5%* and a kinetic one of *3%*, and in case of tall and slim piers is prudent to assume a static value of about *7%*;

– a coefficient, $\psi_i \leq 1$, which reduces the breakaway force of the back saddles because of the elastic behavior of the deck, which, compressed by the thrust force, staggers longitudinally the saddle breakaway. This reduction is influential only in rather long bridges.

Therefore, the thrust force necessary to detach the deck does not increase linearly with the number n of launching saddles because of the formwork friction, whose influence reduces as launch proceeds, and of the lack of simultaneity of breakaway, and it has a trend of the type:

$$F = F_f + \sum_{1}^{n} \psi_i F_i^0 \qquad (4.4)$$

Figure 4.32 shows the thrust forces recorded in the Serio River Bridge[171].

Once the highest theoretical thrust force is defined, when sizing thrust devices it is advisable to introduce a safety factor which covers possible unforeseen events, such as insufficient lubrication of neo-flon plates, wear of contact surfaces, Teflon's tendency to seize to launching saddles during the intervals between two subsequent launches, and so on.

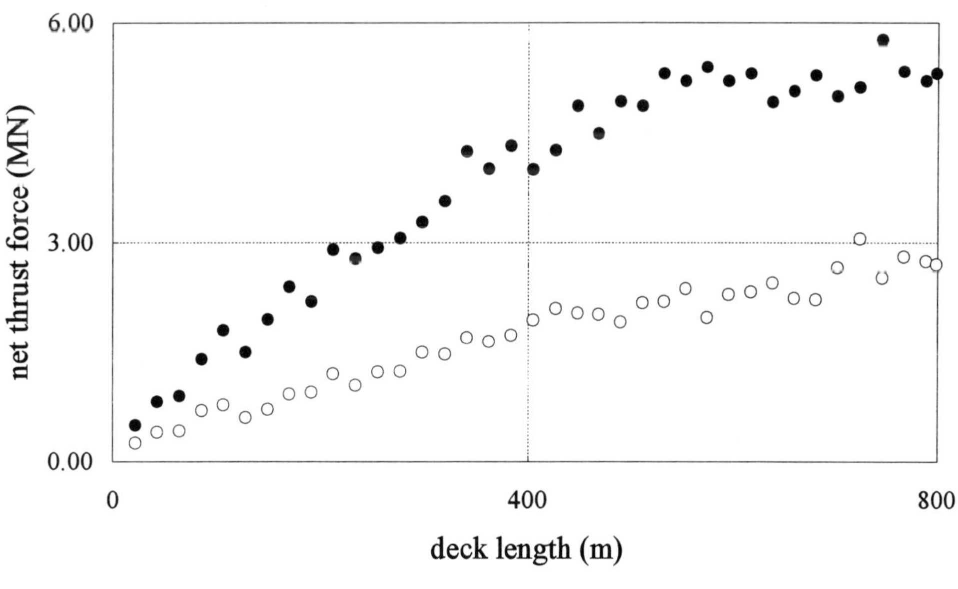

Figure 4.32 - Thrust forces for the Serio River Bridge. Different from the kinetic friction force (white dots), the breakaway force (black dots) does not increase linearly during launch due to the lower influence of the formwork friction, and to the lack of simultaneity of breakaway.

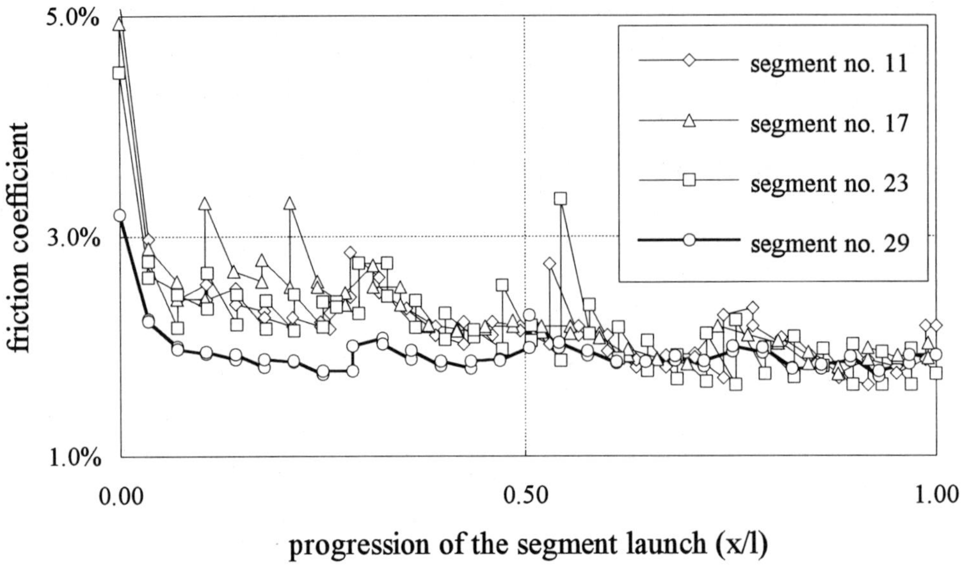

Figure 4.33 - Thrust force evolution during some segment launches (Serio River Bridge).

Monitoring of the thrust force during launch[171] (figure 4.33) always confirms the rapid passage from the breakaway condition to kinetic friction. Therefore, the thrust devices could be sized for kinetic friction, and additional back-thrust jacks could join in supplying the breakaway force.

Anyway, because of possible unforeseen events, it is general practice to determine the thrust force directly with an overall friction coefficient about *4* to *5%*, and to add to this value the longitudinal slope of launching surface.

When launching downgrade, great care must be taken in the determination of the minimum friction. In fact, once the breakaway phase is passed and a good temperature is reached for neo-flon plates, new and well lubricated materials can easily reduce the overall friction coefficient to less than *2%*, which can require braking the deck instead of thrusting it, even on grades about *1.5%*.

Finally, particular care must be taken when sizing the strand tendons necessary to launch the first deck segments, as in these phases the absence of an effective transverse guide can require keeping the formwork close to the deck in order to correct its alignment, which increases noticeably the total friction. In the launch of the first segment of the bridge of figure 4.34, the breakaway friction has reached *20%*.

Figure 4.34 - First segment extraction of the Milan-Palizzi Bridge (author).

Chapter 5

LAUNCH STRESSES AND BRIDGE DESIGN

The design of a bridge built by incremental launching develops through a sequence of operations:

- pre-sizing of the deck "current" zone: aesthetic considerations and the statistic analysis of bridges already built can help in defining the main dimensions of the cross section;
- definition of the most suitable device for launching stress reduction: launching nose, temporary piers, cable-stayed schemes, or a combination of these;
- pre-sizing of the deck front zone by means of approximate models according to the front device adopted;
- analysis of the pre-sized deck for the service load effects;
- exact analysis of launch stresses by taking into account thermal gradients, differential settlement of bearings and time-dependent phenomena;
- exact analysis of service load effects.

BACK "CURRENT" ZONE OF THE DECK

The most characteristic design issue of incremental launching is the presence of temporary stresses in the deck due to the launch support configurations. Each cross section of the deck passes cyclically in mid-span and on the piers, and therefore is subject both to the highest positive moment and to the highest negative moment, and the highest shear force. Even though under dead load only (but, however, in the presence of thermal gradients and differential settlement of bearings which complicates the situation), each cross section of the deck has to stand temporary stresses which are not only high, but also noticeably different from the final ones.

The situation evolves cyclically between the two limit conditions of figure 5.1. Considering two consecutive piers, the condition A is similar to the final position, with the support diaphragms of the deck placed above the piers, while the second position, B, is with the deck advanced by half a span and therefore supported on the sections which, on completed launch, will become mid-span.

In both positions the mid-span sections will be unlikely to turn out critical, as the shear force is low, and a bending moment much lower than the values reached under live loads permits absorption of edge tension by means of an adequate launching prestressing. In addition, box girders are morphologically better suited to resist positive moments, since the wide top slab provides a large compression area and draws the centroid upward.

As regards support sections, in position A launching stresses are only a share of the final ones and the sizing of the cross section for service loads is generally adequate. However, adequate launch prestressing and web reinforcement for the full absorption of the tensile stresses produced by shear force shall be provided, since parabolic prestressing will reduce the shear diagram only after its introduction at the completion of launch. On the contrary, the cross sections which on finished launch will be in mid-span, designed to resist positive moments and reduced shear forces, are subjected in the position B to the highest negative moment and to the highest shear force. Almost always they have to be sized for these temporary stresses, and therefore tend to be oversized compared with service requirements.

Since during launch each cross section passes over the piers, these temporary stresses impose a marked uniformity of the cross sectional properties along the deck, while with other construction techniques the mid-span cross sections are frequently lightened, especially for long spans and constant depth decks. This results in oversizing of the entire deck and in a heavier structure.

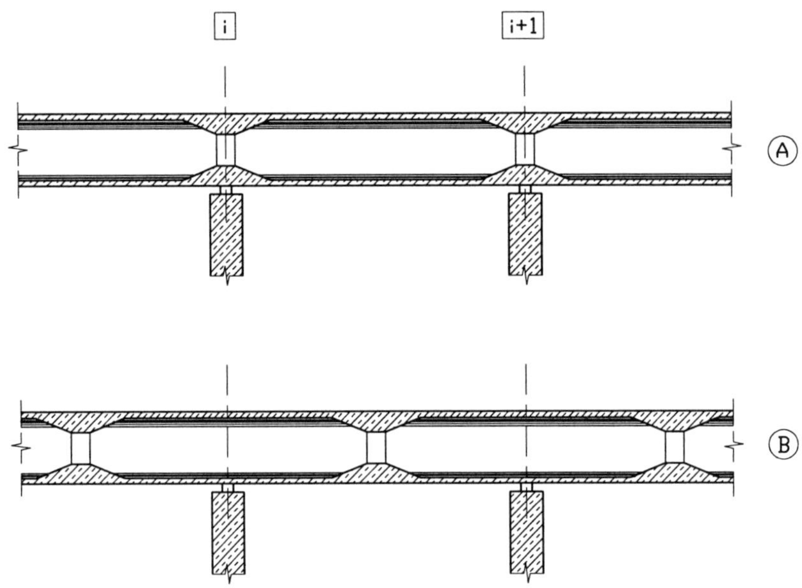

Figure 5.1 - Limit support configurations for the current zone of the deck.

Moreover, the cyclic change of sign of moment and shear does not allow the use of parabolic prestressing, and the edge tension is absorbed by a centroidal launch prestressing which results in uniform compression of all cross sections. At the completion of launching, this uniform compression is not only insufficient to absorb live load tensile stresses, and therefore needs to be integrated by more efficient parabolic tendons, but is even detrimental, since it increases the stresses in the deck edges already compressed by the bending moment final diagram. Consequently, these temporary stresses require numerous checks and a specific type of prestressing that optimizes, among the various aspects, the deck performance during launch as well, reducing the requirement for material at those points where oversizing, in the service phase, would be pointless or, worse, damaging.

A structure which presents so many and different load conditions requires a careful pre-sizing, to avoid excessive stresses in one of the many launch or service stages, but without excessive prudence, which would result in oversized structures. For this purpose, many characteristic aspects of the construction technique must be taken into account:

- during launching, the load effects in the deck current zone are uniform and substantially known, while the front zone stresses can be analyzed with approximate design theories which avoid proceeding by trial end error with sophisticated computer programs;
- the evolution of the concrete strength with time does not create particular problems. The differences of elastic modulus along the deck are almost always negligible, and strength problems generally arise only in the curing zone near the formwork and are solved by reducing the spacing of temporary bearings;
- the stresses caused by thermal gradients, particularly important because of the absence of paving, and by bearing settlement, tend to increase launching prestressing, but have little repercussion on the cross-sectional pre-sizing;
- time-dependent phenomena can be neglected in pre-sizing, as except in cases of long term stoppage of construction they cause only launch prestressing losses and a modest redistribution of the launch bending moment.

As most of the launching stresses depend linearly on the distributed dead load q of the deck, its reduction is of fundamental importance. Besides, set p, the distributed service load, the bridge sizing, and therefore its total cost, depends on $p+q$. The general efficiency, intended as the ratio of the service rendered to the investment made, can be expressed as[169]:

$$\rho_g = \frac{p}{p+q} \qquad (5.1)$$

The reduction of dead load limits the influence of launching stresses and at the same time increases the general efficiency of the bridge by creating a reserve available for service loads.

Since the deck cost is strictly connected to material consumption, and therefore to the area, A, of its cross section, while the moment of inertia, I, is the main parameter in estimating carrying capacity, the efficiency of the cross section can be measured in terms of moment of inertia reached, self-weight being equal – that is, of radius of gyration:

$$r^2 = \frac{I}{A} \qquad (5.2)$$

Area being equal, in a section of depth H, whose centroid is distance z_u and z_l respectively from the upper and lower edge, the highest radius of gyration is obtained by concentrating masses at the edges and results in:

$$r^2_{max} = z_u z_l \qquad (5.3)$$

Therefore, the cross sectional efficiency can be measured in relative terms as structural efficiency:

$$\rho_s = \frac{r^2}{z_u z_l} = \frac{I}{z_u z_l A} \qquad (5.4)$$

which is the better the closer to one. The height of the central core is:

$$H_c = \rho_s H \qquad (5.5)$$

and the highest fluctuation of the launch bending moment is:

$$\Delta M_{max} = K\gamma A l^2 \qquad (5.6)$$

in which γ is the specific weight of the concrete and, approximately, $K = 0.16$ to 0.22 in the deck front zone and $K = 0.12$ to 0.15 in the back zone.

Launch prestressing can avoid edge decompression only when the moment fluctuation is such as to maintain its resulting force inside the central core – that is, the minimum axial prestressing, F_{min}, is:

$$F_{min} = \frac{\Delta M_{max}}{H_c} = \frac{K\gamma A l^2}{\rho_s H} \qquad (5.7)$$

For the same span, F_{min} depends linearly on γ and inversely on the depth of the cross section and on its efficiency. Consequently, a higher efficiency can also permit savings in prestressing. The role of area can be emphasized by expressing the minimum prestressing as:

$$F_{min} = K z_u z_l \gamma \frac{A^2 l^2}{IH} \qquad (5.8)$$

Since F_{min} is a quadratic function of the area, it is imperative to reduce the weight of the cross section as much as possible.

Evidently, in order to improve structural efficiency, it is necessary that the reduction of area does not penalize excessively the moment of inertia, so that the radius of gyration (5.2) can increase on the whole. Since static and technological constraints prevent lightening the cross section excessively, it is almost always necessary to increase at the samc time the moment of inertia. This cannot be obtained by acting on the slab thickness, since this affects the area directly. It is therefore inevitable, and more effective, to intervene on the lever arm of the resisting couple rather than on masses, by increasing directly the depth of the cross section. A larger deck depth thus becomes the aesthetic characteristic of this construction method: the bridges tend to lose the typical slenderness of continuous beam decks, especially for the longest spans, to become similar to the most usual simply supported bridges.

The implications of incremental launching on deck sizing have been studied in statistic analyses of highway bridges. Figure 5.2 shows the relationship between the deck external depth and the span.

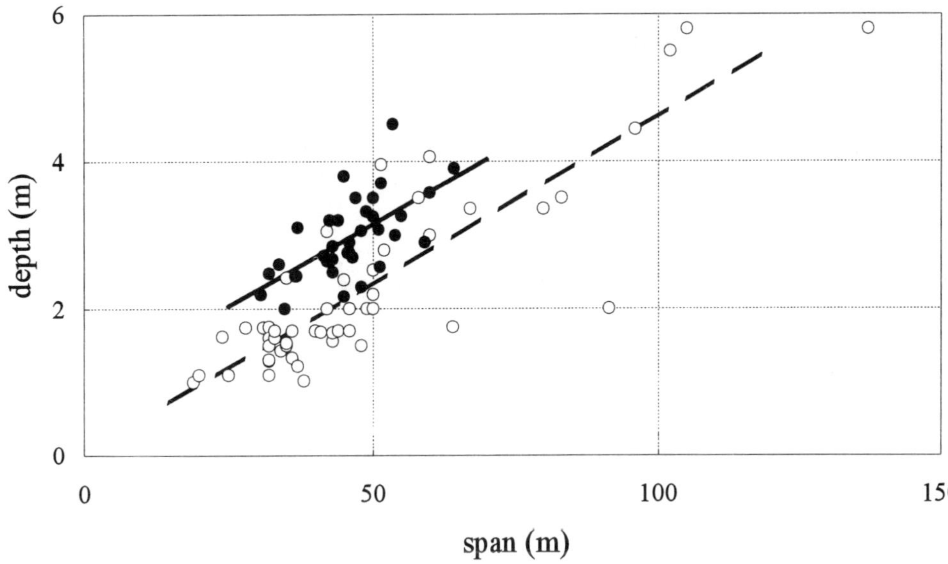

Figure 5.2 - Evolution of the deck depth with the span. Black dots are launched bridges, while white dots are continuous beam bridges built with other techniques.

Launched bridges are evidently less slim: the linear regression for highway bridges is[166]:

$$H = 0.94 + \frac{l}{22.7} \qquad (5.9)$$

compared with $H = 0.04 + l/21.8$ for bridges built using other techniques[166], while in railway bridges it is generally $12 \leq H/l \leq 14$[28]. The influence of the larger depth on the architectural aspect of the bridge can be reduced by inclining webs to avoid large cantilevers and by using surface texture, which is easy and inexpensive to obtain in a fixed formwork supported on the ground. The transverse dimensions of the cross section confirm this trend. In launched bridges, the ratio of the cantilever width to the top slab width is about 0.24 compared with 0.27 for decks built with other techniques, while the ratio of the bottom slab width to the top one is in both cases about *0.44*.

The average slab thickness is similar to that of cast-in-situ continuous beams, depending mainly on the transverse bending under live loads and on the necessity to contain prestressing tendons.

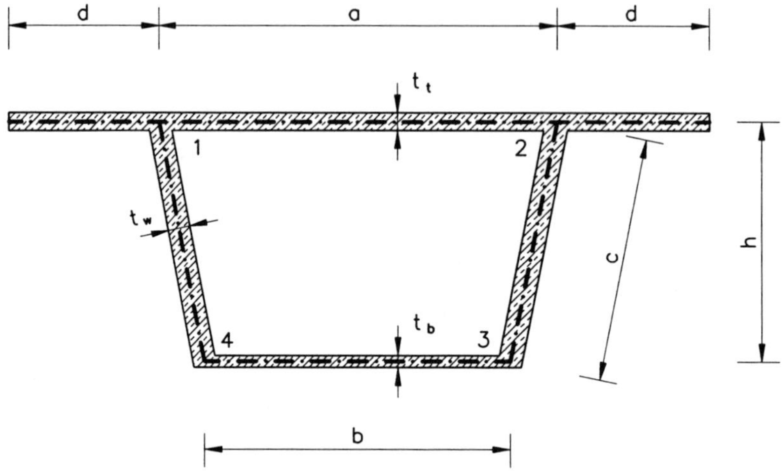

Figure 5.3 - Cross section symbols.

For the top slab of highway bridges, it is (always in meters, and with the symbols of figure 5.3)[28]:

$$\frac{a}{t_t} = 25 \text{ to } 30 \qquad (5.10)$$

with a minimum value of *0.22* to *0.26 m*[28], depending on the distance between webs and the eventual transverse prestressing, and for the bottom slab[166]:

$$t_b = (b - t_w)\left(0.07 - \frac{l}{3700}\right) \qquad (5.11)$$

with minimum values, for highway bridges, of *0.18 m* or *0.24 m* respectively in absence or in presence of longitudinal prestressing tendons[28]. This value generally increases in the support sections to resist the negative bending moment better. In railway bridges, the minimum thickness for both slabs is about *0.30 m*[28].

The mid-span web thickness suffers from the fact that the highest launch shear force is much higher than the shear in the final position of the deck under live loads, and therefore the thickness tends to increase with the span, contrary to what occurs in bridges built with other techniques. In the case of highway bridges with internal parabolic prestressing, the thickness of each web can be estimated, in meters, as[166]:

$$t_w = B\left(\frac{l}{2100} + 0.02\right) \qquad (5.12)$$

and with external prestressing[28]:

$$t_w = \frac{Bl - 500}{2000} + 0.30 \qquad (5.13)$$

with a minimum value of $0.30\ m$. In railway bridges the mid-span thickness is about $0.50\ m$ for the usual spans of 40 to $50\ m$, and in the support sections it generally reaches 0.90 to $1.00\ m$[28].

Evidently, a larger depth of the deck together with webs of constant thickness tends to produce higher material consumption. The average concrete consumption (total volume to deck surface ratio, inclusive of stiffeners and diaphragms) in a highway bridge is, in meters[166]:

$$h_c = 0.25 + \frac{l}{110} \qquad (5.14)$$

and in railway bridges it increases to[28]:

$$h_c = 0.40 + \frac{l}{100} \qquad (5.15)$$

Prestressing steel is about 40 to $60\ kg/m^3$ in highway bridges[166] and 65 to $70\ kg/m^3$ in railway spans[28]. Reinforcement is about 90 to $150\ kg/m^3$ in highway bridges[166] and 140 to $150\ kg/m^3$ in railways span[28].

FRONT ZONE OF THE DECK

In a continuous beam composed of several bays of span l and of constant dead load, far from the ends the support sections remain vertical and the static scheme of each span is that of the perfectly fixed beam. On the contrary, at each approach to a pier, the front zone of the deck overhangs beyond the first support with a cantilever l long. Therefore, the negative moment in the first support section is up to six times higher than the one at the rear supports, and the shear force is double (figure 5.4).

Reasoning necessarily in terms of the cylindrical geometry of a solid to "extrude," and therefore of a constant depth, sizing carried out on the basis of the launching stresses in the front zone would burden the entire structure, while sizing optimized for the rear zone would be inadequate in the front one. As this second alternative is more economical than the first one (the back zone is much longer than the front one) it is necessary to introduce devices which control the stresses in the front zone beyond that critical cantilever, l_{cr}, for which the negative moment at the first support is equal to the perfectly fixed beam moment in the rear zone:

$$\frac{ql_{cr}^2}{2} = \frac{ql^2}{12} \qquad (5.16)$$

and which results in $l_{cr} = 0.41\,l$.

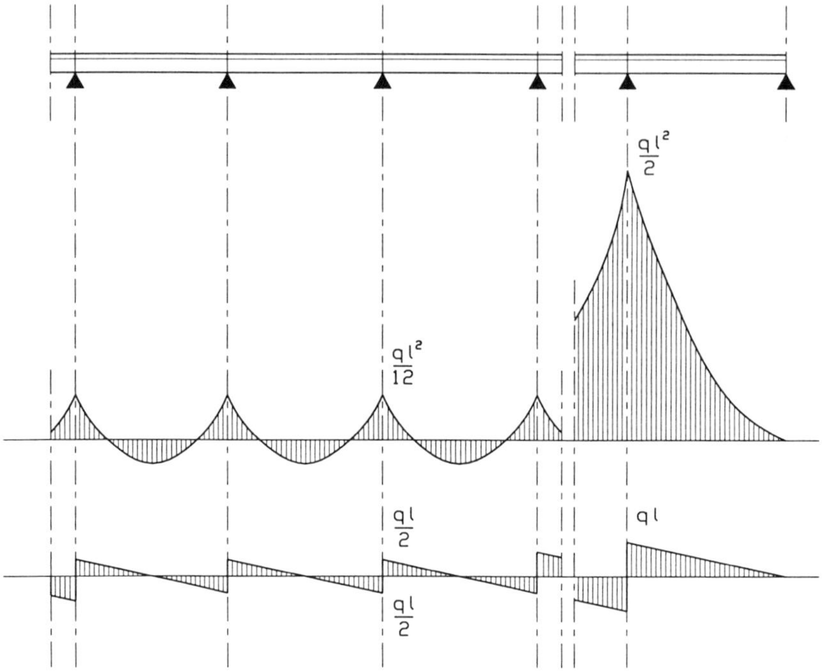

Figure 5.4 - Bending moment and shear force diagrams in the front and in the rear zone of the deck.

There are two possible solutions: to reduce the launching stresses by increasing the number of supports, or to limit the difference between the launching stresses in the front and rear zones by supporting the cantilever or by reducing its weight.

The first applications of incremental launching were carried out by dividing the spans with temporary piers distributed along the entire length of the deck (figure 1.4). In fact, it was observed that using a single temporary pier located under each cantilever span would have involved significant time and cost, acceptable only for spans so long as to make launching stresses reduction convenient even in the rear zone of the deck. It was also observed that the adoption of many temporary piers in each span would not have effectively produced the dramatic reduction of stresses predictable at first sight, because of the different elastic and thermal deformability of the supports, the construction tolerances, and the deck flexural stiffness. This resulted in the general use of a single temporary pier per span.

Nowadays, the cost of temporary piers, although prefabricated and reusable, and of the additional labor they require, is justified only in spans so long or variable as to compensate for the lower consumption of launch prestressing (unless made worthless by a higher final parabolic prestressing) and of reinforcement.

These limitations narrow the field of application of the principle of overall launching stress reduction. Instead, normally the stress differences along the deck are reduced, to allow overcoming the full span with a cantilever.

In the first of these techniques, the cantilever is supported by means of a stayed scheme composed of a tower integral with the deck and strand tendons anchored symmetrically to the tip of the cantilever and two spans behind, to create a sort of external prestressing with great eccentricity. In relation to the position reached by the tower during launch, this scheme requires continuous adjustment of the force in the stays because of the concentrated action exerted by the tower on the deck. Just before contact with the next pier, the tower passes onto the first support and the maximum pull in the stays supports the cantilever and reduces its elastic deflection, but once the bearing has been reached, the pull must be relieved to avoid, during the course of launch, increasing the positive moment in the deck because of the concentrated action of the tower in the front span. Resorting to this technique is discouraged by the longer time required for the continuous adjustments of the stay force and the deflection of the deck, but especially by its intrinsic delicateness and complexity, presenting considerable stress concentrations and serious risks in case of errors.

The second technique provides, instead of a support for the deck end, a limitation of its cantilever weight by means of a lighter extension, the launching nose, which anticipates the deck contact with the pier. The launching nose represents a "passive" solution perhaps less elegant than the stayed scheme, but much simpler, safe, fast, and economic, to the point that its adoption has become virtually standard in full span incremental launching. Even using temporary piers, the reduction of the cantilever weight is advantageous, and essential in the longest spans. The stayed scheme also takes advantage of it. The launching nose thus characterizes practically all the applications of incremental launching.

Launching Nose

In the thirty years of life of incremental launching, launching noses of all types have been used, more or less rigid, trusses or plate girders, made of steel or concrete, with front realignment sledges or hydraulic systems to recover the elastic deflection (figure 5.5). These launching noses were sometimes conceived and sized according to custom and experience, but often were already available, and therefore ready to be reused.

The most frequent type of launching nose consists of two steel plate girders clamped to the deck front end with their bottom flanges aligned with the lower edge of the deck. Acting as an extension of the deck perfectly integral at the joint, the action of the launching nose affects the stresses in the deck (or, more exactly, in the continuous beam composed of the deck and the launching nose itself) and permits their homogenizing in relation to the efficiency of the elastic system composed of the launching nose and the front zone of the deck.

Figure 5.5 - The elastic deflection of the launching nose tip has been recovered by means of hydraulic pistons (author).

The behavior of the nose-deck elastic system is governed by three dimensionless parameters which describe its geometrical and mechanical characteristics:

- the length of the launching nose compared with the span to be overcome, l_n / l;
- the unit weight of the launching nose compared with the unit weight of the deck front zone, q_n / q;
- the flexural stiffness of the launching nose compared with the flexural stiffness of the deck front zone, $E_n I_n / EI$.

Nose-Deck Interaction

In order to develop a theoretical model of the nose-deck interaction, the following assumptions are made[160,170]:

- the launching nose and the deck are of constant stiffness and weight: as generally they vary both in the launching nose and in the deck front zone, it is correct to consider their weighted average;
- the number of spans behind support C (figure 5.6) is assumed to be so large as to assimilate the deck into a continuous beam composed of infinite bays of constant span l ;
- launch prestressing is centroidal also in the deck front zone, so as not to introduce additional redundant unknowns.

Defining the progression of launch with the distance x of the nose-deck joint section from support B, the cantilever configurations assumed before support A is reached vary from the starting position, $x = 0$, to the contact one, $x = l - l_n$, and the dimensionless launching progress $\alpha = x/l$ varies in the range $0 \le \alpha < 1 - l_n/l$.

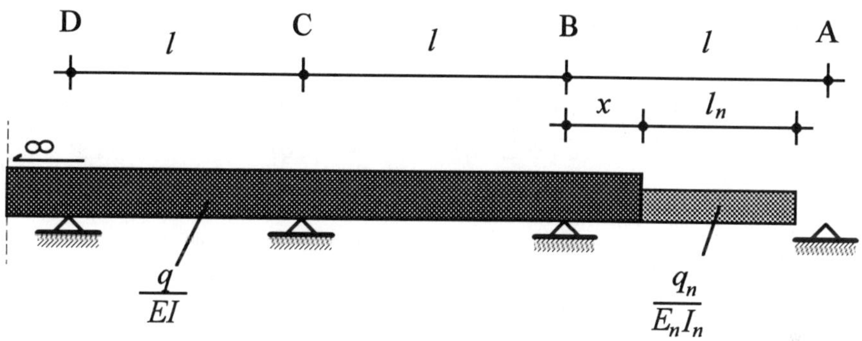

Figure 5.6 - Nose-deck elastic system in the first phase of launch.

At the start of launch, for $\alpha = 0$, the bending moment in the support section B is that of the cantilever launching nose, which in a dimensionless form (hereinafter the moments due to cantilever masses will be marked with an asterisk) is:

$$\frac{M_B^*}{ql^2} = -\frac{q_n}{2q}\left(\frac{l_n}{l}\right)^2 \qquad (5.17)$$

Once launch starts, this increases (in absolute value) with the dimensionless law:

$$\frac{M_B^*}{ql^2} = -\frac{\alpha^2}{2} - \frac{q_n}{q}\frac{l_n}{l}\left(\alpha + \frac{l_n}{2l}\right) \qquad (5.18)$$

and if one sets:

$$C_1 = \frac{l}{3EI} \qquad (5.19)$$

$$C_2 = \frac{l}{6EI} \qquad (5.20)$$

$$C_3 = \frac{l}{2\sqrt{3}EI} \qquad (5.21)$$

$$C_4 = \frac{ql^3}{24EI} \qquad (5.22)$$

$$C_5 = \frac{ql^3}{24\sqrt{3}EI} \qquad (5.23)$$

the moment M_C at the back support C is:

$$\frac{M_C}{ql^2} = -\frac{C_5 + C_4}{C_3 + C_1}\frac{1}{ql^2} - \frac{C_2}{C_3 + C_1}\frac{M_B^*}{ql^2} \qquad (5.24)$$

from which:

$$\frac{R_B}{ql} = \frac{M_C}{ql^2} + \frac{(1+\alpha)^2}{2} + \frac{q_n}{q}\frac{l_n}{l}\left(1 + \alpha + \frac{l_n}{2l}\right) \qquad (5.25)$$

Once support A has been reached, the tip of the launching nose is hoisted until realigned with the theoretical profile of the deck in order to permit the subsequent sliding on launching saddles. The recovery of the elastic deflection creates a positive moment that reduces M_B, and the second launch phase starts (figure 5.7), in which the launching nose slides on the support A until the concrete deck arrives. This phase is defined by $1 - l_n/l < \alpha \leq 1$.

Once one sets (with $\dot{C}_i$ the variable terms during launch and with C_i the constant terms):

$$\dot{C}_6 = \frac{l}{3EI}\left[1 + \left(\frac{EI}{E_n I_n} - 1\right)(1-\alpha)^3\right] \qquad (5.26)$$

$$\dot{C}_7 = \frac{l}{EI}\left[\alpha^2\left(\frac{1}{2} - \frac{\alpha}{3}\right) + \frac{EI}{E_n I_n}\left(\frac{1}{6} - \frac{\alpha^2}{2} + \frac{\alpha^3}{3}\right)\right] \qquad (5.27)$$

and:

$$\dot{C}_8 = \frac{ql^3}{2EI}\left\{\left[\alpha(2-\alpha)+\frac{q_n}{q}(1-\alpha)^2\right]\left(\frac{\alpha^2}{2}-\frac{\alpha^3}{3}\right)-\left(\frac{\alpha^3}{3}-\frac{\alpha^4}{4}\right)\right\}+$$
$$+\frac{ql^3}{24EI}\frac{EI}{E_nI_n}\left[4\alpha^2(1-\alpha)^3+\frac{q_n}{q}\left(4\alpha^5-15\alpha^4+20\alpha^3-10\alpha^2+1\right)\right] \tag{5.28}$$

one obtains:

$$\frac{M_B}{ql^2} = \frac{\left(\frac{C_5+C_4}{C_3+C_1}C_2-C_4-\dot{C}_8\right)\frac{1}{ql^2}+\dot{C}_7\frac{q_n}{2q}\left(\frac{l_n}{l}+\alpha-1\right)^2}{C_1+\dot{C}_6-\frac{C_2^2}{C_3+C_1}} \tag{5.29}$$

from which (5.18):

$$\frac{R_A}{ql} = \frac{M_B}{ql^2}+\frac{\alpha^2}{2}+\frac{q_n}{q}\frac{l_n}{l}\left(\alpha+\frac{l_n}{2l}\right)=\frac{M_B}{ql^2}-\frac{M_B^*}{ql^2} \tag{5.30}$$

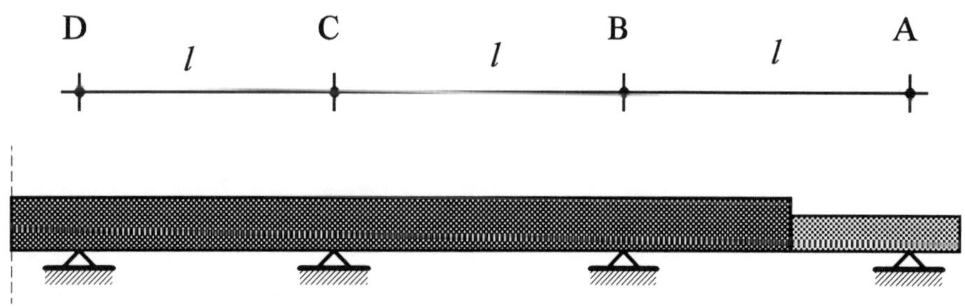

Figure 5.7 - Nose-deck elastic system in the second phase of launch.

Returning to M_B, to prevent application of different formulae in relation to the position of the deck, (5.18) in the cantilever configuration and (5.29) after realignment, and to better evaluate the effect of limiting the cantilever moment M_B^* exerted by R_A, it is preferable, once one calculates R_A with (5.30), to define M_B as:

$$\frac{M_B}{ql^2} = \frac{M_B^*}{ql^2}+\frac{R_A}{ql} \tag{5.31}$$

where M_B^* is the moment in the cantilever configuration (5.18) and the reaction R_A acts only for $1 - l_n/l \le \alpha \le 1$.

To be effective, a launching nose must at the same time reduce the cantilever moment M_B^* and assure high values of R_A, and therefore must satisfy two a priori contradictory requirements:

- on the one hand it must be light, to reduce M_B^*, and so long that on contact with the front pier the portion of cantilevered concrete deck, much heavier, is as short as possible;
- on the other hand it must be rigid, so that recovery of the elastic deflection at support A produces an initially high reaction R_A which immediately increases, as launch proceeds, to balance the increment of M_B^* deriving from the progressive lengthening of the concrete segment overhanging the first span.

For a long launching nose, $l_n/l = 0.80$, of relative weight $q_n/q = 0.10$, the progression of M_B with launch in relation to the relative stiffness of the system is described in figure 5.8. Until contact with support A, for $\alpha < 0.20$, M_B^* grows with (5.18). Once the support is reached, the positive moment generated by the recovery of the elastic deflection reduces M_B^*, and as launch continues M_B depends on the relative stiffness and tends to an end-of-launch (EOL) value which, set $\alpha = 1$ in (5.29), results in:

$$\frac{M_B^{EOL}}{ql^2} = 0.134\frac{q_n}{q}\left(\frac{l_n}{l}\right)^2 - 0.106 \qquad (5.32)$$

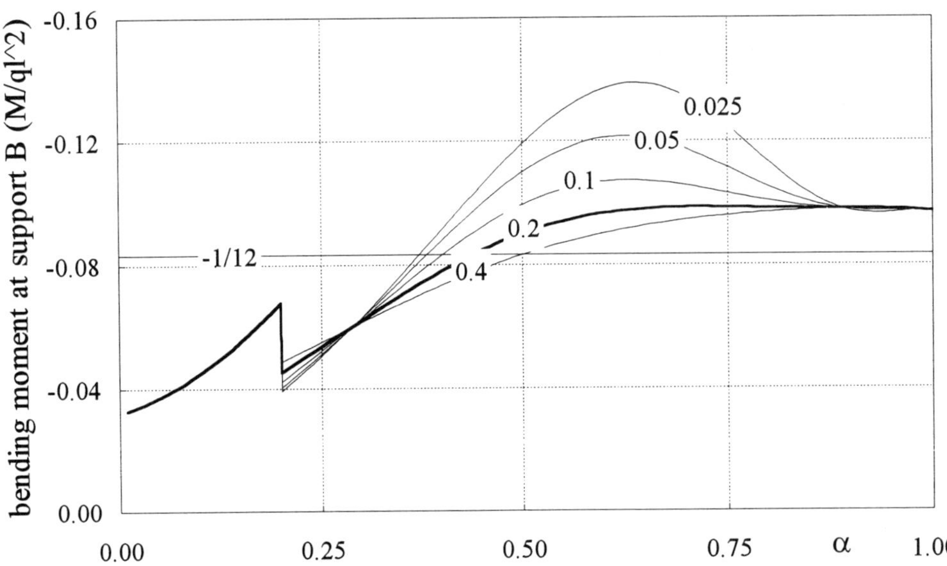

Figure 5.8 - Progression of M_B with launch for $l_n/l = 0.80$ and $q_n/q = 0.10$ in relation to the relative flexural stiffness, E_nI_n/EI. In the case of a long launching nose, the maximum cantilever moment is lower than that the end of launch.

As in the cantilever configuration, for $\alpha < 0.20$, also in this final stage of launch, for $\alpha > 0.90$, M_B does not depend on E_nI_n/EI. Therefore, the rigidity of the launching nose can be used only to ensure that the launching moments after contact with support A and until the stabilization of the EOL value ($0.20 < \alpha < 0.90$) do not exceed the moments at contact and at EOL, since deck oversizing is generally more expensive than stiffening the launching nose. Optimum rigidity, in this case, turns out to be around 0.200 (thicker line). Greater rigidity is pointless, since it cannot prevent reaching the EOL moment, and less rigidity increases the negative design moment of the deck. Moreover, the maximum moment reached in the cantilever configuration is much lower than the EOL one: contact with the pier is therefore uselessly anticipated – that is, the launching nose is excessively long, and it is economically convenient to shorten it.

On the other hand, in the case of a short nose (equal relative weight but $l_n/l = 0.50$) the progression of M_B is shown in figure 5.9.

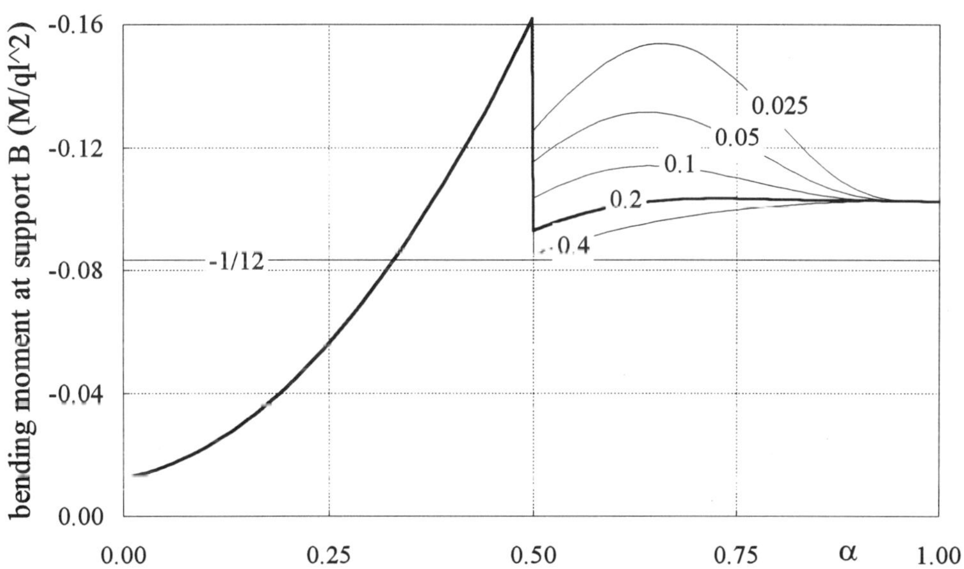

Figure 5.9 - Progression of M_B with launch for $l_n/l = 0.50$ and $q_n/q = 0.10$ in relation to the relative flexural stiffness, E_nI_n/EI. In the case of a short launching nose, the maximum cantilever moment is higher than that at the end of launch.

Until the contact with support A it is favorably affected by the lower weight of the launching nose but unfavorably by the greater length of the cantilever concrete segment. Once A is reached, M_B^* is reduced more effectively, and at the EOL it stabilizes on the value of (5.32). The effects of the relative stiffness are comparable as well, and also in this case optimum rigidity should be around 0.200 (thicker line). However, keeping M_B below the EOL value during the intermediate stages of launching loses significance since this threshold is substantially exceeded in the previous cantilever configuration. Contrary to what occurs for the long launching nose, in the case of a short nose, the maximum negative cantilever moment is greater than the EOL moment, and it is therefore the first phase of launch to affect the sizing of the deck.

Contact with the pier occurs therefore too late – that is, the launching nose is too short and should be extended. Proceeding by trial and error, the optimum length of the launching nose is obtained in figure 5.10. With $E_n I_n / EI = 0.200$, the moment at contact is equal to the EOL moment, and in the intermediate stages it is lower.

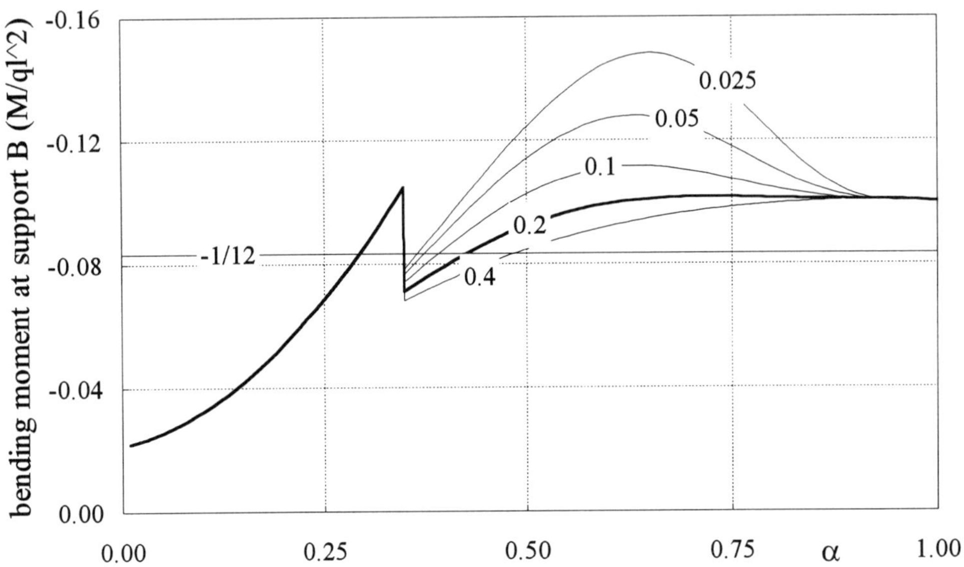

Figure 5.10 - The optimum nose length for $q_n / q = 0.10$ is obtained by trial and error: with $l_n / l = 0.65$, the maximum cantilever moment is equal to that at the end of launch.

The diagrams examined so far analyze the behavior of the nose-deck elastic system for a given relative weight. Figure 5.11 compares the diagrams of M_B calculated for the optimum parameters obtained for $q_n / q = 0.10$, as a function of this ratio. M_B depends significantly on the relative weight of the launching nose only during the first phase of launch. In the second phase, it is limited to a family of curves contained in *15* to *20%* of the EOL value, which twists because of the effect of M_A^*.

The different relationship between M_B and q_n / q in the two consecutive phases of launch means that, once the system has been optimized for a given relative weight, an increase in q_n / q modifies the diagram of M_B towards the type of figure 5.9 ("short" launching nose) and a reduction modifies the diagram towards the type of figure 5.8 ("long" launching nose). To force the value of M_B^* on contact toward the EOL moment, it is then necessary to modify the length of the launching nose, shortening the cantilevered concrete segment by lengthening the launching nose as the relative weight of the latter increases.

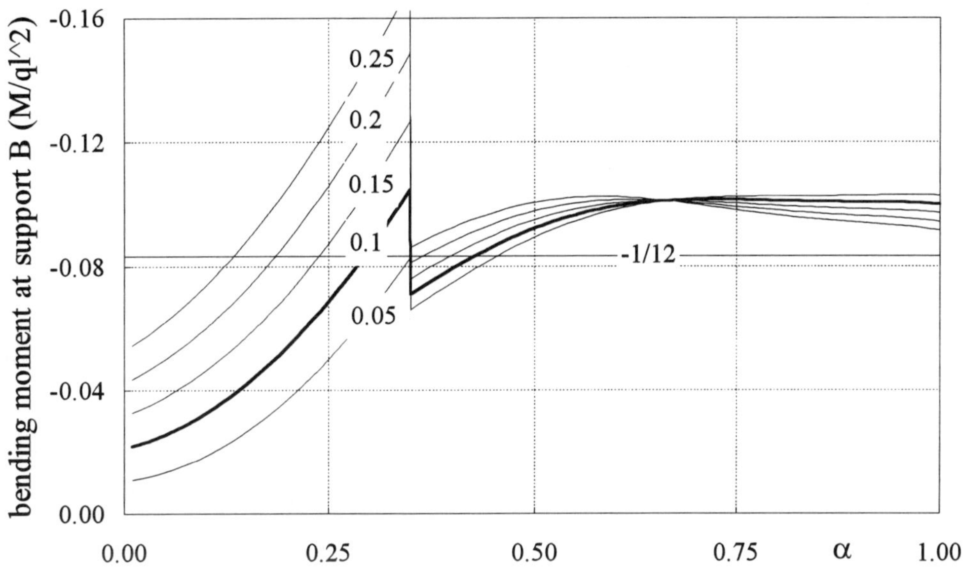

Figure 5.11 - Progression of M_B with launch for $l_n / l = 0.65$ and $E_n I_n / EI = 0.200$ in relation to the relative weight q_n / q. The thicker line is the optimum diagram of figure 5.10, for $q_n / q = 0.10$.

Figure 5.12 compares the diagrams of M_B that restore this condition: starting from $q_n / q = 0.10$, as the relative weight increases, equalization is obtained with longer launching noses and for progressively lower values of M_B which tend to the perfectly fixed beam moment of the current zone of the deck. Proceeding by trial and error again, the curve of figure 5.13 is obtained defining, as q_n / q changes and for the optimum flexural stiffness, the values of l_n / l which equalize the negative moment on contact with the EOL moment. The diagram allows one to determine easily the optimum length of the launching nose in relation to its relative weight.

In conclusion, therefore, a correct sizing of the nose-deck system can be developed starting from a length l_n equivalent to about two thirds of the typical span and from a first attempt unit dead load q_n obtained by statistical formulae such as[28]:

$$q_n = K\, l_n^2 \qquad (5.33)$$

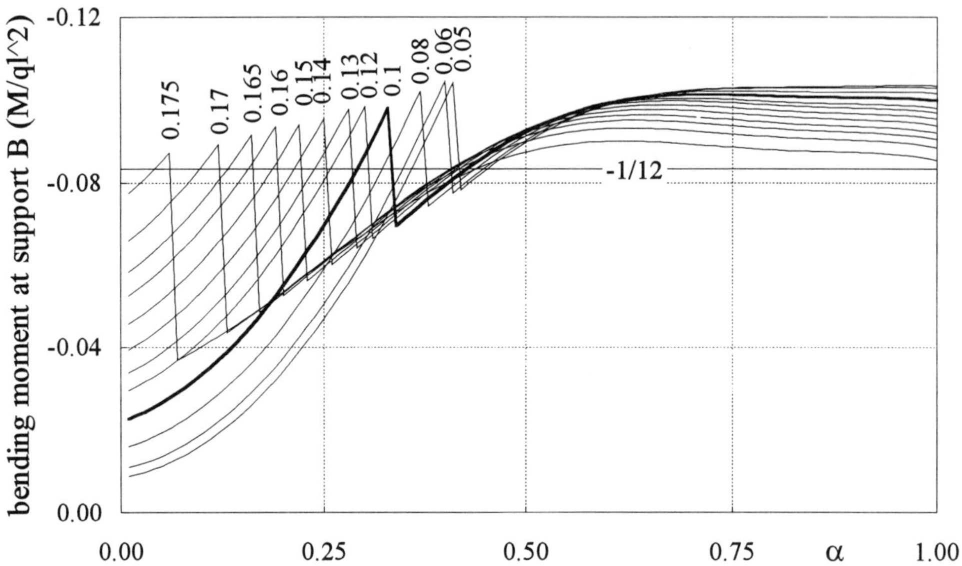

Figure 5.12 - As q_n/q changes, the equalization of bending moment on contact and at the end of launch is obtained by adjusting the nose length.

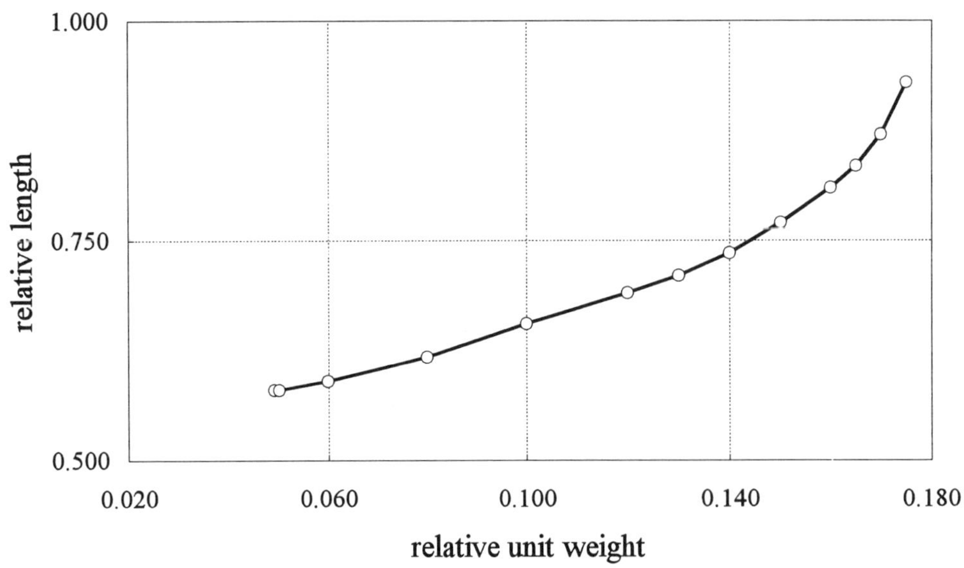

Figure 5.13 - Evolution of the value of l_n l, which equalizes the negative moment on contact with the end-of-launch moment as q_n q changes (for the optimum value of $E_n I_n$ EI).

in which, in kN/m and meters, $0.012 \leq K \leq 0.020$ for highway bridges and $0.018 \leq K \leq 0.030$ for railway spans, with progressively higher values in case of wide or heavy superstructures, as in both cases dead load increases. The values of l_n and q_n so defined will be refined until the equalization of the contact and EOL moments is obtained. Then, the minimum flexural stiffness necessary to contain the moment in the intermediate launch stages under this common threshold can be defined.

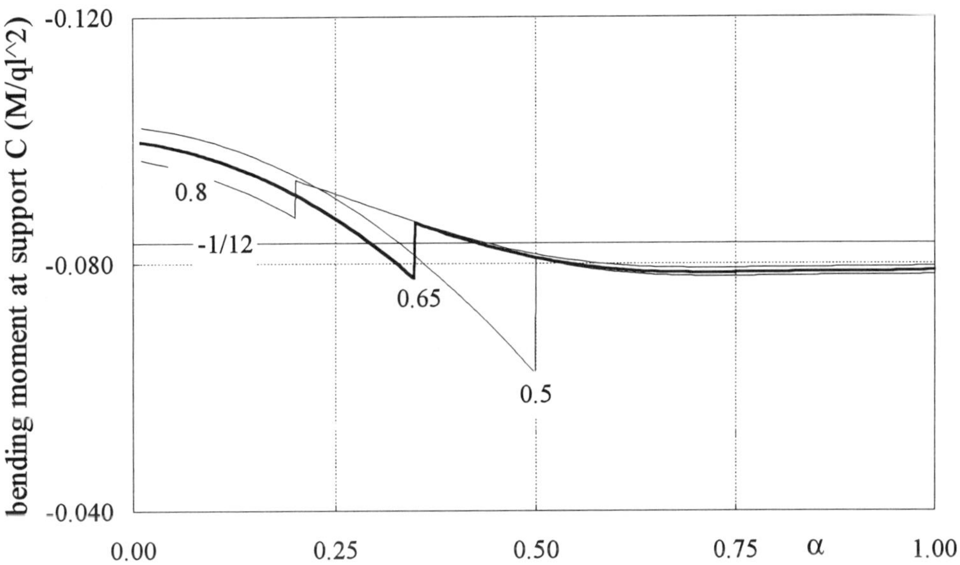

Figure 5.14 - Evolution of M_C with launch for different values of l_n/l. From $\alpha = 0.60$ onwards, M_C is practically no longer affected by the front elastic system, and the deck current zone begins.

In figure 5.14, the progression of M_C (5.24) shows that, independently from the length of the nose, the recovery of its elastic deflection forces M_C towards a single curve which stabilizes for $\alpha = 0.60$ on a slightly lower value than the perfectly fixed beam moment. From this point onwards, except for irrelevant additional fluctuations upon the contact with the following piers, the support moment is no longer affected by the front elastic system, and the deck current zone begins.

In short, as regards negative moment, it may be concluded that adopting launching noses of sufficient flexural stiffness and adequate length for the relative weight, the deck zone which feels the effects of the nose-deck elastic system is limited to a length of about $1.5\ l$, increasing to $2\ l$ in case of very deformable launching noses.

The highest positive moment in the front span is:

$$\frac{M_{A-B}^{max}}{ql^2} = \frac{R_A}{ql}\left(1-\alpha-\frac{q_n}{q}\frac{l_n}{l}\right)+\frac{1}{2}\left(\frac{R_A}{ql}\right)^2+\frac{q_n}{2q}\left(\frac{l_n}{l}\right)^2\left(\frac{q_n}{q}-1\right) \tag{5.34}$$

and figure 5.15 illustrates its evolution in case of a long launching nose. Contrary to what happens to M_B, the relative flexural stiffness of the launching nose affects only the progression of M_{A-B}^{max} towards the maximum value, which is reached around $\alpha = 0.90$ and then starts to decrease because of the effect of M_A^*.

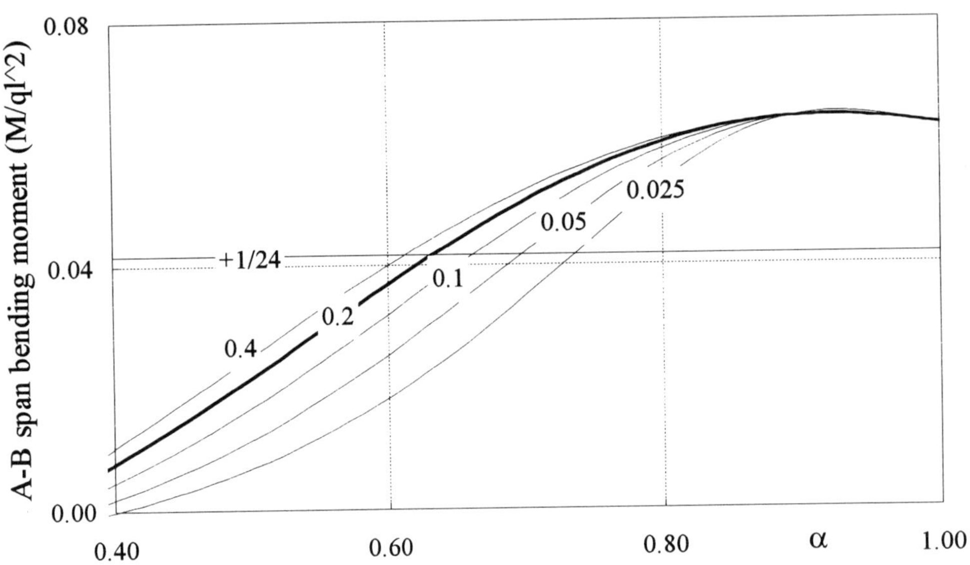

Figure 5.15 - Evolution of the maximum positive moment, M_{A-B}^{max}, in the front span for $l_n/l = 0.80$ and $q_n/q = 0.10$ in relation to E_nI_n/EI. The relative flexural stiffness affects only the evolution of M_{A-B}^{max}, but not its maximum value.

By shortening the launching nose, the progression of M_{A-B}^{max} is substantially similar (figure 5.16). The maximum value is obtained a little further on in the launch, for $\alpha = 0.95$, and is slightly higher because of the smaller contribution of M_A^*.

Therefore, the flexural stiffness of the launching nose cannot govern the positive moment behavior of the system, which depends essentially on the continuous beam static scheme of the concrete deck and can be modified only by means of M_A^*.

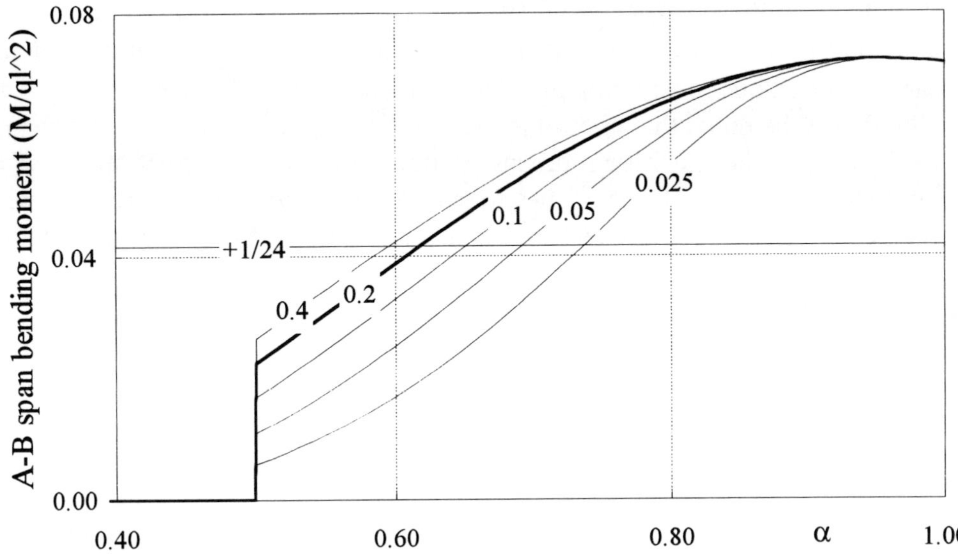

Figure 5.16 - Evolution of the maximum positive moment, M^{max}_{A-B}, in the front span for $l_n/l = 0.50$ and $q_n/q = 0.10$.

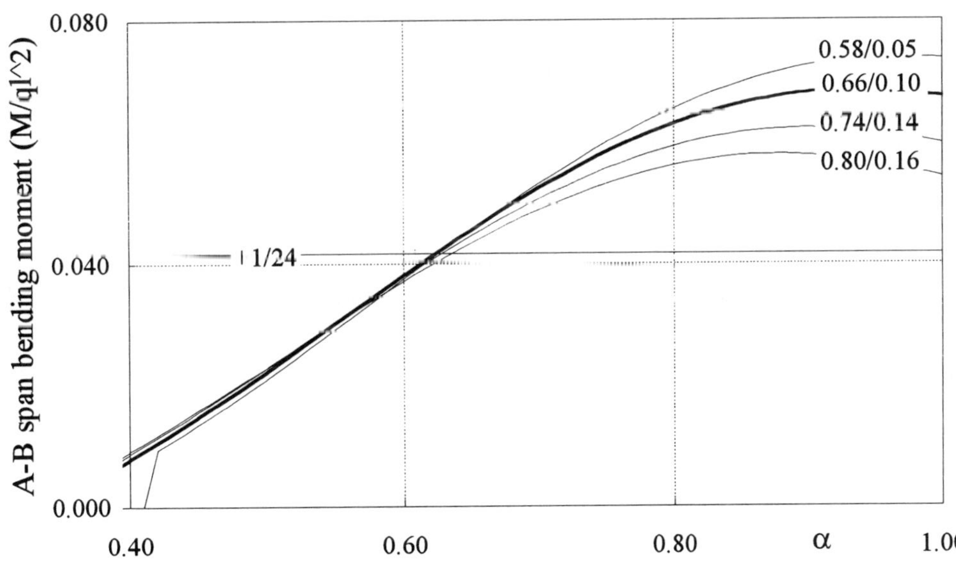

Figure 5.17 - Evolution of M^{max}_{A-B} for the optimum combinations of l_n/l and q_n/q defined in figure 5.13.

A synergistic action of increasing the length and the weight of the launching nose, carried out with the optimum values of figure 5.13, can reduce at the same time the negative equalized moment (figure 5.12) and the maximum positive moment (figure 5.17). This can be advantageous if the greater cost of the launching nose is more than compensated for by lower structural costs. Remaining within reasonable dimensions, the correction is effective, more for M^{max}_{A-B} than for M_B, and the values reached with $l_n/l = 0.65$ and $q_n/q = 0.10$ decrease respectively *15%* and *4%* with $l_n/l = 0.80$ and $q_n/q = 0.16$. Consequently, the choice among available launching noses can be carried out considering also these savings.

Finally, it shall be noted that assuming that the first bay of the continuous beam is always subject to higher positive moments is deceiving, since correct sizing of the bridge involves shorter end spans just to compensate for these increases, while during launch the end span deck section has to overcome all the longer intermediate spans.

For the B-C span one obtains:

$$\frac{M_{B-C}^{max}}{ql^2} = \frac{1}{2}\left(\frac{M_C}{ql^2} - \frac{M_B}{ql^2} + \frac{1}{2}\right)^2 + \frac{M_B}{ql^2} \qquad (5.35)$$

and the relative diagram, compared with the progression of M_C, confirms that the current zone of the deck begins with reasonable approximation at a distance of *1.5 l* from its front end.

Sizing of the Launching Nose

The dimensions and the mechanical characteristics of the launching nose depend on many factors, including economic considerations. In fact, the convenience of amortizing the investment over several bridges dictates designing launching noses to be as versatile as possible. If one thinks that incremental launching construction can reach free spans of about sixty meters, it can be convenient to design the launching nose directly for the longest spans, even in the case of initial use on shorter spans, and to adapt it to the lower depth of the deck. The probable need to reduce the length of the launching nose can be addressed by subdividing it into joined elements, while its greater stiffness and weight would further reduce stresses in the deck front zone. Therefore, the role of the launching nose, originally aimed at homogenizing launch negative bending moments, can extend to the control of the front span positive moment. The advantage of the higher investment required for this purpose has to be verified case by case on the basis of the amortization expected and of the smaller oversizing of the deck front zone.

In most cases, the launching nose is composed of two twin steel girders braced horizontally and in the cross section. Prestressed concrete launching noses have been adopted only in the case of impossible reuse or of difficult transport, to be demolished after completion of the launch.

If there are no limitations in the depth, the search for the greatest cross sectional efficiency of the launching nose can be carried out by adopting truss schemes. In these, however, the support reaction migration produces important local stresses in the bottom chords which require their sizing for the whole shear force and the highest local bending moment, and this, together with the large quantity of manual welding, makes this solution uncompetitive. Therefore, the general trend is toward the use of steel plate girders composed of transportable elements connected by vertical bolted joints (figure 5.18), and in the case of long spans and therefore of deep girders, even by horizontal joints (figure 5.19).

The bracing system is generally specific for each bridge (that is, to be changed each time) in order to align the steel girders of the launching nose with the deck webs and use the same launching saddles. The use of special wider saddles could avoid the replacement of the bracing, but the cost of the saddles and the stresses in the deck front diaphragm caused by the transverse eccentricity of the shear forces transferred by the nose make this solution undesirable.

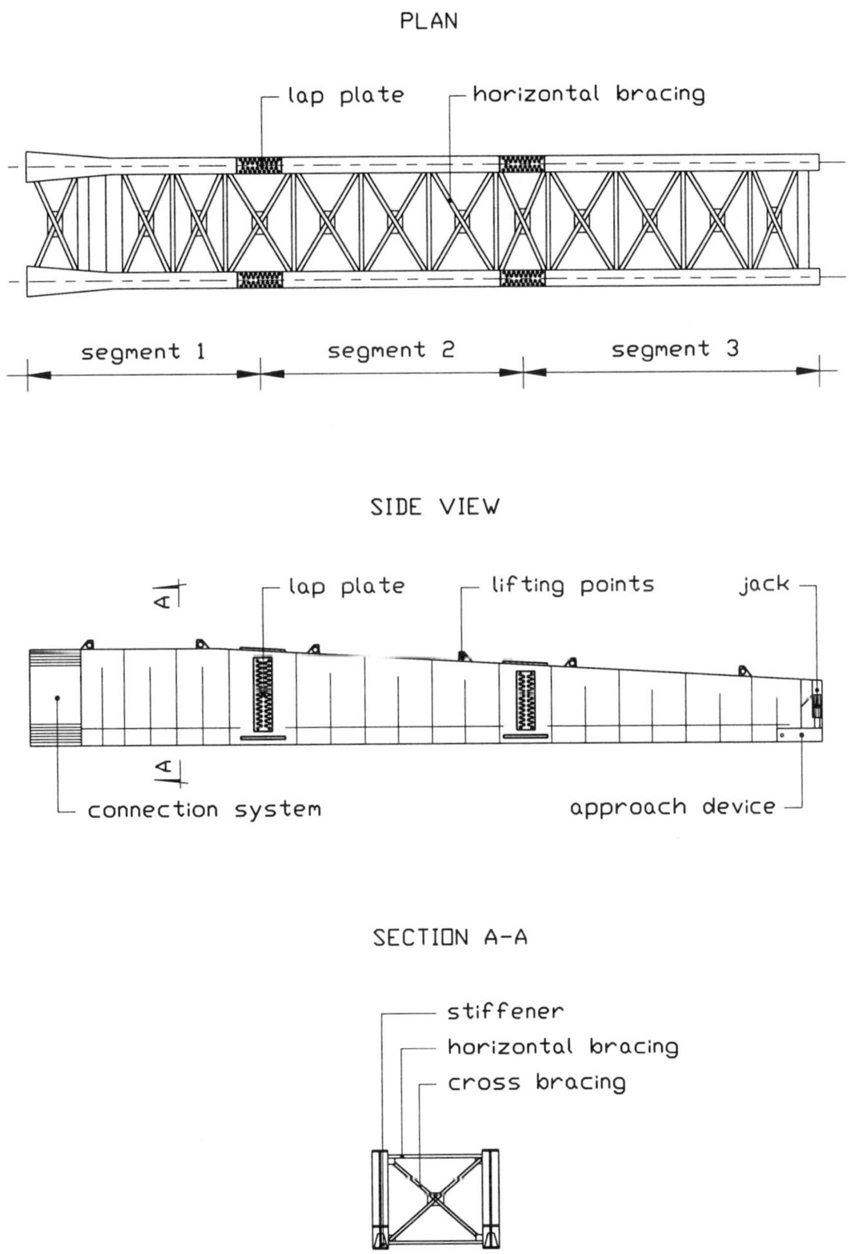

Figure 5.18 - Construction scheme of the launching nose.

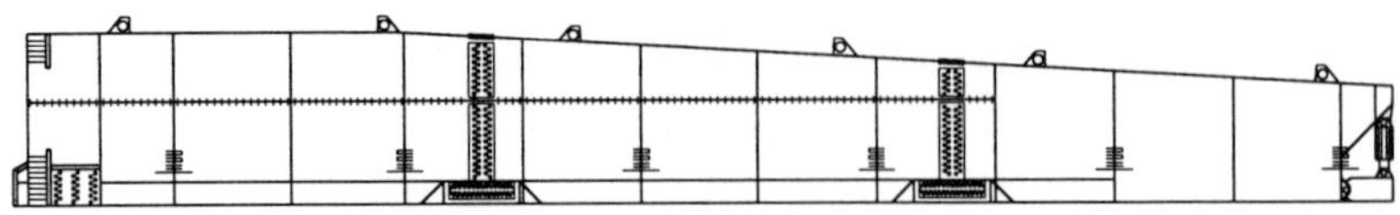

Figure 5.19 - In the case of deep girders, horizontal joints facilitate transport.

The external edges of the launching nose bottom flanges are often aligned with the lower corners of the deck in order to be guided in the transverse plane by the same anti-drift devices. Clearly, this is aimed only at centering R_A on the vertical steel plates, as deck alignment correction by acting on the launching nose would cause excessive stresses in its bracing.

When sized for a specific bridge, the highest depth of the launching nose is generally equal to the depth of the deck, and is maintained for few meters only from the back joint section. In fact, the diagrams of M_B show that the whole flexural stiffness is necessary only in the last launch phases, and the front zone can be lightened by reducing both the plate thickness and the girder depth, with savings in material and in cantilevered masses.

Design stresses should be kept low because of load differences between the girders caused by construction and assembly tolerances in joints, by the surface imperfections of the bottom flanges, and by misplacement of launching bearings. These load differences can be very high in the case of launching noses with closely spaced girders.

Starting from the moment envelope in the launching nose, once the depth and the allowable stresses are defined, calculation of the moment of inertia is immediate, and from it the area of the two horizontal flanges. The bottom flange width can be chosen so as to use the same neo-flon plates as the deck. Defining the width of a neo-flon plate in the transverse direction with b_{nf}, its length with l_{nf}, and the total length of the plates placed on the launching saddle with Σl_{nf} (figure 4.24), the calculation of the contact pressure, and then of the flange thickness, t_b, is immediate.

The transverse flexural stresses produced by the support reaction, R_A, can be calculated by assuming that the neo-flon plates are perfectly centered and the support reaction is equally distributed between the two girders:

$$\sigma_t = \frac{3R_A b_{nf}}{8t_b^{\,2}\Sigma l_{nf}} \qquad (5.36)$$

but, more conservatively, a certain eccentricity could be taken into account.

Once the area of the bottom flanges is defined, the calculation of the area of the top flanges is immediate. The ratio of their width, b_t, to their thickness, t_t, should be such as to prevent instability under the compression produced by R_A, and, for instance, it should be[28]:

$$\frac{b_t}{t_t} \leq 0.9\sqrt{\frac{E}{f_y}} \qquad (5.37)$$

where f_y is the yield stress of the steel. As an alternative, for AASHTO Standard 92, it should be:

$$\frac{b_t}{t_t} \leq \frac{270}{\sqrt{f_b}} \qquad (5.38)$$

where f_b is the computed compressive bending stress in the flange, with $b_t/t_t \leq 24$.

Using the shear force envelope, the web thickness, t_w, is chosen by limiting shear stresses so as to reduce the number of the vertical ribs of the girder. Once their spacing is defined, the formulae for web panel buckling permit definition of the minimum web thickness, which should be at least 10 to $12\,mm$ in order to handle plates easily in the workshop and to obtain an adequate overall stiffness. Clearly, the web thickness can vary along the launching nose. Local buckling under the thrust of the compressed top flange should also be checked. For instance, with Eurocode 3, it should be[28]:

$$t_c \geq \frac{h\sqrt{1+\frac{h}{3r}\frac{E}{f_y}}}{0.55\frac{E}{f_y}\sqrt{\frac{A_w}{A_t}}} \qquad (5.39)$$

where r is the radius of plan curvature of the web, h is the net depth of the panel, A_t is the area of the top flange, and A_w the area of the web plate. Finally, the local vertical compression in the web is:

$$\sigma_v = \frac{R_A}{2t_w \Sigma l_{nf}} \qquad (5.40)$$

If the bottom flange transverse bending (5.36) or the web vertical compression (5.40) are excessive, closely-spaced vertical stiffeners can create further support lines in the flange and permit a better distribution of the vertical compression (figure 5.20). As an alternative, continuous longitudinal plates can create two triangular cells (figure 5.21).

In proximity of the nose-deck joint section, the stresses in the launching nose lower nodes are rather high because of the simultaneous longitudinal tension, vertical compression, and shear stress. By assuming the stress state is bi-axial, principal stresses should be calculated in order to verify the safety factor with respect to the yield strength of the steel. In this zone of the launching nose, however, it is convenient to increase the number of the bottom flange stiffeners (figure 5.22), or to adopt directly the triangular cellular scheme, in spite of the geometric restraints due to the clearance required for the tensioning jacks for the strands or the bars which clamp the launching nose to the deck.

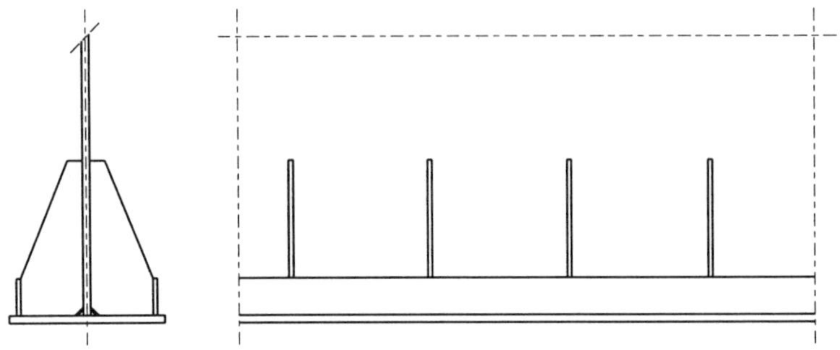

Figure 5.20 - Closely spaced vertical stiffeners create additional support lines in the bottom flange and improve the distribution of the vertical compression.

Figure 5.21 - Two symmetrical longitudinal plates create a cellular bottom flange (author).

Figure 5.22 - Vertical stiffeners in the nose-deck joint zone (author).

Bracing

Launching nose twin girders are connected to each other by several cross diaphragms and by horizontal bracing.

The purpose of the diaphragms is to join the twin girders for the absorption of asymmetrical loads, assuring their distribution in relation to the stiffness of their members; besides, when the launching nose is twisted, they oppose distortion.

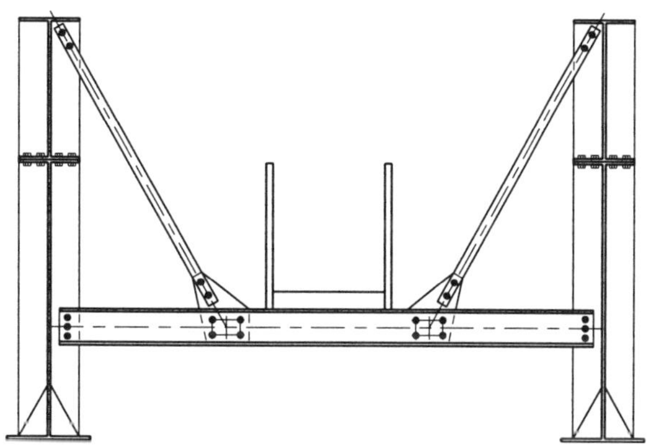

Figure 5.23 - Frame diaphragm with a rigid connection between the web vertical stiffeners and the horizontal cross beam. Additional diagonals improve the compression top flange stability.

Diaphragms can be designed as crosses, or as frames with a rigid connection between the vertical web stiffeners and the horizontal cross beams placed in the plane of the horizontal bracing, with diagonals which further prevent the instability of the compressed top flange (figure 5.23). The cross beams usually also support the walkway to the front work platform. In narrow noses the longitudinal spacing of cross diaphragms is often such that the panels of the horizontal bracing system are square; in wider noses the rectangular scheme is more common.

The role of the horizontal bracing is to stiffen the main girders in resisting the wind loads and the transverse forces produced by the anti-drift devices which center the nose on the launching saddles. They are composed of cross or K schemes, where the verticals are the cross beams and the chords are the main girders.

Coupling to the Deck

The nose-deck coupling must guarantee the transfer of the bending moment and shear force by providing perfect structural continuity, since an internal hinge would compromise the correct behavior of the nose-deck elastic system. At the same time, upon completion of the launch, it must permit an easy disassembly of the launching nose.

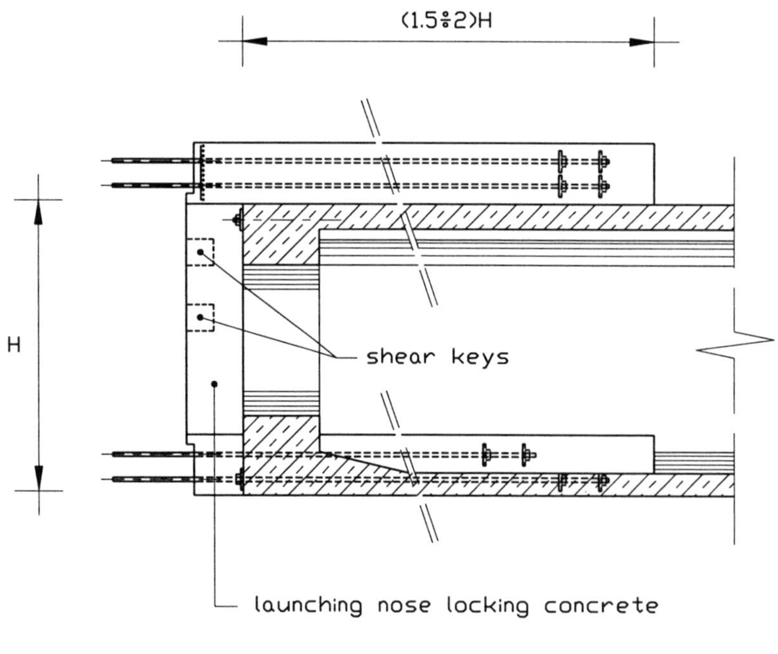

Figure 5.24 - Side view of the launching nose anchor devices. When the nose is deeper than the deck, the top bars are anchored in concrete blocks cast-in-situ over the deck. A special shimming concrete can be cast between the nosc gusset plates and the deck front end so as not to damage it with the nose shear keys, or in order to join the nose to the deck after the first segment casting.

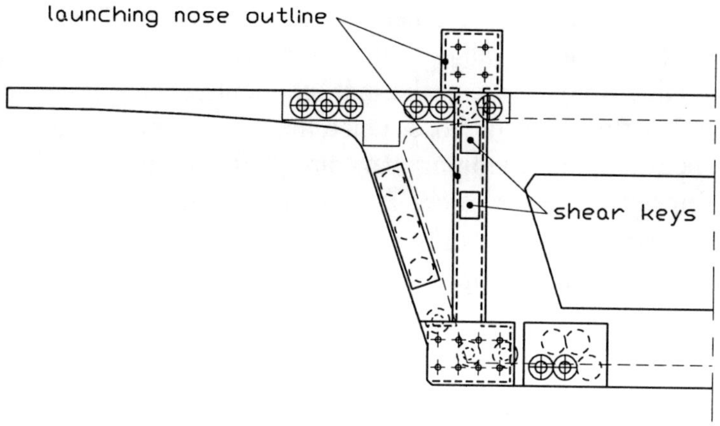

Figure 5.25 - Front view of the anchor scheme of figure 5.24. When the deck webs are sloped, the external anchor blocks do not affect the anchorages of the launch prestressing tendons.

The solution usually adopted for transferring the bending moment consists in prestressing the joint by means of bars. The use of strands is less frequent because of the difficulty in relieving them for maintenance or monitoring.

The bar length should be almost double the deck depth, in order to obtain an anchorage as long as the whole deck depth both for the bars and for the launch prestressing tendons (figures 5.24 and 5.25).

It shall be noted that the presence of the launch prestressing anchorages in the top slab or the availability of a nose deeper than the deck can require joining the nose top flanges to temporary blocks.

Since the bar position is defined by the launching nose geometry, the layout of the launch prestressing tendons must be modified to avoid the bar anchorages and to introduce bars deeply between them to make the stress transmission as direct as possible. This may cause sharp curvatures and losses of prestress. In particular cases, and in the case of truss noses which tend to concentrate the couple transferred more, the launch prestressing of the front segment can be obtained with several small tendons or with prestressing bars (figure 5.26).

The complication of tendon layout increases in decks with sloped webs, as the nose girders maintain a vertical position and their top flanges are distant from the deck webs (figure 5.25). This can require a mild transverse prestressing in the deck front diaphragm to absorb the effects of this eccentricity.

The maximum negative moment transferred to the deck at the joint section is given by (5.17). The positive moment is:

$$\frac{M^+}{ql^2} = \frac{M_B}{ql^2}(1-\alpha) + \frac{\alpha^2}{2}(1-\alpha) + \frac{q_n}{q}\frac{l_n}{l}\left(1-\alpha-\frac{l_n}{2l}\right)\alpha \qquad (5.41)$$

and its maximum value is obtained by trial and error. This is an approximate formula, and the definitive value results from enveloping the moment diagrams obtained by taking into account thermal gradients, bearing settlement, and the varying flexural stiffness of the nose. Once the maximum moments transferred and the distance between the clamping bars are defined, the calculation of the two prestressing forces (the lower force, F_l, and the upper one, F_u) is immediate. The bars will be oversized in order to absorb the differences of load between the nose girders and to avoid joint decompression and bar overload.

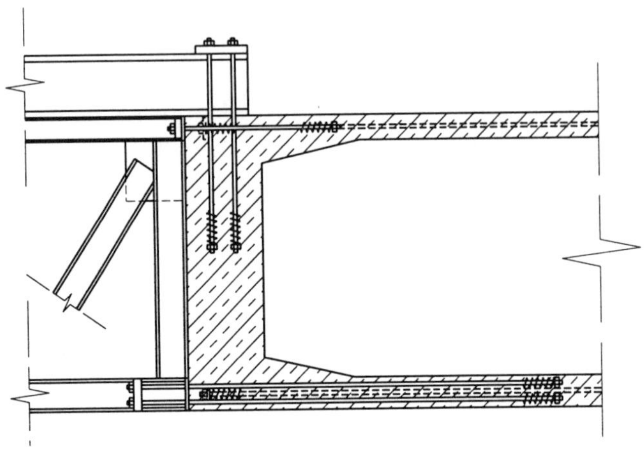

Figure 5.26 - A truss launching nose has smaller gusset plates, and therefore closely spaced anchor bars. The deck may be prestressed by means of high strength bars instead of strands.

In bar sizing, the reentry of anchorages, friction losses, elastic shortening of the compressed concrete, and (only in the case of long bridges whose construction requires a long time) even the losses of prestress due to time-dependent phenomena will be taken into account. If needed, however, the bars can always be retensioned.

Once bars are sized and their position is defined according to the launching nose geometry, the joint surface, of area A_j and moment of inertia I_j (both excluding bar ducts), is sized by trial and error by assuming prestressing forces as centroidal in the respective rectangles and a linear distribution of the contact pressure along the joint surface (figure 5.27). For the maximum positive moment the extreme fiber stresses:

$$\begin{cases} \sigma_u = \dfrac{F_u + F_l}{A_j} + \dfrac{M^+_{max} + F_u e_u - F_l e_l}{I_j} z_u \\[2ex] \sigma_l = \dfrac{F_u + F_l}{A_j} - \dfrac{M^+_{max} + F_u e_u - F_l e_l}{I_j} z_l \end{cases} \qquad (5.42)$$

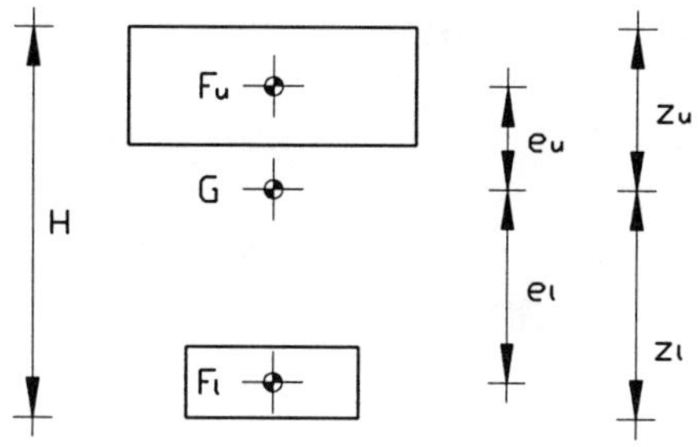

Figure 5.27 - Nose-deck joint cross section.

shall be compatible respectively with the lowest safety joint compression and with the allowable concrete compressive stress. The same verification will be carried out for the negative moment produced by the nose dead load and by the eccentricity of the joint prestressing.

Shear force, V, is transferred to the deck by steel blocks (figure 5.19) or by shear keys distributed along the contact plates. Figure 5.24 shows a solution which eliminates carving the deck with shear keys. In spite of these shear transfer devices, it is a good practice to check that the prestressing force guarantee the joint stability by friction, by adopting a steel-concrete friction coefficient of *0.4* and a safety factor equal to *2*:

$$\frac{V}{F_u + F_l} \leq 0.2 \qquad (5.43)$$

As already mentioned, when the launching nose is deeper than the deck, it is anchored by means of temporary steel blocks (figure 5.22) or of concrete blocks cast directly above the deck. In the case of concrete blocks, the solution with reinforcement crossing the horizontal joint between the block and the deck has been abandoned with the years, as upon completion of the launch it requires demolishing the block, cutting the joint reinforcement, and patching the deck surface. Instead, blocks are connected to the deck by friction (figure 5.28).Once the horizontal force transferred by the launching nose to the portion of the contact surface, A_j, affecting the upper block is calculated (a force which in the block-deck joint acts as a shear force, V), the minimum compression, N_{min}, to apply by means of vertical bars shall be such as to guarantee[134]:

$$\frac{V}{N_{min}} \leq 0.3 \qquad (5.44)$$

in order to reach a safety factor equal to *2* with respect to a friction coefficient between the two concrete surfaces of about *0.6*. This friction can be increased by introducing distributed shear keys, obtained by pressing corrugated sheet metal on the fresh concrete of the deck.

Figure 5.28 - In the Skye Bridge, the concrete anchor blocks were clamped to the deck by vertical high strength bars (reproduced with permission, Dywidag Systems International).

Recovery of the Elastic Deflection

When approaching the pier, the nose tip is generally distant from the theoretical profile of the bridge because of the elastic deflection of the nose-deck system.

In the case of a regular sequence of spans and of piers with similar elasticity, the nose tip deflection at contact is always downward, and it can easily exceed a decimeter. In steel or composite bridges, this deflection is much larger, and the need to recover it in order to permit the continuation of launch is often solved by means of a "hoisting sledge" – that is, by inclining the bottom flange of the nose so that its advance on the pier forces its progressive realignment (figure 5.29).

In prestressed concrete bridges, this device produces high horizontal forces, which together with friction cause excessive bending stresses in the piers. Therefore, realignment is always obtained by means of hydraulic devices. The simplest method consists in placing on the pier cap, immediately before the launching saddles, a couple of jacks topped by stainless steel plates. Once the launching nose has arrived above these jacks, neo-flon plates are inserted between them and the nose, the jacking permits realignment of the nose with the theoretical deck profile, and the neo-flon plates permit the launch to continue until the nose arrives on the launching saddles.

Figure 5.29 - The front “hoisting sledge” of the Schrotetal Bridge (author).

Figure 5.30 - Elastic deflection of the launching nose tip (author).

The need to move the jacks to subsequent piers often makes it convenient to use steel extensions hinged to the nose and moved by hydraulic pistons (figure 5.30): once they reach the launching saddles, these extensions are lowered until the nose bottom flanges are realigned with the deck soffit profile.

Sizing of the Deck Front Zone

The load effects in the deck front zone depend on its interaction with the launching nose, and therefore on the geometric and mechanical parameters of both of them.

In the presence of a regular sequence of spans and of piers with similar elasticity, the pre-sizing of the deck current zone can be carried out according to the equations from (5.9) to (5.15). Then, from its unit dead load, the optimum length and weight of the launching nose can be evaluated.

In the front zone of the deck, the maximum negative support moment, M_B, can be evaluated with (5.32) in the case of a long or optimized launching nose. With a short nose – that is, of a cantilevered configuration with higher moments than during the second phase of launch, it is obtained by (5.18):

$$\frac{M_B^*}{ql^2} = \left(1 - \frac{q_n}{q}\right)\left(1 - \frac{l_n}{2l}\right)\frac{l_n}{l} - \frac{1}{2} \qquad (5.45)$$

With these values and a rough estimate of the effects of thermal gradients and bearing settlement, the axial prestressing of the deck front zone can be defined, and then the increase to apply to the launch prestressing of the back current zone can be derived. The maximum positive moment can be calculated with (5.34) in order to evaluate the need to further refine the launch prestressing.

When there is an excessively low margin between the actual and allowable stresses at both the extreme fibers, it can be convenient to increase the slab thickness locally and to optimize again the launching nose and local prestressing.

Cable-Stayed Scheme

A second solution to the need to reduce stresses deriving from cantilever configurations consists in supporting the deck front end by means of its own back zone, with a temporary cable-stayed device composed of a tower hinged to the deck and strand tendons fixed at the tower head and fan-anchored symmetrically with respect to it, to obtain an adjustable external prestressing system with high eccentricity (figure 5.31).

In the initial phases of construction, when the first two spans of the deck are still in the curing zone (figure 5.32), a short launching nose which makes the approach operations easier and a steel tower are anchored to the deck. The latter is positioned so that when the nose tip is near the pier $n+1$ the tower is some meters behind the pier n. Transversely, the two legs of the tower are supported on the deck by means of long-stroke hydraulic jacks which permit modification of the vertical force transferred by the tower to the deck, and then simultaneous adjustment of the pull of stays. In fact, for the device to be effective in all launch phases, it is necessary to modify continuously the pull of stays because of the concentrated load that the tower transfers to the deck. When the tower approaches the pier the force is the highest and the system is efficient (figure 5.33), but when it advances to mid-span the same force would increase the positive moment, and it must be reduced by loosening stays.

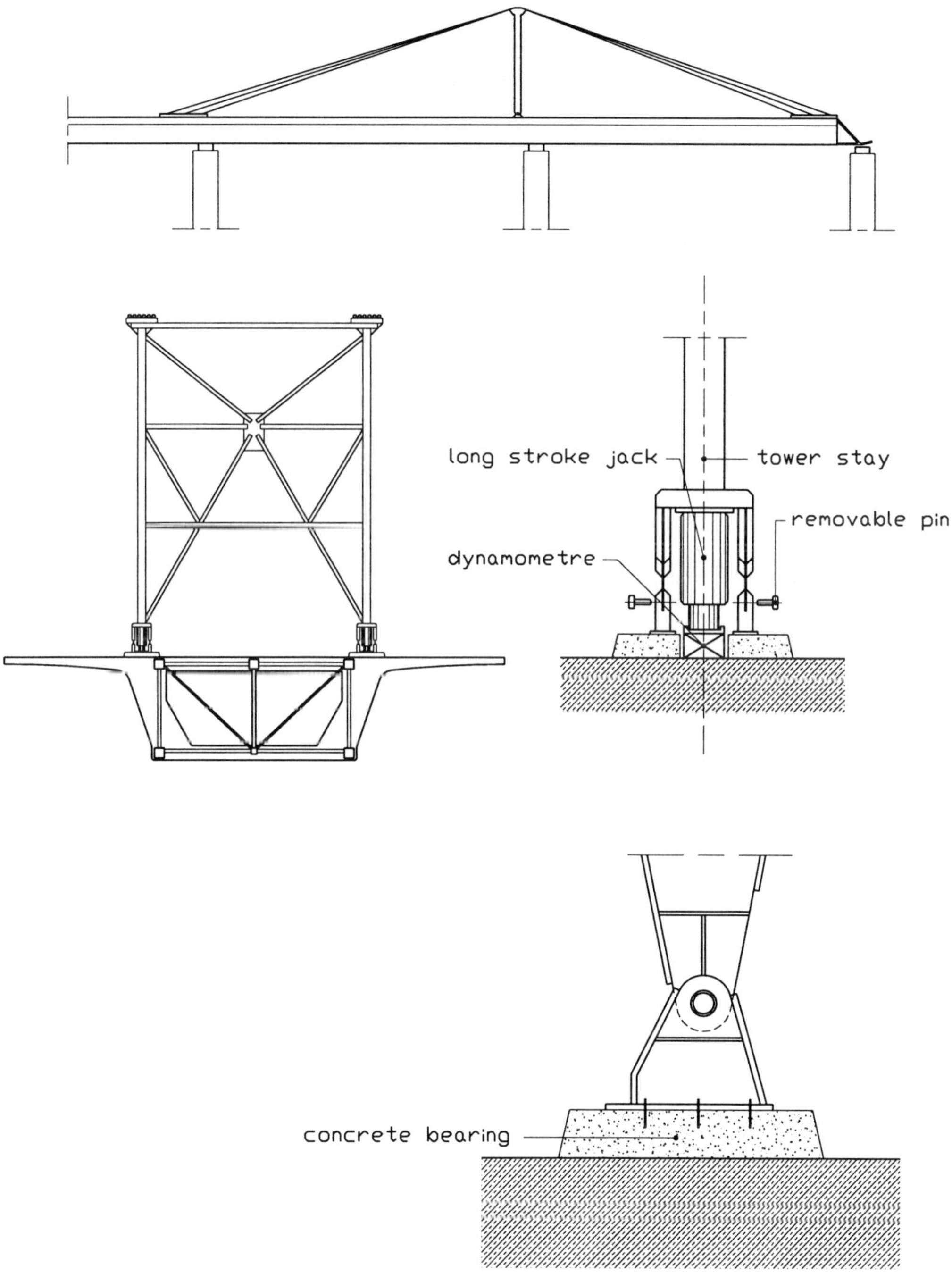

Figure 5.31 - Cable-stayed front scheme.

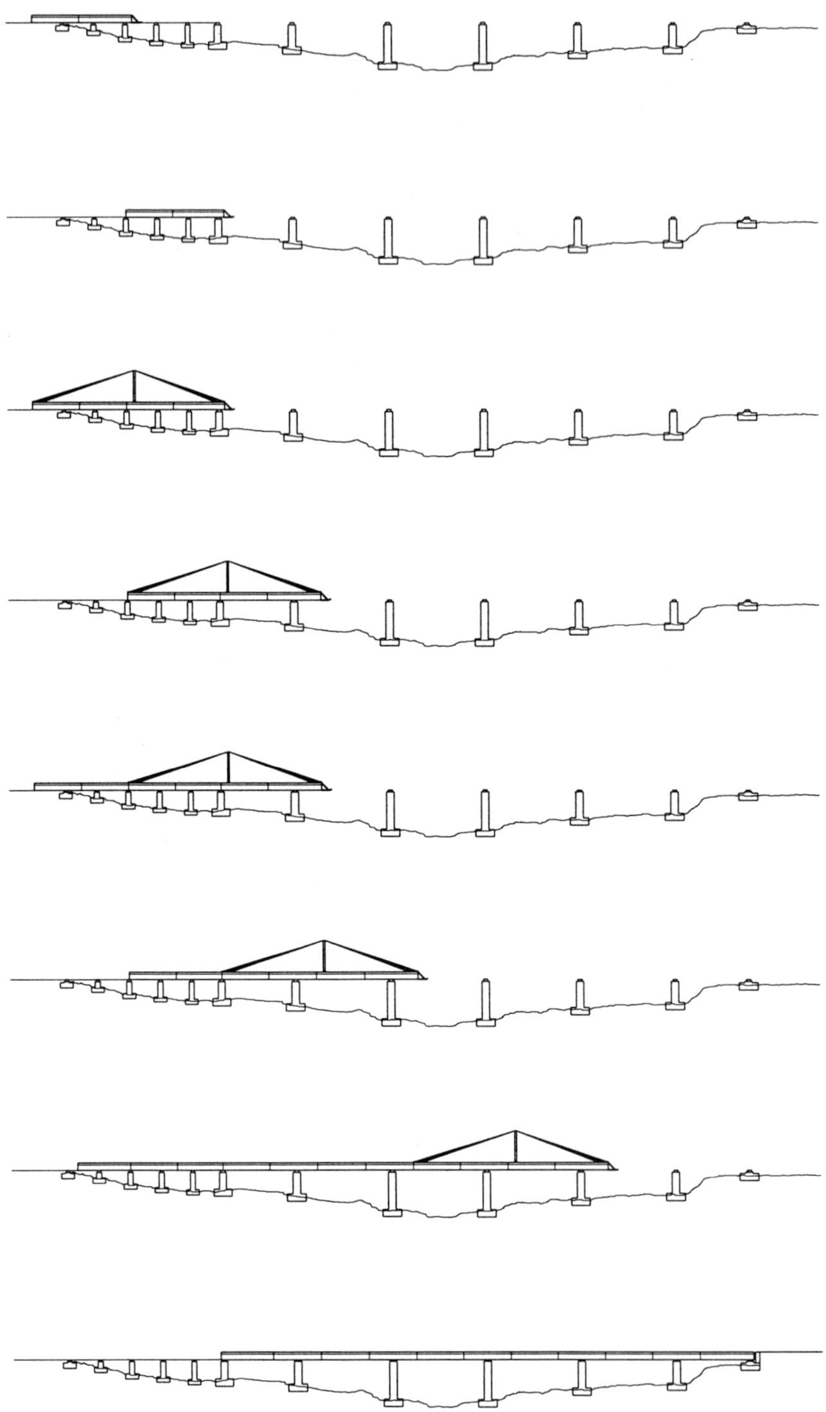

Figure 5.32 - Launch phases with the cable-stayed scheme. In case of a shortest launching nose, each launch is as long as a whole span.

Figure 5.33 - Cable-stayed scheme for the Boivre Bridge in the position of maximum pull in the stays (courtesy Spic Batignolles[16]).

In practice, many factors limit the use of this support scheme to the longest spans, and always in combination with a launching nose to reduce as much as possible the cantilevered weight (figure 5.35) and with it the required capacity of the stays:

- the need for a long curing yard (about two spans)[16]. When launching on the first span, a back deck segment cured enough to anchor stays and long enough to guarantee adequate safety against tilting is necessary, unless a temporary pier is added in the first span. At the same time, in the previous phases the absence of stays does not permit long cantilevers beyond the abutment;
- the complexity of calculation, since the tension in the stays depends on the elastic deflection of the deck-stays-nose system during launch, on the adjustable action introduced by the jacks of the tower, on the pier deformation, the bearing misplacement, and the thermal gradients both in the deck and in the stays;
- the need to control and to modify continuously the stresses in the tower and in the stays;
- the load concentration in the deck zone under the tower, which in spite of its specific stiffening often suffers cracking problems;
- the intrinsic delicateness of the solution, the risks involved in case of error, and the need always to reach the pier at the end of launch to avoid long stops with stressed stays, a fact that in the case of short launching noses adds further restraints on the length of the curing zone;

- the irreversible deck deformations produced by creep in cases of launches as long as the whole span, which suggest ending each launch with a front cantilever able to reduce the positive moment in the first span, or to adopt a longer launching nose, $0.4 \leq l_n / l \leq 0.6$, which permits halving the length of launch;
- compared with the use of the launching nose only, the greater length of the deck front zone that needs higher axial prestressing.

Figure 5.34 - Detail of the stay anchorage (courtesy Spie Batignolles[16]).

The combined use of a relatively short launching nose[28],

$$0.4 \leq \frac{l_n}{l} \leq 0.6 \qquad (5.46)$$

and of a cable-stayed scheme with a tower of height h_t:

$$0.40 \leq \frac{h_t}{l - l_n} \leq 0.44 \qquad (5.47)$$

is then generally limited to wide and heavy bridges and to 60 to $70\,m$ spans, and in this case it permits use of launching noses sized for spans 10 to 40% shorter. Over these spans the use of temporary piers is the most convenient solution.

This scheme is effective also in the case of a long central span obtained by launching symmetrically from the opposite abutments, since the cable-stayed support of the two cantilevers permits undertaking central spans well over a hundred meters long, as in the case of the Mainflingen Bridge[33].

Figure 5.35 - Charix Viaduct, launched on spans *64 m* long by means of a short launching nose and a cable-stayed front scheme (reproduced with permission, SCETAROUTE - Berenguier).

Compared with the use of a launching nose only, the action of stays modifies the evolution of launch stresses, since before the contact with the pier they absorb most of the excess of negative moment (compared to the end-of-launch value, figure 5.9) due to the use of a short launching nose. Once the support is reached, the stays are relieved and launch continues with the nose only.

The optimum flexural stiffness of a launching nose (5.46) with dead load (5.33) is computed based on the already mentioned criteria for a conventional nose, since it acts only after the pier has been reached and the stays are relieved. Since it is unavoidable to reach the EOL support moment M_B^{EOL} (5.32), it is convenient to design the deck for this value, and the action of stays so as not to exceed it in the cantilever configurations. At the beginning of launch the stays are unstressed, and they are not necessary until a critical advance l_{cr} which produces a negative moment in the front support equal to M_B^{EOL}, and which results in:

$$\alpha_{cr} = \frac{l_{cr}}{l} = -\frac{q_n}{q}\frac{l_n}{l} + \sqrt{\left(\frac{q_n}{q}\frac{l_n}{l}\right)^2 - 1.268\frac{q_n}{q}\left(\frac{l_n}{l}\right)^2 + 0.212} \qquad (5.48)$$

Beyond α_{cr}, the action of stays should limit the support moment below or at the EOL value until the contact with the pier – that is, for $\alpha_{cr} \le \alpha < 1 - l_n/l$. By imposing that the excess of M_B^* (5.18) over M_B^{EOL} be absorbed by the stays, if β is their inclination with respect to the horizontal, the following equation for the adjustment of the force F_s is obtained:

$$\frac{F_s}{ql} = \frac{1}{sin\beta}\left[\frac{\alpha}{2} + \frac{q_n}{q}\frac{l_n}{l} + \frac{0.634}{\alpha}\frac{q_n}{q}\left(\frac{l_n}{l}\right)^2 - \frac{0.106}{\alpha}\right] \tag{5.49}$$

This value is only preliminary, since the concentrated load transferred from the tower can produce excessive stresses in the span $B-C$, and therefore can require one to exceed M_B^{EOL} in the cantilever phase. During the course of launch, in each position assumed by the deck the pull in the stays must be higher than a minimum value which depends on the allowable tensile stress at the upper edge of the front support section, but lower than a maximum value which depends on the allowable stresses both in the back span and in the front one, suspended at its tip. Then, it is possible to define the diagram of the minimum and maximum tension in the stays as the deck position changes (figure 5.36 shows the case of a very short launching nose) and the actual procedure of tension adjustment must reside between these two curves[16]. If they are excessively close or if they intersect it is necessary to increase axial prestressing in the deck front zone, provided that this higher force, added to the axial component $F_s \cos\beta$ of the tension in the stays, is not excessive in terms of allowable compression.

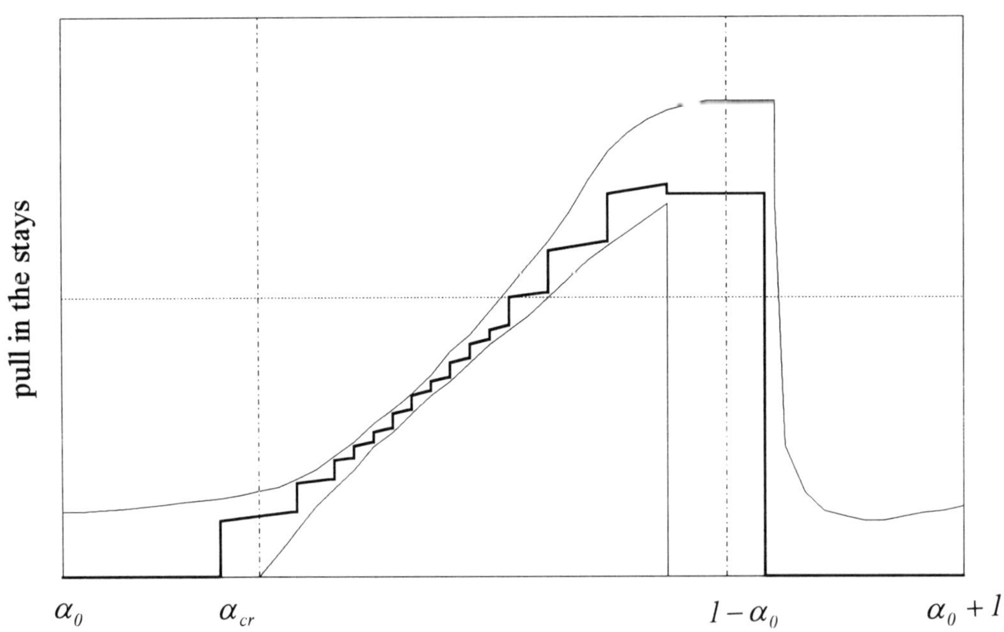

Figure 5.36 - Minimum and maximum pull in the stays as the deck position changes in case of a very short launching nose. The actual procedure of pull adjustment, thicker line, must fall between these two curves.

This diagram of tension is discontinuous, since adjustment occurs during the stops between the subsequent launches. The calculation of the vertical load on the tower is then immediate, and this value is used in adjusting hoisting jacks. In case of a narrow tension adjustment range, and therefore of very delicate operations, jacks can be placed on dynamometers[16] (figure 5.31).

Finally, once the pier is reached, the realignment of the nose front end is obtained by means of jacks or extensions with hydraulic pistons, to avoid excessive stresses in the stays and in the tower.

Temporary Piers

The first experiences in incremental launching were carried out by halving the spans by means of temporary piers. As years went by this solution has been adopted more and more rarely and only in bridges composed of very long or varying spans, in which the economies obtained by reducing the launch stresses can cover the cost of the piers.

Figure 5.37 - Temporary piers composed of precast reinforced concrete modules (courtesy Leonhardt[108]).

By adopting one temporary pier in each span, its relative settlement, η, produces a positive moment in the deck section over the pier and a negative moment in the support sections on the adjacent final piers, which in a deck with constant section and final spans l are both:

$$M \cong \pm \frac{24EI}{l^2}\eta \qquad (5.50)$$

The tensile stresses in the extreme fibers of these sections, to be absorbed by means of an equal additional launch prestressing, are:

$$\sigma_{u,l} \cong -\frac{24E\eta}{l^2} z_{u,l} \qquad (5.51)$$

They do not depend on the moment of inertia – that is, all decks are equally sensitive to differential settlement. On the contrary, deck flexural stiffness and construction tolerances reduce the actual possibility of obtaining satisfactory support reactions at all the temporary piers, and the stress reduction predictable in theory with a large number of piers is illusory because of the loss of contact at the supports.

Figure 5.38 - The Steine Bridge temporary piers were placed on transverse foundations, which allowed their shifting from one side to the other one (courtesy Greisch[6]).

Figure 5.39 - Shifting of the main temporary pier of the Steine Bridge, *80 m* high (courtesy Greisch[6]).

The use of only one temporary pier in each span permits use of shorter and lighter launching noses and some savings both in axial prestressing (although a larger number of parabolic tendons is often necessary) and in reinforcement, because of the lower influence of launch shear stresses. But the piers, their foundations, and the additional labor produce higher costs.

The use of more than one temporary pier per span is rare. In the Kufstein Bridge (figure 1.8), the three *102 m* interior spans were reduced to less than *35 m* with two temporary piers, while the *59 m* end spans required only one pier. So it was possible

Figure 5.40 - The temporary piers of the Steine Bridge were shifted by pulling strand cables (courtesy Greisch[6]).

to absorb edge tensile stresses only with reinforcement avoiding axial prestressing, and to use a light launching nose.

For the longest spans, temporary piers permit increase in the competitive threshold of incremental launching versus free cantilevering to more than *100 m*. The *168 m* central span of the Worth Bridge[74] was launched on three temporary supports. In case of varying spans, temporary piers permit standardization of the sequence of launching supports. In the Brohltal Bridge[98], sized for launching on a *50 m* span, the two *70 m* central spans were built by inserting two temporary piers and by leaving the final pier between them unloaded, so as to obtain three consecutive spans shorter than the design span.

The first temporary piers were made of concrete (some cast-in-situ as the final piers, others obtained by filling steel caissons or pipes) and were demolished at the end of launch. Later, modular precast towers were used, composed of reinforced concrete pillars or plates (figure 5.37) positioned under the deck webs, and braced transversally by steel members. Concrete modular elements are reusable, less deformable than steel towers, less sensitive to thermal variations, and generally less expensive under *10 m* height. On the other hand, steel towers are light (figure 5.38), and can be easily transferred from one side to the other in case of parallel decks, to halve the investment (figures 5.39 and 5.40).

Figure 5.41 - The temporary piers for the Milan-Palizzi Bridge were hinged at the base and cable-stayed at the cap. The symmetrical stays were contained in self-supporting steel pipes (coated with polyethylene to prevent electric arcs) to avoid the risk of a broken strand falling on the electric lines. The rotation of the pier was allowed by thick neoprene bottom plates (author).

Steel towers are deformable and can require the absorption of the horizontal force transferred by the deck by means of steel tension rods anchored to the final piers behind them (figure 5.38), or of strand stays anchored to the ground. In this case the stay tension may be adjusted by means of jacks controlled by dynamometers to avoid applying backward forces in kinetic friction conditions, once the static friction has been overcome. In the particular case of figure 5.41, the static scheme of the temporary pier was the one of a rod hinged at the base and stayed symmetrically at the cap, to reduce flexural stresses in the foundation. Even their vertical deformation can be high, and in the case of tall piers their best position in the launch static scheme is not in mid-span but slightly before it[6] (figure 5.42), as thus the increase of the end-of-launch moment caused by their incapacity to produce the reaction of a rigid support is reduced. At the same time, this position allows loading the four corners of the pier from the very realignment[6] (figure 5.38) in order to take advantage of the deck flexural stiffness in restraining the pier cap – that is, in

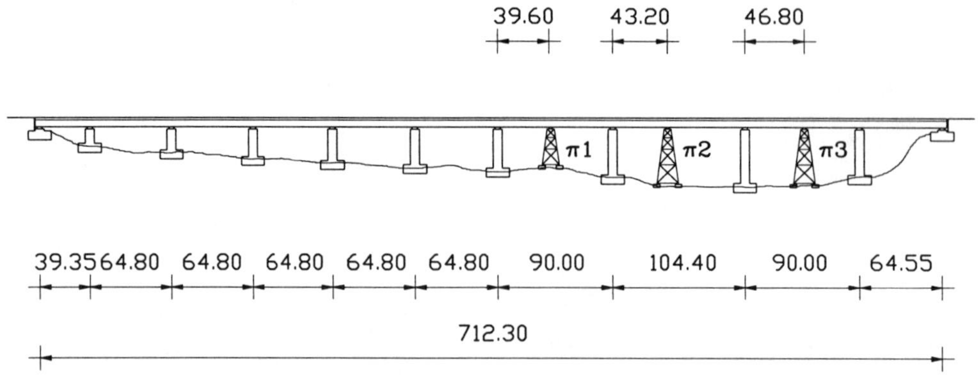

Figure 5.42 - Position of the temporary piers in the static scheme of the Steine Bridge (Figure 5.38).

Figure 5.43 - In the Hamburg Bridge, temporary piers on both sides of the permanent piers allowed launch of a bridge with permanent bearings narrower than the webs (courtesy Dywidag Systems International[5]).

reducing the effective length of the pier in the longitudinal direction to improve its stability.

Finally, steel temporary piers have a thermal inertia much lower than concrete ones and suffer sudden dimensional variations in case of changes of temperature. For this reason, in the most delicate cases, steel pipes are filled with concrete.

To conclude, temporary piers can be used successfully only in particular cases: low height of the piers, direct foundations, accessible areas, spans longer than *60 m* or very variable, bridges composed of two parallel decks built subsequently so as to transfer piers from one side to the other. Other than in these cases and on long spans, it can be convenient to reduce the deck slenderness (for instance[108] to $12 \leq H/l \leq 15$)

and to use long launching noses or short noses together with cable-stayed schemes, to avoid the use of temporary supports.

Finally, some particular uses of temporary piers: on the sides of a narrow pier with closely spaced permanent bearings[5], to widen the pier cap during launch and support the webs directly (figure 5.43); in a case of decks with varying width or radius of curvature (figure 4.27); and under the closure segment in decks launched symmetrically from both the abutments to guarantee its adequate curing.

Specific aspects of the design of temporary piers are described in Chapter 8.

SOLUTION OF THE CONTINUOUS BEAM

The construction of a bridge by incremental launching produces temporary stresses which are different from service stresses and which require optimization of the structural behavior of the deck even for launching stages.

For this purpose, the movement of the deck above fixed bearings would require the examination of an infinite number of static schemes, since the front segment of the deck overhangs beyond the first pier until reaching the next one, and the growth of the negative moment over the first support propagates backwards in the continuous beam. Even discretizing the problem – that is, checking a finite number of closely spaced support configurations, decks on the order of several hundreds meters long require calculation of hundreds of bending moment and shear force diagrams, each one calculated in the worst combination of thermal gradients, differential bearing settlement, and permanent deformations due to creep.

Such a ponderous task requires the use of a computer, by means of programs which simulate the launch and iteratively recall the continuous beam solver subroutine for the successive support configurations.

The calculation of an indeterminate system can be carried out in many ways, and in most cases (apart from artifices of symmetry or wise choice of redundant unknowns) there is a substantial equivalence between them all, since they are based on the same differential equations and require the same number of integrations. The practical interest of a method can therefore be found in the use of the same type of integrals, easily collectable in libraries, or in the application of an agile and repetitive algorithm which minimizes the risk of mistakes and can be easily implemented in a computer program.

The reduced-transfer-matrix (RTM) method[168] meets both these requisites, provides an exact, repetitive, and elegant solution, and greatly benefits from the repetitiveness of the segments composing launched bridges.

Let us consider the general case of a continuous beam contained in the $x-y$ plane, the local frame of reference with its x-axis on the centroidal line of the beam and the orthogonal y-axis, with any constraints, constant or variable cross section, and its principal axis of inertia contained in the plane. In each cross section, the state of stress is defined by three forces, N, V_y, and V_z, and three moments, M_x, M_y, and M_z, and the state of deformation by the three components of the displacement, η, and the three components of the rotation, φ.

Assuming only external forces orthogonal to the $x-y$ plane and external moments about in-plane axes, in each cross section the six load components reduce to three: orthogonal shear force $V = V_z$, bending moment $M = M_y$, and torsional moment $T = M_z$. Their internal work results from the orthogonal displacement $\eta = \eta_z$, the flexural rotation $\varphi = \varphi_y$, and the torsional rotation $\vartheta = \varphi_x$ of the section.

Since launched bridges are in most cases straight or moderately curved and the adoption of box girders (whose high torsional rigidity limits tangential stresses even when the torsional moment is relatively high) is nearly a rule, the straight beam theory is generally sufficient. Then, assuming only external forces applied along the x-axis, unknown deformations are only η and φ, and unknown load effects are only M and V.

To these components one has to add the action of prestressing, computed by applying a system of equivalent forces composed of orthogonal distributed loads along the curved parts of the tendons and of concentrated loads applied at their anchorages producing shear and axial components. The axial component on an anchorage is balanced by the same component on the opposite tendon anchorage and by the axial resultant of distributed friction, and this system of forces does not introduce secondary effects, since launching bearings, neglecting friction, do not prevent axial deformations in the deck. Therefore, the action of prestressing can be computed by considering separately the uniform compression directly obtainable from the axial force diagram and the flexural effects of the equivalent system: couples produced by the eccentricity of the anchorages, and orthogonal forces resulting from concentrated and distributed loads.

So, leaving axial force and the corresponding deformation out of consideration, in each cross section of the beam the elastic system can be described by a *"state array"* $\{S\}$:

$$\{S\} = \begin{Bmatrix} \eta \\ \varphi \\ M \\ V \end{Bmatrix} \qquad (5.52)$$

and the linear relation between the state-arrays of two subsequent sections by a matrix, $[T]$, which, "transferring" the definition of the state array, is called *"transfer matrix."*[103] Between an origin section, 0, and the section 1 of the beam it is:

$$\{S\}_1 = [T]_0^1 \times \{S\}_0 \qquad (5.53)$$

between the two ends of the subsequent beam segment it is still:

$$\{S\}_2 = [T]_1^2 \times \{S\}_1 \qquad (5.54)$$

that is:

$$\{S\}_2 = [T]_1^2 \times [T]_0^1 \times \{S\}_0 \qquad (5.55)$$

and in a beam composed of n segments the relation between the state arrays of its end sections is obtained by successively multiplying the transfer matrices of the single segments:

$$\{S\}_n = [T]_{n-1}^n \times [T]_{n-2}^{n-1} \times \ldots\ldots \times [T]_1^2 \times [T]_0^1 \times \{S\}_0 \qquad (5.56)$$

and can be summarized in just one transfer matrix:

$$\{S\}_n = [T]_0^n \times \{S\}_0 \qquad (5.57)$$

obtained by multiplying:

$$[T]_0^n = [T]_{n-1}^n \times [T]_{n-2}^{n-1} \times \ldots\ldots \times [T]_1^2 \times [T]_0^1 \qquad (5.58)$$

Hereinafter, the generic cross section will be defined with a number, and the support sections with a capital letter. The terms of the transfer matrix, $[T]_0^1$, between the ends of the beam segment oriented $0 \rightarrow 1$, of length, l, and constant EI, are obtained from the differential equations of the elastic beam. These are solved by means of four successive integrations, and integrals depend on four constants, which, if the state array of the origin section 0 is defined, are η_0, φ_0, M_0, and V_0, on the mechanical properties of the beam segment, and on the loads applied to it.

Neglecting shear deformation (a hypothesis generally acceptable for prestressed concrete box girders but excessively approximate for the prestressed composite sections of Chapter 9), integrals are:

$$\begin{cases} \eta_1 = \eta_0 + l\varphi_0 - \dfrac{l^2}{2EI} M_0 - \dfrac{l^3}{6EI} V_0 + f_1(p, l, E, I) \\ \varphi_1 = \varphi_0 - \dfrac{l}{EI} M_0 - \dfrac{l^2}{2EI} V_0 + f_2(p, l, E, I) \\ M_1 = M_0 + lV_0 - f_3(p, l) \\ V_1 = V_0 - f_4(p, l) \end{cases} \qquad (5.59)$$

and can be expressed in matrix form as:

$$\{S\}_1 = \begin{Bmatrix} \eta \\ \varphi \\ M \\ V \\ 1 \end{Bmatrix}_1 = \begin{bmatrix} 1 & l & -\frac{l^2}{2EI} & -\frac{l^3}{6EI} & f_1(p,l,E,I) \\ 0 & 1 & -\frac{l}{EI} & -\frac{l^2}{2EI} & f_2(p,l,E,I) \\ 0 & 0 & 1 & l & -f_3(p,l) \\ 0 & 0 & 0 & 1 & -f_4(p,l) \\ 0 & 0 & 0 & 0 & 1 \end{bmatrix}_0^1 \times \begin{Bmatrix} \eta \\ \varphi \\ M \\ V \\ 1 \end{Bmatrix}_0 = [T]_0^1 \times \{S\}_0 \quad (5.60)$$

where the fifth row has been introduced to keep the transfer matrix square, all the terms of its first four columns depend only on the mechanical and geometrical properties of the beam segment, and only the terms of the fifth column depend also on external loads.

In launched bridges, few load types characterize construction and service stages, and a complete library of cases is easy to create. For a distributed load, p, applied uniformly to the whole $0 \rightarrow 1$ segment, by integrating one obtains:

$$T_{15} = \frac{pl^4}{24EI} \quad T_{25} = \frac{pl^3}{6EI} \quad T_{35} = -\frac{pl^2}{2} \quad T_{45} = -pl \qquad (5.61)$$

For a distributed load whose density varies linearly from p_0 in section 0 to p_1 in section 1:

$$\begin{aligned} T_{15} &= \frac{4p_0 + p_1}{120EI} l^4 \quad T_{25} = \frac{3p_0 + p_1}{24EI} l^3 \\ T_{35} &= -\frac{2p_0 + p_1}{6} l^2 \quad T_{45} = -\frac{p_0 + p_1}{2} l \end{aligned} \qquad (5.62)$$

and the generalization for any type of distributed load is immediate by observing that the terms T_{i5} are the successive integrals of the $-p(x)$ function, divided by $-EI$ in the first two rows. The case of distributed loads applied only to some portions of the beam segment, or of variable EI, can always be solved by integration, but it is generally convenient to divide the beam segment into more sub-segments with full distributed load or approximately constant EI, to obtain the transfer matrix by their multiplication. For a prestressing force, N, whose eccentricity varies linearly from e_0 in section 0 to e_1 in section 1 (parabolic prestressing or axial prestressing but variable position of the centroid), the beam segment is loaded by a moment which varies from $M_0 = Ne_0$ to $M_1 = Ne_1$, and it is:

$$T_{15} = -\frac{2M_0 + M_1}{6EI} l^2 \quad T_{25} = -\frac{M_0 + M_1}{2EI} l \quad T_{35} = 0 \quad T_{45} = 0 \qquad (5.63)$$

For a linear thermal gradient between a warmer top surface and a colder bottom one, the constant value of the moment:

$$M_0 = M_1 = EI\frac{\alpha \Delta T}{H} \qquad (5.64)$$

should be introduced in (5.63). In the case of concentrated loads or distortions applied to a section at distance x from the local origin 0, an infinitesimal sub-segment is introduced. Once the $0 \rightarrow 1$ segment is divided into three sub-segments $0 \rightarrow l$, $l \rightarrow r$ and $r \rightarrow 1$, with the central one limited by the cross sections l immediately on the left and r immediately on the right of the section where the concentrated action is applied, the transfer matrix $[T]_0^1$ is again obtained by multiplying:

$$\{S\}_1 = [T]_0^1 \times \{S\}_0 = [T]_r^1 \times [T]_l^r \times [T]_0^l \times \{S\}_0 \qquad (5.65)$$

with:

$$[T]_0^l = \begin{bmatrix} 1 & x & -\dfrac{x^2}{2EI} & -\dfrac{x^3}{6EI} & 0 \\ 0 & 1 & -\dfrac{x}{EI} & \dfrac{x^2}{2EI} & 0 \\ 0 & 0 & 1 & x & 0 \\ 0 & 0 & 0 & 1 & 0 \\ 0 & 0 & 0 & 0 & 1 \end{bmatrix} \qquad (5.66)$$

$$[T]_l^r = \begin{bmatrix} 1 & 0 & 0 & 0 & \Delta\eta \\ 0 & 1 & 0 & 0 & \Delta\varphi \\ 0 & 0 & 1 & 0 & \Delta M \\ 0 & 0 & 0 & 1 & \Delta V \\ 0 & 0 & 0 & 0 & 1 \end{bmatrix} \qquad (5.67)$$

$$[T]_r^1 = \begin{bmatrix} 1 & (l-x) & -\dfrac{(l-x)^2}{2EI} & -\dfrac{(l-x)^3}{6EI} & 0 \\ 0 & 1 & -\dfrac{(l-x)}{EI} & -\dfrac{(l-x)^2}{2EI} & 0 \\ 0 & 0 & 1 & (l-x) & 0 \\ 0 & 0 & 0 & 1 & 0 \\ 0 & 0 & 0 & 0 & 1 \end{bmatrix} \qquad (5.68)$$

If the concentrated action comes from an elastic support (piers with different height or temporary piers alternated with the stiffer final ones) and therefore is:

$$R = \Delta V = k\eta \qquad (5.69)$$

the $l \rightarrow r$ transfer-matrix becomes:

$$[T]_l^r = \begin{bmatrix} 1 & 0 & 0 & 0 & 0 \\ 0 & 1 & 0 & 0 & 0 \\ 0 & 0 & 1 & 0 & 0 \\ k & 0 & 0 & 1 & 0 \\ 0 & 0 & 0 & 0 & 1 \end{bmatrix} \qquad (5.70)$$

Usually, the solution of the continuous beam takes place by assuming support reactions as redundant unknowns[103].

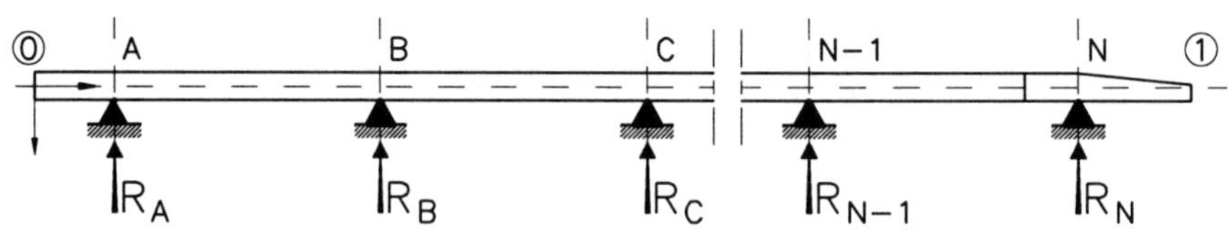

Figure 5.44 - Static scheme of the continuous beam.

Starting from the free back end 0 of the deck (figure 5.44), once the transfer matrix $[T]_0^A$ between the ends of the back cantilever is defined, it is:

$$\{S\}_A = [T]_0^A \times \{S\}_0 \qquad (5.71)$$

In the same way, once the transfer matrix $[T]_A^B$ between the ends of the first span is defined, it is:

$$\{S\}_B = [T]_A^B \times \{S\}_A \qquad (5.72)$$

where $[T]_A^B$ also includes the unknown support reaction R_A, and consequently:

$$\{S\}_B = [T]_A^B \times [T]_0^A \times \{S\}_0 \qquad (5.73)$$

Proceeding so, after the last support, N, the free front end of the launching nose, section 1, is reached. The relation between the state arrays of the beam end sections is:

$$\{S\}_1 = [T]_N^1 \times [T]_{N-1}^N \times \ldots\ldots \times [T]_A^B \times [T]_0^A \times \{S\}_0 \qquad (5.74)$$

in which each span-by-span transfer-matrix except for $[T]_0^A$ contains one redundant unknown, and can be summarized in:

$$\{S\}_1 = [T]_0^1 \times \{S\}_0 \qquad (5.75)$$

Without these redundant unknowns, (5.75) would consist of four linear equations with eight unknowns, and the additional four equations should be obtained by imposing the boundary conditions in the state arrays of the two end sections. The presence of redundant unknowns requires the definition of an equal number of additional equations, which can be obtained by calculating the support section state arrays and by imposing in them zero deflection, or a definite one in case of settlement. For this reason, during the launch, the number of equations progressively increases because of the introduction of new supports, and their solution requires an iterative algorithm, generally slower than the direct solution. But, above all, the method itself loses simplicity and repetitivity.

Since additional unknowns are concentrated in the support section state arrays, where they constitute the only possible discontinuity, it is convenient to limit the analysis to the continuous terms of these arrays (the rotation, φ, for compatibility and the bending moment, M, for equilibrium) and to calculate the remaining ones later. Then, considering two subsequent support sections, J and K, of the beam, which limit its generic span, $J \rightarrow K$:

$$\{S\}_K = [T]_J^K \times \{S\}_J \qquad (5.76)$$

that is:

$$\begin{cases} \eta_K = \eta_J + T_{12}\varphi_J + T_{13}M_J + T_{14}V_J + T_{15} \\ \varphi_K = \varphi_J + T_{23}M_J + T_{24}V_J + T_{25} \\ M_K = M_J + T_{34}V_J + T_{35} \\ V_K = V_J + T_{45} \end{cases} \qquad (5.77)$$

in which the coefficients are the terms of the $[T]_J^K$ transfer matrix. Since J and K are two support sections, their vertical deflections are null, or constant when settlements are introduced to simulate the movements imposed on the deck in order to replace neo-flon plates or to remove launching bearings. Therefore, once set:

$$\eta_J = \dot{\eta}_J = \text{constant} \qquad \eta_K = \dot{\eta}_K = \text{constant} \qquad (5.78)$$

by extracting V_J from the first equation of (5.77) and substituting it in the other three, a direct relation between rotations and bending moments in the two support sections can be obtained:

$$\begin{cases} \dot{\eta}_K = \text{constant} \\ V_K = -\dfrac{T_{12}}{T_{14}}\varphi_J - \dfrac{T_{13}}{T_{14}}M_J + \dfrac{1}{T_{14}}\left(\dot{\eta}_K - \dot{\eta}_J - T_{15}\right) + T_{45} \end{cases}$$

$$\begin{cases} \varphi_K = \left(1 - T_{24}\dfrac{T_{12}}{T_{14}}\right)\varphi_J + \left(T_{23} - T_{24}\dfrac{T_{13}}{T_{14}}\right)M_J + \dfrac{T_{24}}{T_{14}}\left(\dot{\eta}_K - \dot{\eta}_J - T_{15}\right) + T_{25} \\ M_K = -T_{34}\dfrac{T_{12}}{T_{14}}\varphi_J + \left(1 - T_{34}\dfrac{T_{13}}{T_{14}}\right)M_J + \dfrac{T_{34}}{T_{14}}\left(\dot{\eta}_K - \dot{\eta}_J - T_{15}\right) + T_{35} \end{cases} \qquad (5.79)$$

Once the *"reduced state array"* of the support section is defined[168]:

$$\{S_R\} = \begin{Bmatrix} \varphi \\ M \\ 1 \end{Bmatrix} \qquad (5.80)$$

the relative *"reduced transfer matrix"*, RTM, is obtained from the third and fourth equation of (5.79):

$$[T_R]_J^K = \begin{bmatrix} 1 - \dfrac{T_{24}T_{12}}{T_{14}} & T_{23} - \dfrac{T_{24}T_{13}}{T_{14}} & \dfrac{T_{24}}{T_{14}}\left(\dot{\eta}_K - \dot{\eta}_J - T_{15}\right) + T_{25} \\ -\dfrac{T_{34}T_{12}}{T_{14}} & 1 - \dfrac{T_{34}T_{13}}{T_{14}} & \dfrac{T_{34}}{T_{14}}\left(\dot{\eta}_K - \dot{\eta}_J - T_{15}\right) + T_{35} \\ 0 & 0 & 1 \end{bmatrix} \qquad (5.81)$$

In the continuous beam of figure 5.44, the relation between the reduced state arrays of the end support sections is:

$$\{S_R\}_N = [T_R]_{N-1}^N \times \ldots\ldots \times [T_R]_A^B \times \{S_R\}_A = [T_R]_A^N \times \{S_R\}_A \quad (5.82)$$

that is, two equations with four unknowns. Since bending moments, M_A and M_N, are immediately calculable, unknown terms are only the two rotations, φ_A and φ_N, and (5.82) can be immediately solved:

$$\varphi_A = \frac{1}{T_{R21}}\left(M_N - T_{R22} M_A - T_{R23}\right) \qquad (5.83)$$

Knowing φ_A, and therefore the reduced state array $\{S_R\}_A$, with (5.82) one can define the rotation and the bending moment in each support section of the deck, and with the second equation of (5.79), the shear force. Once the definition of the support section state arrays is completed ($\dot{\eta}_I$ is assumed constant), shear, bending moment, deflection, and rotation in each point of the deck can be obtained by means of the transfer matrix between the section at issue and the nearest support section.

A small spreadsheet program has been developed in order to show the advantages of the RTM method compared with more consolidated methods (equilibrium and compatibility, iterative, finite differences, finite elements) which all suffer the progressive increase of the grade of redundancy in the course of launch. The use of the RTM method in a spreadsheet permits taking advantage of all the features of the software package (graphics, file management, friendly environment) and to obtain an exact solution which is perfectly comparable with the one obtained by other methods.

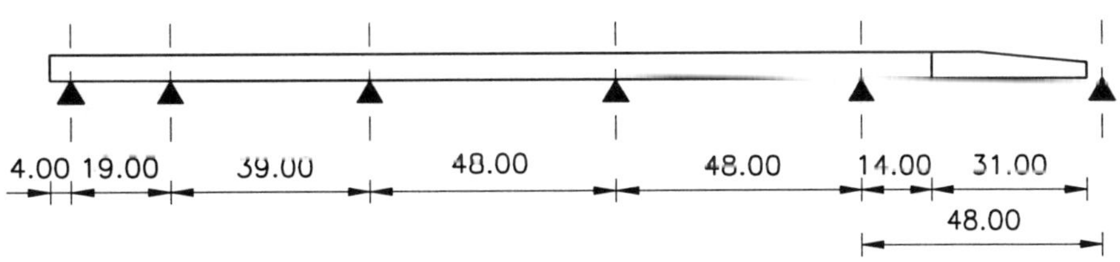

Figure 5.45 - Continuous beam analyzed by the RTM method and a spreadsheet, composed of a concrete deck with $q = 0.207\ MN/m$, $E = 37000\ MPa$, and $I = 6.8m^4$, and of a steel launching nose with $q_n = 0.024\ MN/m$, $E_n = 210000\ MPa$, and $I_n = 0.24m^4$.

To analyze the sample structure of figure 5.45, the first step is the calculation of the span-by-span transfer matrices (figure 5.46), and then of the reduced ones (figure 5.47). By successively multiplying the latest (figure 5.48), one obtains the RTM $[T_R]_A^E$ between the two end supports of the continuous beam, and then, with (5.82), the support moments.

	$[T]_A^B$	$[T]_B^C$	$[T]_C^D$ and $[T]_D^E$
T_{12}	19	39	48
T_{13}	-0.000717	-0.003023	-0.004579
T_{14}	-0.004544	-0.039295	-0.073259
T_{15}	0.004467	0.079306	0.181976
T_{23}	-0.000076	-0.000155	-0.000191
T_{24}	-0.000717	-0.003023	-0.004579
T_{25}	0.000941	0.008134	0.015165
T_{34}	19	39	48
T_{35}	-37.3635	-157.4235	-238.464
T_{45}	-3.933	-8.073	-9.936

Figure 5.46 - Terms of the span-by-span transfer-matrices for the continuous beam of Figure 5.45 under only dead load (neither thermal gradients nor support settlement have been taken into account).

	$[T_R]_A^B$	$[T_R]_B^C$	$[T_R]_C^D$ and $[T_R]_D^E$
T_{R11}	-2	-2	-2
T_{R12}	0.000038	0.000078	0.000095
T_{R13}	0.000235	0.002033	0.003791
T_{R21}	79452.63	38707.69	31450
T_{R22}	-2	-2	-2
T_{R23}	-18.68175	-78.71175	-119.232

Figure 5.47 - Terms of the reduced-transfer-matrices.

The course of launch in the $E \rightarrow F$ span is analyzed in two subsequent phases. The first one applies the static scheme of figure 5.45 until the contact with pier F (figure 5.49). The second one consists of the recovery of the elastic deflection of the launching nose at pier F and the subsequent launch stages until the arrival of the

$$[T_R]_A^C = [T_R]_B^C \times [T_R]_A^B = \begin{bmatrix} -2 & 0.000078 & 0.002633 \\ 38707.69 & -2 & -78.71175 \\ 0 & 0 & 1 \end{bmatrix}_B^C \times \begin{bmatrix} -2 & 0.000038 & 0.000235 \\ 79452.63 & -2 & -18.68175 \\ 0 & 0 & 1 \end{bmatrix}_A^B = \begin{bmatrix} 10.157895 & -0.000231 & 0.000115 \\ -236320.65 & 5.461538 & -32.246885 \\ 0 & 0 & 1 \end{bmatrix}_A^C$$

$$[T_R]_A^D = [T_R]_C^D \times [T_R]_A^C = \begin{bmatrix} -2 & 0.000095 & 0.003791 \\ 31450 & -2 & -119.232 \\ 0 & 0 & 1 \end{bmatrix}_C^D \times \begin{bmatrix} 10.157895 & -0.000231 & 0.000115 \\ -236320.65 & 5.461538 & -32.246885 \\ 0 & 0 & 1 \end{bmatrix}_A^C = \begin{bmatrix} -42.8583 & 0.000982 & 0.000485 \\ 792107.09 & -18.173077 & -51.111418 \\ 0 & 0 & 1 \end{bmatrix}_A^D$$

$$[T_R]_A^E = [T_R]_D^E \times [T_R]_A^D = \begin{bmatrix} -2 & 0.000095 & 0.003791 \\ 31450 & -2 & -119.232 \\ 0 & 0 & 1 \end{bmatrix}_D^E \times \begin{bmatrix} -42.8583 & 0.000982 & 0.000485 \\ 792107.09 & -18.173077 & -51.111418 \\ 0 & 0 & 1 \end{bmatrix}_A^D = \begin{bmatrix} 161.275304 & -0.003698 & -0.002053 \\ -2932107.69 & 67.230769 & -1.771442 \\ 0 & 0 & 1 \end{bmatrix}_A^E$$

Figure 5.48 - Calculation of the reduced-transfer-matrix, $[T_R]_A^E$, between the two end supports of the continuous beam.

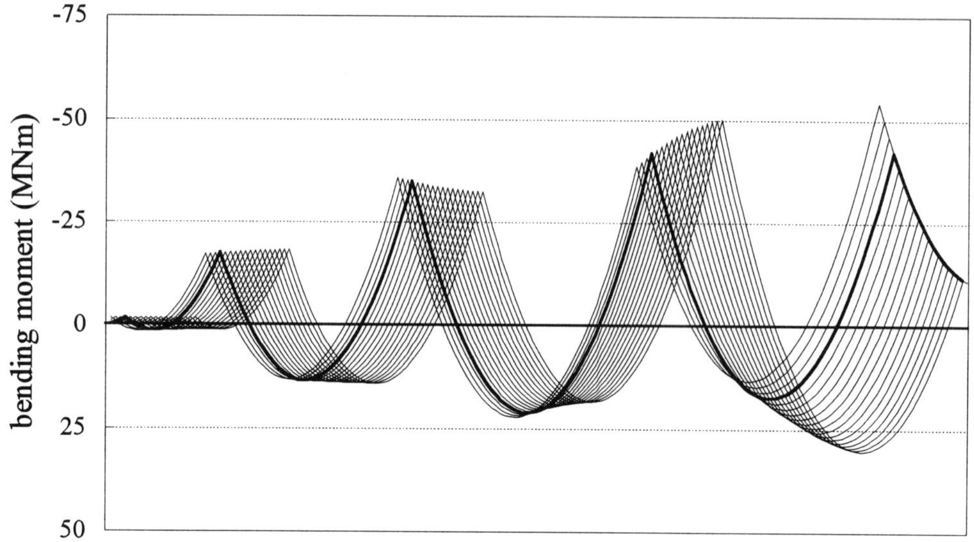

Figure 5.49 - Dead load bending moments (aligned supports, no thermal gradients) with the course of launch, starting from the position with the nose-deck joint section on pier E, and until the contact of the nose tip with pier F. The thicker line is the moment diagram for the support scheme of figure 5.45.

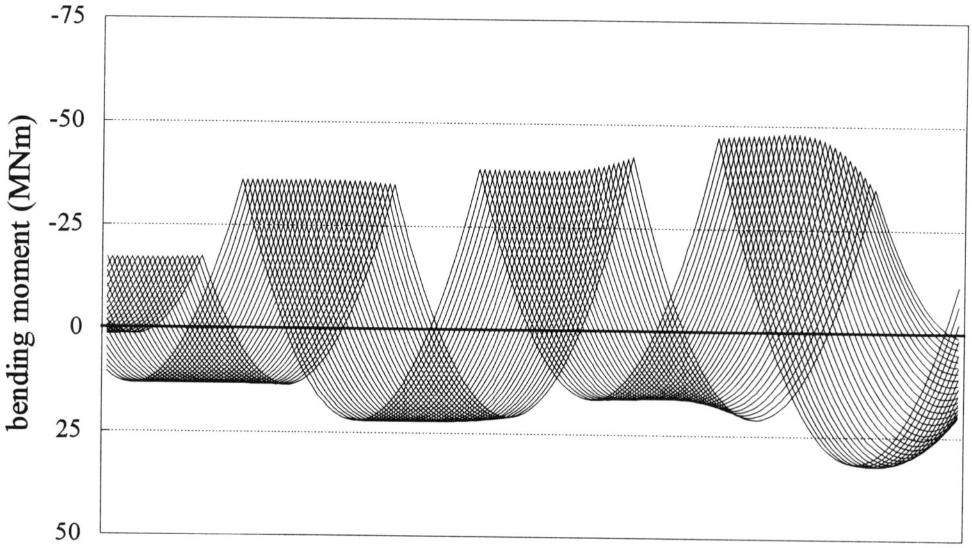

Figure 5.50 - Once pier F has been reached by the nose, the new transfer-matrix, $[T]_E^F$ (whose terms vary during launch until the nose-deck joint section reaches pier F) and the relative RTM, $\left[T_R\right]_E^F$, permit one to define $\left[T_R\right]_A^F$ and to analyze the deflection recovery and the course of launch until the subsequent contact.

concrete deck (figure 5.50). After contact, it is necessary to analyze a new static scheme which includes the new support F, and this requires a new transfer matrix, $[T]_E^F$, whose terms vary with the course of launch according to the different lengths of the concrete and steel segments in the $E \to F$ span. Once the new RTM $\left[T_R\right]_E^F$ has been defined, $\left[T_R\right]_A^F$ is obtained with a further multiplication.

Figure 5.51 compares the evolution of M_E, analyzed by the model of the nose-deck elastic system examined in this chapter, by finite elements, and by the RTM method.

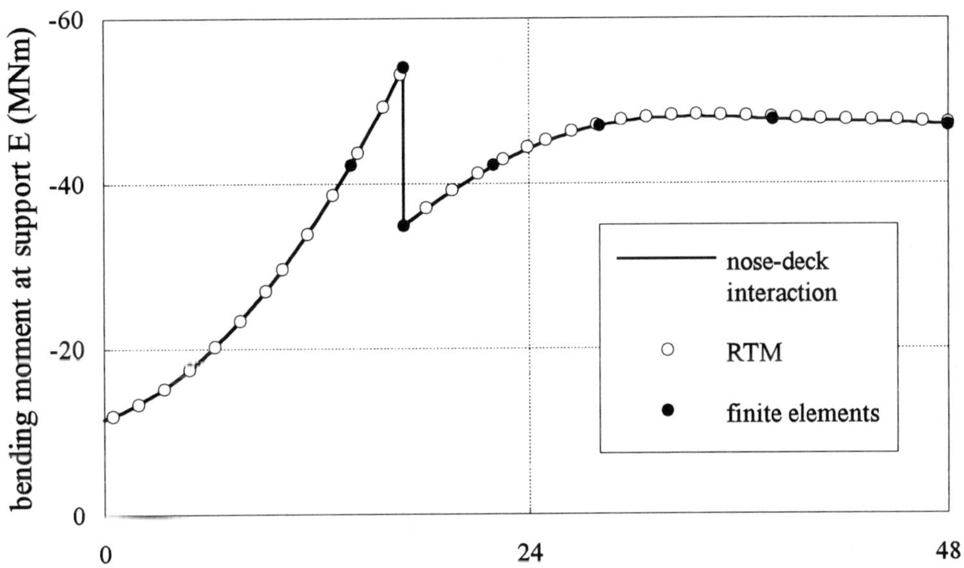

Figure 5.51 - Evolution of M_E analyzed by the model of the nose-deck interaction (continuous line), by finite elements (black dots), and by the RTM method (white dots). In the case of constant dead load and span length, results are practically equal, but both the RTM and finite element methods permit analysis of any static scheme and load condition.

Since all these methods are based on the same differential equations, results are practically equal. As finite elements, the RTM method allows one to analyze continuous beams with varying loads, flexural stiffness, and span, while the assumptions of the nose-deck model are more restrictive.

The superimposition of the diagrams of figures 5.49 and 5.50 is shown in figure 5.52, and permits to draw immediately the launching moment envelope (figure 5.53).

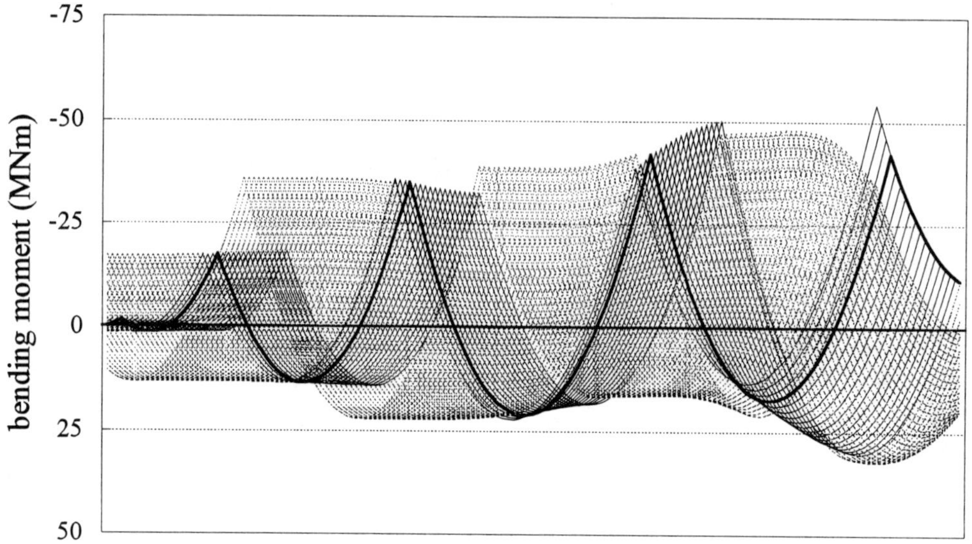

Figure 5.52 - The superimposition of the bending moment diagrams for the two launch phases (figures 5.49 and 5.50) completes the flexural stress analysis.

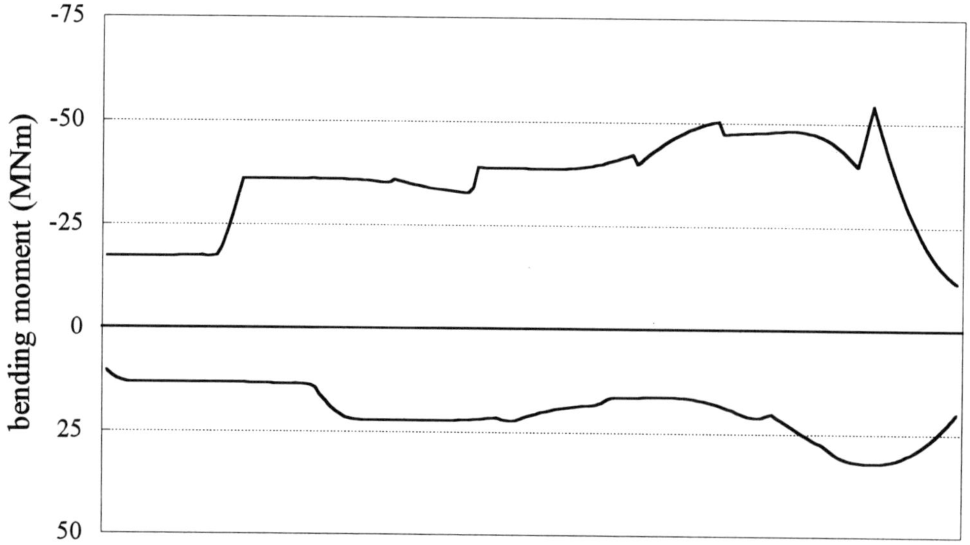

Figure 5.53 - Envelope of the launch bending moment diagrams (aligned supports and no thermal gradients) obtained by means of the @MIN and @MAX functions of the spreadsheet.

By modifying the length of the launching nose (in figure 5.54, the case of $l_n/l = 0.40$), the recalculation is immediate, and the optimization of the nose-deck elastic system is possible by trial and error.

The same results can certainly be obtained by finite elements, but through a long hand manipulation of outputs and a lot of subsequent iterations for the different static schemes, while the RTM method permits a fast iterative analysis.

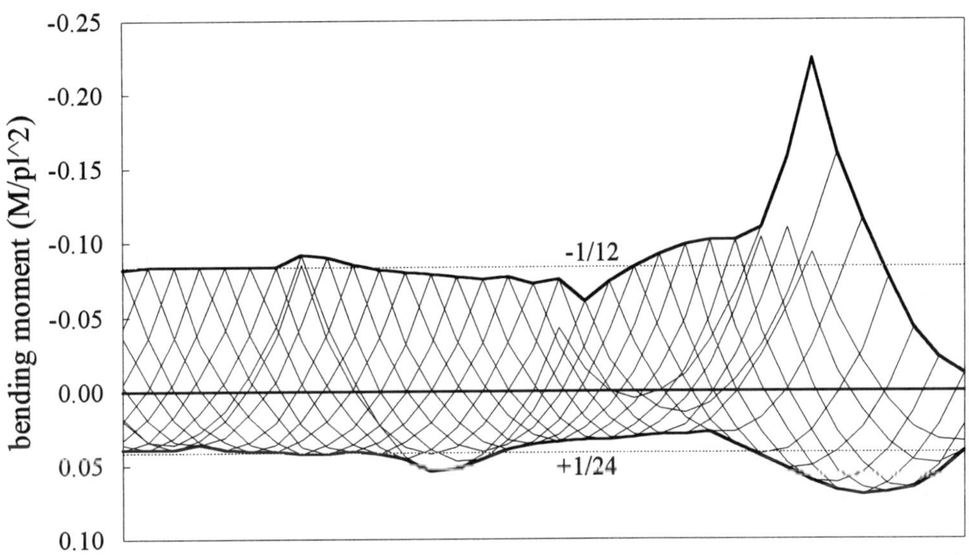

Figure 5.54 - By modifying the nose length (in this case $l_n/l = 0.40$) the moment envelope in the two front spans changes significantly (for research purposes, it can be immediately expressed in dimensionless form).

SHEAR AND BENDING MOMENT ENVELOPE

The use of the envelopes of bending and shear diagrams, and from these of the envelopes of extreme fiber stresses, instead of hundreds and hundreds of printout pages, is definitely advantageous. In addition to convenience, envelopes permit one to visually summarize the evolution of launch stresses and to limit local strengthening to the zones in which it is strictly necessary.

In the case of constant dead load, equal spans, and optimized launching nose, the bending moment envelope[188] looks like figure 5.55. Figure 5.56 shows the bending moment envelope in the three front spans of the Amiens Viaduct[83]. Thanks to external prestressing, this bridge (two *41.5 m* end spans and seventeen *50.5 m* interior ones) is characterized by the light cross section of figure 5.57.

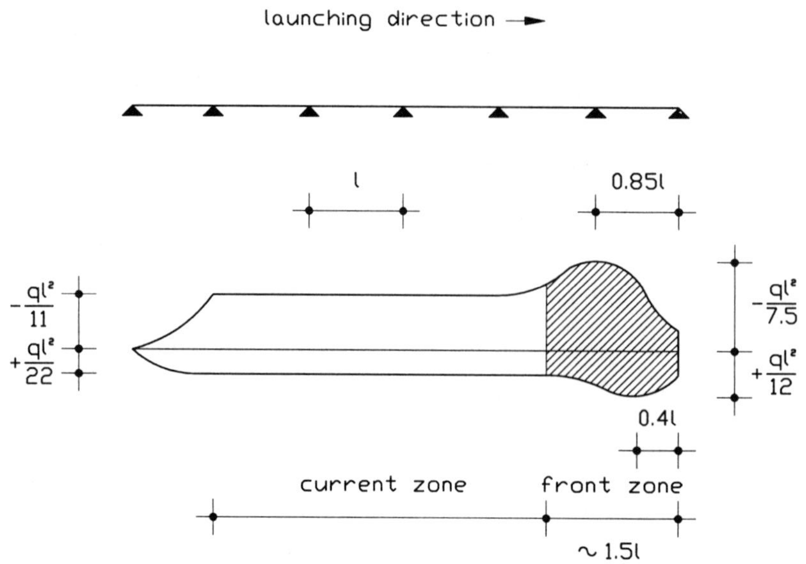

Figure 5.55 - Typical bending moment envelope for constant dead load and span length, and optimized launching nose (aligned supports and no thermal gradients).

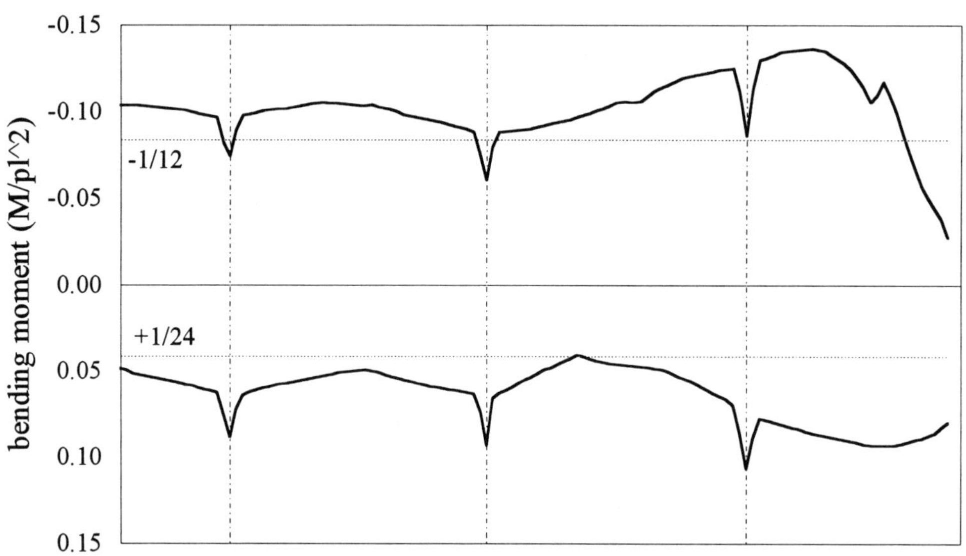

Figure 5.56 - Dimensionless moment envelope in the three front spans of the Amiens Viaduct ($l_n / l = 0.69$).

The curve presents aspects common to most launched bridges and takes into account both dead load and the equivalent system of prestressing, since the different heights of the centroids and the eccentric local tendons influence the stress diagrams:

- in the back current zone, the migration of the support diaphragms produces regular undulations in the envelope with a period equal to the span length, since the highest moments are reached when these heavier sections are in mid-span;
- the envelope of positive moment has its maximum at the support sections, and the envelope of the negative one at the mid-span sections, on which the deck is supported when diaphragms are in mid-span;
- ΔQ being the concentrated load corresponding to the increase in weight in the support sections, the amplitude of these undulations is approximately:

$$\Delta M \cong \pm \Delta Q \frac{l}{8} \qquad (5.84)$$

- both for the positive and negative moments. Since full or thick diaphragms cause oversizing of axial prestressing it can be convenient to build them only after finished launch;
- in the stiffened support sections, the eccentricity of the prestressing force (generally balanced on the centroidal axis of the current section) from a definitely lower centroid causes a positive peak, important even with low eccentricities (in the Amiens Viaduct of only *0.14 m*).

The envelopes often leave the effects of thermal gradients and differential settlement out of consideration, since the first are immediately superimposable and the second produce a high number of load combinations, and therefore are calculated for some typical situations and then added. In case of temporary piers, the envelopes are affected by the higher deformability of some bearings, which is made worse, in the case of steel piers, by further imposed deformations produced by varying thermal inertia.

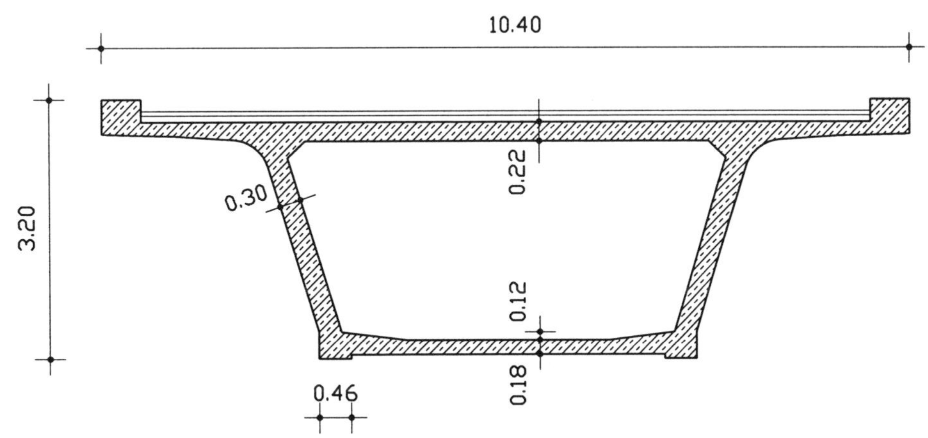

Figure 5.57 Cross section of the Amiens Viaduct. Thanks to external prestressing, the web thickness is only *0.30 m* for *50.5 m* long spans.

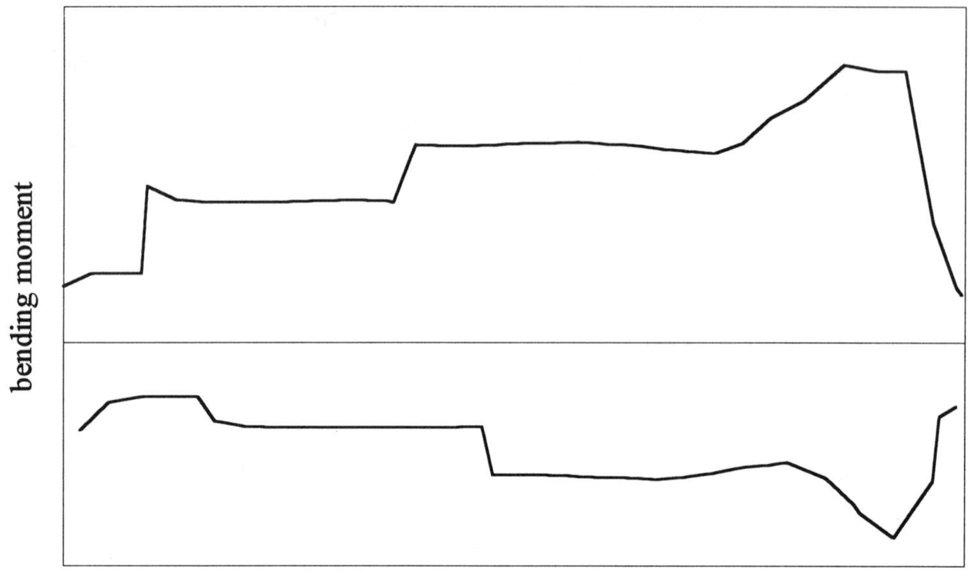

Figure 5.58 - Launch moment envelope in the presence of a central *58 m* long span in a regular sequence of *48 m* long spans. The whole deck segment which passes above this span needs oversizing.

Finally, important anomalies are produced by irregular sequences of spans. The presence of a longer span in a regular spacing influences the moment envelope of the whole deck section which passes above this span. In this case the envelope presents horizontal steps aligned on the subsequent spans, as in figure 5.58. In the case of irregular spans, therefore, the use of homogenizing temporary piers or the temporary elimination of some of the final piers can be convenient. Otherwise, in the presence of symmetry and of a long central span, one can launch from both the abutments with mid-span closure.

EDGE STRESSES

Even the extreme fiber stress-envelopes generally represent only the effects of dead load and the prestressing equivalent system, and the effects of differential settlement and thermal gradients are added later.

For the bending moment envelope of figure 5.56, the envelopes of the longitudinal stresses respectively in the upper and the lower extreme fibers of the deck current zone[83] are shown in figures 5.59 and 5.60. While the upper edge has adequate margins from decompression, the lower edge is decompressed by the dead load alone, and the tensile stresses produced by settlement and thermal gradients (a *5°C* gradient and settlement of *0.01 m* produced tensile stresses of *1.5 MPa*) are resisted by bottom slab reinforcement.

In figures 5.61 and 5.62, the envelopes of the longitudinal stresses in the deck front zone are drawn[83]. In the upper edge, the maximum compression is not excessive, even in cases of thermal gradients, and the minimum is similar that in the current zone. In the lower edge, the low tension is acceptable (in the end span of the continuous beam, the effects of thermal gradients and bearing settlement are smaller) and the maximum compression can be reduced by locally stiffening the deck, or absorbed by increasing the concrete strength.

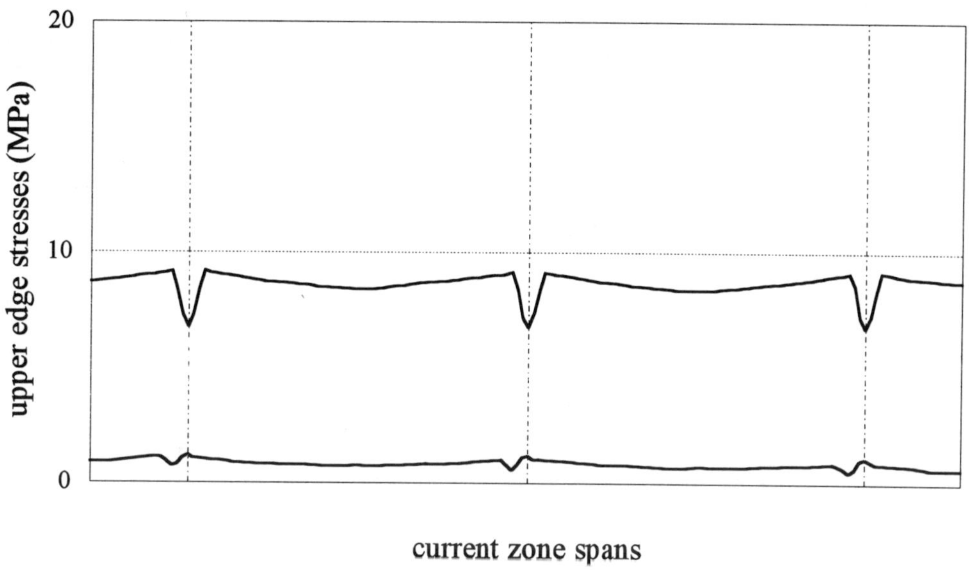

Figure 5.59 - Deck current zone. Envelope of the upper edge stresses due to the moment envelope of figure 5.56.

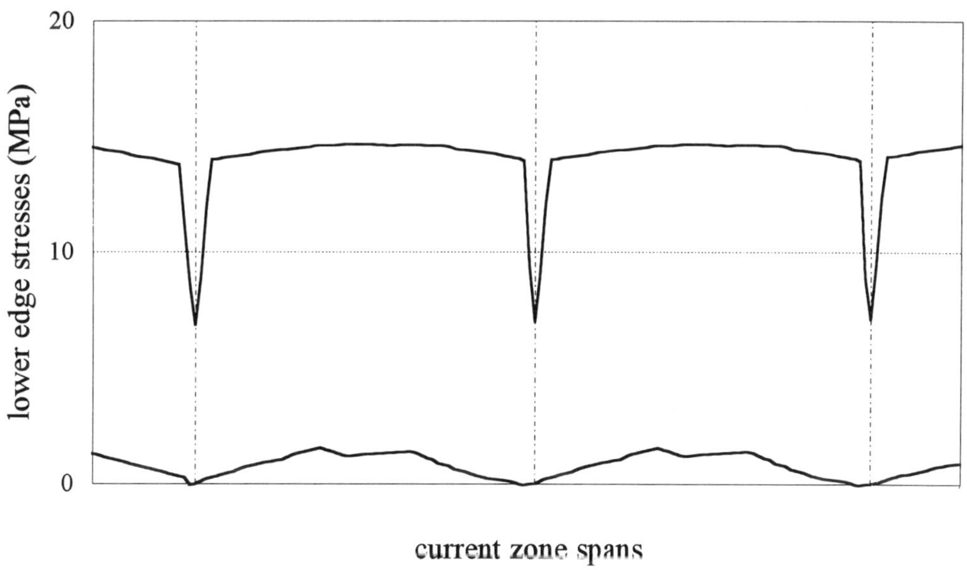

Figure 5.60 - Deck current zone. Envelope of the lower edge stresses due to the moment envelope of figure 5.56.

Another representation of the stress state in the deck compares the envelopes of the extreme fiber tensile stresses with the diagram of axial compression (figure 5.63 … developed based on the envelope of figure 5.54) to emphasize the tension zones where reinforcement must be added or a local complementary prestressing must be introduced.

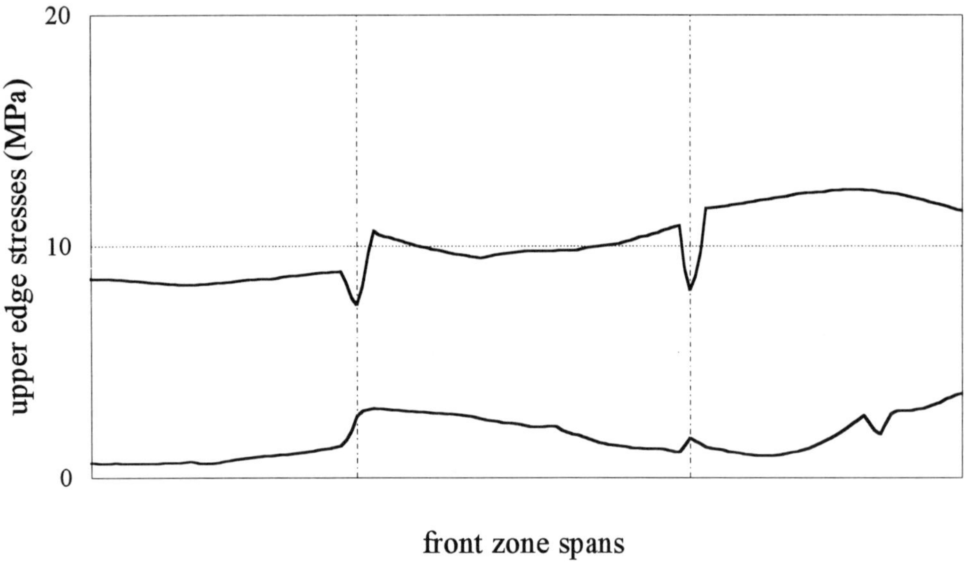

Figure 5.61 - Deck front zone. Envelope of the upper edge stresses due to the moment envelope of figure 5.56.

Figure 5.62 - Deck front zone. Envelope of the lower edge stresses due to the moment envelope of figure 5.56.

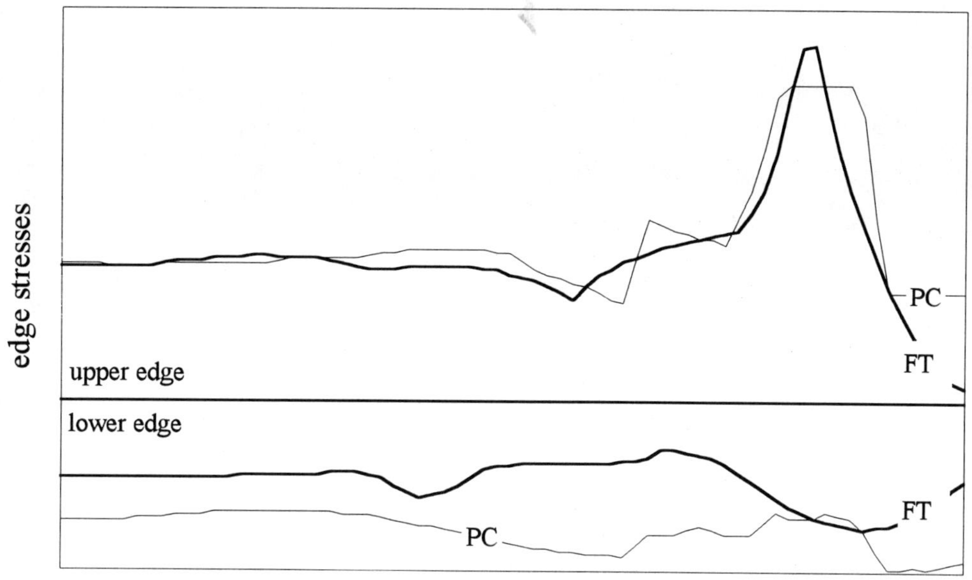

Figure 5.63 - Extreme fiber flexural tensile stresses (FT) are compared with prestressing compressive stresses (PC).

These local anomalies are frequent near the support diaphragms because of the combination of the eccentricity of launch prestressing and the larger area of the cross section, which reduces its average compression.

Chapter 6

ADDITIONAL STRESSES

The stress diagrams of the previous chapters describe the effects of dead load and launch prestressing for the subsequent support configurations of the continuous beam. In turn, these basic diagrams are modified by additional stresses deriving from:

- thermal gradients produced by radiation and ventilation, whose evolution is rather sudden because of the absence of the stabilizing effect of the wearing surface;
- differential settlement of bearings, both in the casting yard, often on direct foundations, and on the piers. The breaking of launching saddles, the accidental slippage of neo-flon plates, or the need to lift the deck produce, in turn, the same effects;
- misplacement of launching bearings due to insufficiently trained labor or topographic errors;
- general geometry defects caused by formwork misplacement, particularly insidious since they can affect the whole deck;
- non-linear effects due to time-dependent phenomena. In the stops between the subsequent launches the deck deforms irreversibly due to creep of the concrete, and the launch of a deformed continuous beam on fixed supports causes effects quite similar to the differential settlement of bearings in an undeformed beam. On the contrary, shrinkage does not produce overall stresses, but deck shortening, if not compensated, causes local stresses due to the eccentricity of the theoretical support axes from the center line of the piers;
- local effects produced by friction and by the grade of the launching surface, which modify the distribution of longitudinal stresses in the support sections and compromise the conservation of plane sections.

THERMAL STRESSES

Most of the standards define the design criteria for the dimensional variations of structural elements subjected to thermal changes.

The effects of these deformations can be analyzed by separating the general effects of dimensional variations from the effects of thermal gradients. The first problem is solved by assuming all the parts of the structure are isothermal, and is limited to the calculation of the changes of dimension of the deck and the piers and the possible consequences, such as bending of piers or sliding of joints and bearings. But, in reality, thermal excursions and the different exposure to the sun and the wind produce non-uniform thermal distributions along the deck and inside its cross section. These thermal gradients produce strain gradients, and since their free distribution is prevented by the conservation of plane sections and by redundancy (longitudinal of the deck and transverse of its cross section), they result in bending and shear stresses.

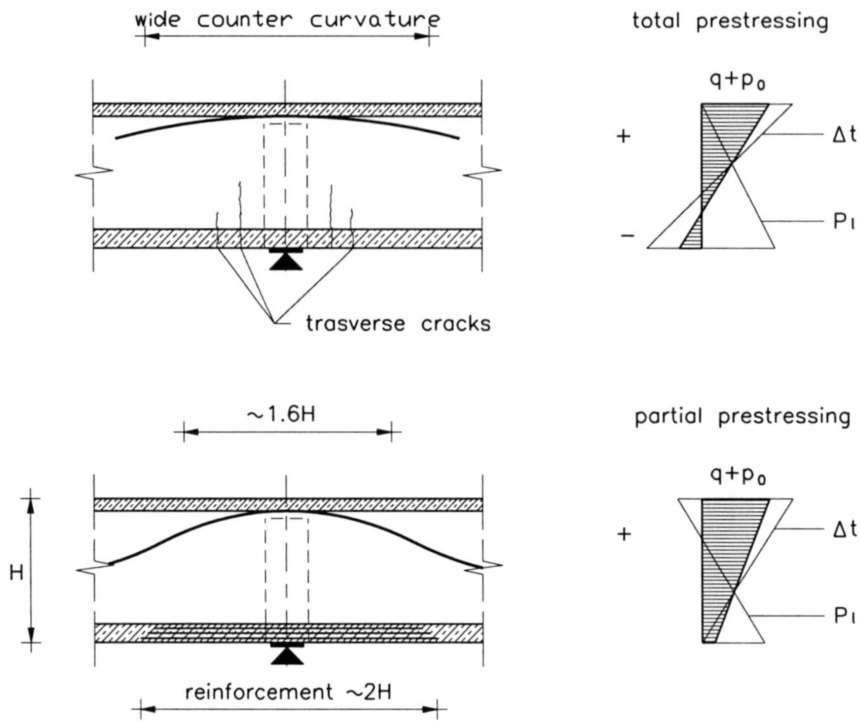

Figure 6.1 - Upon completion of the launch, the positive moments at supports produced by parabolic prestressing and thermal gradients tend to decompress the lower edge and to produce vertical cracks. In case of total prestressing and final release of launch tendons, the curvature of parabolic tendons should be increased, and additional reinforcement may be necessary in the bottom slab.

During launch, the solar radiation on the top slab causes its relative warming, only partially reduced by the thermal inertia of the air trapped inside the box cell. The continuous beam tends to camber upward, and this produces positive bending moments.

Upon completion of the launch, the introduction of parabolic prestressing produces negative moments in mid-span and positive moments on supports, which reduce the lower edge compression. The additional thermal positive moments tend to emphasize this effect until decompressing the lower edge and causing vertical cracks[112] (figure 6.1). When the curvature of parabolic tendons is not sufficient to direct the distributed deviation forces toward the support point of the deck, the cracked zone can extend longitudinally. In the case of total prestressing and final removal of most axial launch tendons, it is prudent to move the inflection points of parabolic tendons to increase their curvature, and to increase reinforcement in the bottom slab to about *1.2* to *2%* for as long as twice the deck depth[112]. But, because of the cost of removing axial launch tendons, it may be less expensive to leave some of them permanently in the structure.

A careful analysis of thermal stresses is essential, although the need to take them into account when sizing structure and prestressing is still an open question. For some authors, adding thermal stresses to the other design stresses and absorbing all these tensile stresses with prestressing is a mistake, since thermal stresses do not reduce the carrying capacity of the structure, while excess prestressing, cost apart, makes its ultimate behavior less ductile. In fact, proceeding towards the ultimate limit state, thermal tensile stresses are annulled by the first cracks and compression falls in the plastic dominion and is detrimental only in case of instability. Then, thermal stresses can be considered only in terms of crack width control[112].

From a theoretical point of view, thermal stress analysis presents two unknowns: the difference in temperature between the surfaces of the cross section, and its internal distribution. The difference in temperature can be very high, since sun radiates upon the concrete directly: in the Kishwaukee River Bridge[131], gradients of $+8°C$ and $-11°C$ have been observed, and the thermal distribution inside the cross section was absolutely not linear.

The problem can hardly be solved experimentally because of the many interacting factors which govern it, and it is convenient to refer directly to those standards which allow for the absence of a wearing surface.

Whatever the gradient distribution may be, the calculation of its effects on the continuous beam is carried out by hinging the beam at the supports, calculating the rotations of the simply supported beams at these sections, and defining the moments necessary to make them zero.

For a linear thermal gradient, this moment is:

$$M = EI \frac{\alpha \Delta T}{H} \qquad (6.1)$$

In the case of non-linear gradients along the cross section depth, stresses arise even in the determinate structure and must be added to those necessary to restore compatibility in the hinged support sections.

In a cross section subject to any thermal diagram, $T(z)$, if all the fibers could expand freely, its deformation would be:

$$\varepsilon(z) = \alpha T(z) \qquad (6.2)$$

and the law of conservation of plane sections would not be respected. In fact, the latest requires a linear expansion law:

$$\varepsilon(z) = \varepsilon_0 + \varphi z \qquad (6.3)$$

where ε_0 is the expansion of the fiber of reference and φ is the cross sectional rotation about it, both unknown. The difference between the free (6.2) and the actual (6.3) expansion generates longitudinal stresses:

$$\sigma(z) = E\left[\varepsilon_0 + \varphi z - \alpha T(z)\right] \qquad (6.4)$$

which produce an axial force, N, and a bending moment, M, about the centroidal axis. By making them both zero (the determinate structure does not prevent elongation and rotation), one obtains:

$$\begin{cases} \varepsilon_0 = -\varphi z_G + \dfrac{\alpha}{A}\displaystyle\int_0^H T(z) b\,dz \\ \varphi = \dfrac{\alpha}{I}\displaystyle\int_0^H T(z)\left(z - z_G\right) b\,dz \end{cases} \qquad (6.5)$$

where z_G is the distance of the centroid from the fiber of reference. Once ε_0 and φ are defined, with (6.4) one obtains the stresses in the determinate structure, to be added to the effects of the bending moment obtained by substituting the rotation, φ, to $\alpha \Delta T / H$ in (6.1).

In addition to longitudinal effects, thermal gradients produce transverse bending in a box girder. Under the sun's radiation, the air trapped in the cell tends to warm up, and with it the internal surfaces of the cross section. The sudden external cooling due to wind creates thermal gradients, with tensile stresses in the external surfaces (longitudinal cracks have been observed many times at the lower corners of the cross section) and compression in the internal surfaces. Similar effects are produced by thermal gradients in the top slab only, and even by the cement hydration heat in sections built in only one casting phase, which have a low mechanical strength during the first curing phases.

The need to limit thermal gradients should not be underestimated, and measures such as water irrigation of the top slab and opening of the deck end diaphragms to improve the cell ventilation can produce good results with insignificant costs.

MISPLACEMENT OF LAUNCHING BEARINGS

In the continuous beam, vertical misplacement of bearings causes secondary stresses.

In the long term, mistakes of *10* to *20 mm* are not very significant, since creep of the concrete redistributes the stresses they produce. But in the short term they cannot be neglected in the structure in service, and even less so in a deck being launched.

Therefore, it is necessary to position the launching saddles with very strict tolerances, about *1* to *2 mm*, to avoid excessive differences with the design assumptions from the very beginning, and to define precise operational sequences for deck lifting aimed at reducing the resulting stresses as much as possible.

The problems induced by differences in elevation of launching bearings are different longitudinally and transversely and cause important stresses. They can occur so rapidly as to prevent progressive shimming (exit of a neo-flon plate) and can affect any cross section of the deck and not only the final support ones, which are generally stiffened and able to mobilize the high torsional stiffness of the box girder without suffering distortional phenomena.

Vertical Misplacement in the Longitudinal Direction

In the longitudinal direction, it is always necessary to provide for vertical movements of launching bearings even in operational conditions, and not only in emergency situations.

In fact, the deck must be liftable at any support (while fixed at all the other supports to avoid prohibitive costs) in order to remove broken launching saddles or damaged neo-flon plates, and upon completion of the launch it is lifted to remove the saddles and install the permanent bearings. But while this last operation occurs only upon completion of launch, when the deck is supported on its stiffened support sections and is often completely prestressed, other situations that require lifting arise in quite random positions, and in presence of only the centroidal prestressing. Therefore, each cross section of the deck must be checked for the stresses produced by an estimated misplacement and the foreseen lifting, and in detail, for instance, by imposing:

- vertical displacement of each launching saddle: a reasonable value is $\pm 5\ mm$ alternated;
- vertical displacement of $\pm 10\ mm$ in only one saddle, assuming all the others are perfectly aligned;
- vertical displacement of $\pm 10\ mm$ at the abutment, when using friction launchers;
- vertical displacement of $\pm 5\ mm$ of the formwork front head because of the elastic settlement of this support in the final stages of extraction of the segment.

As launch prestressing is often sized to absorb tensile stresses in the mid-span lower edges, the moment variations caused by settlement affect this sizing value only marginally. At the same time, the variation of negative moment in the support section subject to a vertical movement, η_m:

$$\Delta M = -K\frac{EI}{l^2}\eta_m \qquad (6.6)$$

with K ranging from 3 to 6, is reduced in quadratic ratio by a generally long span. In both cases, therefore, support settlement is often tolerable without increasing prestressing.

Vertical Misplacement in the Transverse Direction

In a prestressed concrete box girder bridge, eccentric live loads produce relatively modest effects. In fact:

- the cross section torsional stiffness is generally so high as to produce low stresses even under high torsional moments;
- the in-plane flexural stiffness of the plates which compose the cross section, together with its frame-stiffness, rapidly damp distortion;

– live loads are only a component of the design load of a bridge, which is much more affected by dead and superimposed dead loads.

In most cases, therefore, important distortional and torsional phenomena can arise only in the presence of concentrated eccentric loads.

Since the highest concentrated loads which affect a bridge are the support reactions, and in a support section their resulting force becomes transversally eccentric in the case of differential settlement of bearings in the transverse plane, this can produce large in-plane deformations of the cross section.

This possibility is generally disregarded in design, since a pier rotation is critical long before creating local problems to the deck, and a bearing failure is improbable. Moreover, in both cases, the transverse eccentricity of the support reaction affects a stiffened cross section. This makes its distortion negligible, and the problem becomes one of torsional overload in a structural scheme that can easily absorb it.

In launched bridges, diaphragms and cross stiffenings are generally limited to the permanent support sections. Even when their construction is not postponed until the end of the deck launch (this simplifies production and reduces launch prestressing), they cannot prevent distortion, as, for this purpose, their spacing would have to be so reduced to become uneconomical. But differences in elevation between launching saddles occur in practically every case because of their positional tolerances, their failure, or operative errors. In addition, the general geometrical irregularities caused by formwork positional tolerances can produce the same effects as bearing misplacements, making even their most careful positioning completely useless.

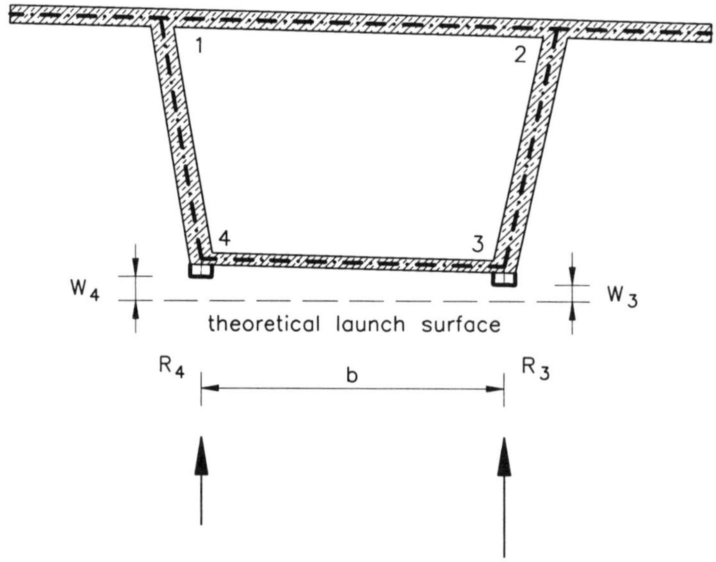

Figure 6.2 - In a deck support section subject to a vertical misplacement of launching bearings, the distances, $W_3 \neq W_4$, of nodes 3 and 4 from the theoretical launch surface unbalance the support reactions, so $R_3 \neq R_4$.

In the generic support section of figure 6.2, defining W_3 and W_4 as the vertical distance of the deck soffit under nodes *3* and *4* from the theoretical launch surface, the average support misplacement is:

$$W_m = \frac{W_3 + W_4}{2} \qquad (6.7)$$

It does not produce transverse effects, and its longitudinal effects in the continuous beam can be analyzed with (6.6). To this average displacement one has to superimpose displacements of the cross section lower nodes equal and opposite in sign:

$$W_h = \frac{W_4 - W_3}{2} \qquad (6.8)$$

which do not produce longitudinal effects in the continuous beam (secondary stresses apart), but produce transverse effects, because of the rotation of the bottom slab, equal to:

$$\varphi = \frac{2W_h}{b} \qquad (6.9)$$

Generally, only this second component of the displacement of the support section lower nodes can cause important stresses. In fact, the cross section is normally not prestressed in the transverse direction (or, if it is, it is limited to the top slab in wide decks, which suffer the phenomenon less thanks to the wider bottom slab) and the distance between bearings is much smaller than in the longitudinal direction – that is, rotations are amplified.

Large rotations and high torsional stiffness produce high torsional moments, and in turn, because of the short transverse distance between launching saddles, large differences of load between them that lead to the rupture of the more loaded saddle. In fact, a correct design of the collapse sequence of the system composed of deck, neo-flon plate, and launching saddle, cannot but entrust to the saddle the role of "fuse" against excessive stresses in the deck.

The time and costs associated with their replacement and with the higher wear of neo-flon plates compel one to assign to these problems the importance they deserve.

Finite-elements analysis requires sophisticated and laborious models and long processing times, and does not distinguish the different phenomena involved in the deck deformation. On the contrary, the model described hereinafter[167] permits an agile and fast analysis of the interaction between torsion and distortion with a simple spreadsheet.

Analytical Model

In a straight prestressed concrete box girder, the effects of a difference in height between the two support points of the cross section deviate from Bredt's theory, since none of its assumptions (rigid in-plane rotation, torsional moment applied through the theoretical distribution of tangential stresses, and free warping) is really respected.

Thus, its deformation results from the superimposition of a rigid rotation and a distortion, effects which depend respectively on the torsional stiffness of the girder, which limits its capacity to rotate about its axis, and on the interaction between the frame stiffness of the cross section and the in-plane flexural stiffness of its plates, which together oppose distortion.

Defining with $\eta_d(x)$ as the vertical displacement of node 3 of the cross section (figure 6.2) due to distortion, and with $\eta_t(x)$ being the displacement due to torsional rotation, in a support section subject to a difference in height between launching saddles (the section assumed hereinafter as the origin of the frame of reference), (6.8):

$$W_h = \eta_d(0) + \eta_t(0) \qquad (6.10)$$

The actual deformation therefore falls between two limit cases: the case of a rigid cross section, which can be analyzed with the theory of torsion, and the case of a section composed of plates totally lacking in transverse flexural stiffness, and which for this reason can be assumed as hinged to each other, which equilibrate the loads by bending in their planes.

The distortion, $\eta_d(x)$, of a single-cell constant-section box girder, neglecting shear strain, is governed by a differential equation identical in form to that for beams on elastic foundations (BEF):

$$EC_d \frac{d^4\eta_d}{dx^4} + k\eta_d = p_h \qquad (6.11)$$

in which C_d is a cross-sectional property related to warping, k is the deformational stiffness of a unit length of the box cell, and p_h is any hemisymmetrical unit load.

The property of the cross section which corresponds to the BEF moment of inertia C_d can be expressed as[167]:

$$C_d = \frac{b}{4} C_8 \left(C_2 C_7 + C_3 C_8 + 2C_4 C_9 \right) \qquad (6.12)$$

with:

$$C_1 = \frac{b^2 t_b + c(a+2b)t_w}{\frac{(a+2d)^3}{a} t_t + c(2a+b)t_w} \tag{6.13}$$

$$C_2 = a\left[\frac{(a+2d)^3}{6a^2} C_1 t_t - \frac{b}{4} t_b - \frac{c}{2}(1 - C_1)t_w\right] \tag{6.14}$$

$$C_3 = -\frac{b^2}{12} t_b \tag{6.15}$$

$$C_4 = -c\left[\frac{b}{4} t_b + \frac{c}{6}(2 - C_1)t_w\right] \tag{6.16}$$

$$C_5 = -a\left[\frac{(a+2d)^3}{6ab} t_t + \frac{b}{4} t_b + \frac{c}{2}\left(1 + \frac{a}{b}\right)t_w\right] \tag{6.17}$$

$$C_6 = -c\left[\frac{b}{4} t_b + \frac{c}{6}\left(2 + \frac{a}{b}\right)t_w\right] \tag{6.18}$$

$$C_7 = 2h\left(\frac{1}{a} + \frac{\frac{C_6}{c} - \frac{C_5}{a}}{C_5 - C_3 - \frac{a-b}{c} C_6}\right) \tag{6.19}$$

$$C_8 = -2h \frac{\frac{C_6}{c} - \frac{C_5}{a}}{C_5 - C_3 - \frac{a-b}{c} C_6} \tag{6.20}$$

$$C_9 = -\frac{h}{c}\left[1 + (a-b)\frac{\frac{C_6}{c} - \frac{C_5}{a}}{C_5 - C_3 - \frac{a-b}{c} C_6}\right] \tag{6.21}$$

The vertical shear force, V, produced in the center of the bottom slab by the cell frame action for a unit hemisymmetrical load is:

$$V = \frac{\dfrac{a^4}{t_t^3} + \dfrac{b^2 c}{t_w^3}(2a+b)}{(a+b)\left[\left(\dfrac{a}{t_t}\right)^3 + \left(\dfrac{b}{t_b}\right)^3 + \dfrac{2c}{t_w^3}\left(a^2+ab+b^2\right)\right]} \qquad (6.22)$$

and the property of the cross section which corresponds to the BEF modulus k is:

$$k = \frac{2(a+b)E}{\left(1-\upsilon^2\right)\left\{\dfrac{a^4}{t_t^3}\left(\dfrac{a}{a+b} - V\right) + \dfrac{cb^2}{t_w^3}\left[\dfrac{2b^2}{a+b} - (2a+b)V\right]\right\}} \qquad (6.23)$$

BEF is solved for the infinite beam scheme. The subsequent support sections of the deck cannot restrain BEF, since, except when a diaphragm passes above them, the deck cross section can distort even in the presence of a perfect transverse alignment of launching bearings. As diaphragms are effective only if so close as to be excessively expensive, hereinafter their presence will be neglected.

In the origin section, the difference in height between launching saddles unbalances the two support reactions R_3 and R_4 (figure 6.2), and the concentrated load to apply to the BEF is their hemisymmetrical component:

$$R_h = \frac{R_4 - R_3}{2} \qquad (6.24)$$

If the cross section of the deck were rigid and the spans constant, the torsional moment produced by this component:

$$T = R_h b \qquad (6.25)$$

would be balanced by two couples equal to $\pm T\ 2$ in the two adjacent support sections, A, of the deck (whose bearings are assumed to be perfectly aligned) and torsion would remain confined within these two spans. But even in the adjacent support sections the balancing torsional moments are produced by hemisymmetrical components, R_{hA}, of the support reactions, which in turn distort the cross section. Rigid rotations occur to compensate for distortion and maintain contact with the launching saddles, and torsion partially overflows toward external spans.

From a theoretical point of view, this overflow affects the whole deck. In reality, high values of C_d and k produce such a rapid damping that the phenomenon is limited to the first two spans on each side of the origin cross section, and the BEF static scheme is that of figure 6.3.

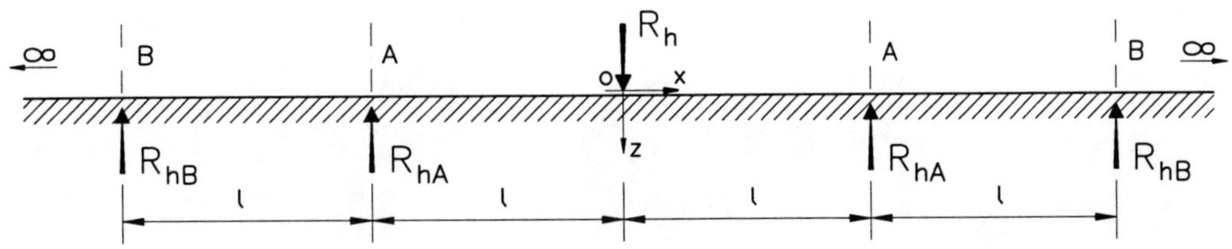

Figure 6.3 - Static scheme of the beam on elastic foundation (BEF). The forces loading the BEF are the hemisymmetrical components of the support reactions.

The equation of the right branch of the BEF elastic line produced by a concentrated load, R_h, applied to its origin is:

$$\eta_d(x) = \frac{R_h \lambda}{2k} e^{-\lambda x}(\cos \lambda x + \sin \lambda x) \qquad (6.26)$$

with:

$$\lambda = \sqrt[4]{\frac{k}{4EC_d}} \qquad (6.27)$$

Once the torsional stiffness of the cross section is defined ($m = 1/\nu$ where ν is Poisson's ratio):

$$C_t = \frac{\frac{m}{2(m+1)} E[(a+b)h]^2}{\frac{a}{t_t} + \frac{b}{t_b} + 2\frac{c}{t_w}} \qquad (6.28)$$

By imposing compatibility of rotations in the support sections one obtains:

$$R_{hA} = \frac{\frac{\lambda}{bk} + \frac{b}{C_t} l}{\frac{4\lambda}{bk} + \frac{2b}{C_t} l} R_h \qquad (6.29)$$

and:

$$R_{hB} = \frac{\dfrac{\lambda}{bk}}{\dfrac{4\lambda}{bk} + \dfrac{2b}{C_t} l} R_h \qquad (6.30)$$

As the effects of R_{hA} and R_{hB} fade within distances of about half a span and therefore do not superimpose on each other, the distortion of any cross section as a function of its abscissa $x \geq 0$ is:

$$\eta_d(x) = \eta_0(x) - \eta_A(x) - \eta_B(x) + \frac{\lambda}{2k} R_{hB} \qquad (6.31)$$

where the different components ($|x|$ is the absolute value of x) are:

$$\eta_0(x) = \frac{R_h \lambda}{2k} e^{-\lambda x}(\cos \lambda x + \sin \lambda x) \qquad (6.32)$$

$$\eta_A(x) = \frac{R_{hA} \lambda}{2k} e^{-\lambda|l-x|}(\cos \lambda|l-x| + \sin \lambda|l-x|) \qquad (6.33)$$

$$\eta_B(x) = \frac{R_{hB} \lambda}{2k} e^{-\lambda|2l-x|}(\cos \lambda|2l-x| + \sin \lambda|2l-x|) \qquad (6.34)$$

and the last addend corrects the frame of reference to take into account the rigid rotation in B. Finally, the deflection, $\eta_t(x)$, produced by torsional rotations is, in the two spans:

$$\eta_t^{0-A}(x) = \frac{b^2}{2C_t}\left[\frac{1}{2}(l-x)R_h + R_{hB} l\right] \qquad (6.35)$$

$$\eta_t^{A-B}(x) = \frac{b^2}{2C_t}(2l-x)R_{hB} \qquad (6.36)$$

Once $\eta_d(0)$ and $\eta_t(0)$ are defined for any couple of hemisymmetrical forces, R_h, the calculation of the hemisymmetrical component of the support reaction (6.24) based on the assumed misplacement of launching saddles is immediate, by linear proportion (6.10). Clearly, in the absence of a bilateral constraint, this force cannot exceed the reaction produced by each launching saddle in balanced conditions:

$$R_m = \frac{R_3 + R_4}{2} \qquad (6.37)$$

that is, *50%* of the continuous beam support reaction. Once this value is reached, only one bearing carries the whole reaction, and the deck rises from the second one, avoiding further deformations.

The local deformation of the launching surface causes important load asymmetries between the two launching saddles. Monitoring of Amiens Viaduct showed support reaction perturbations of about *20%*, while the monitoring of the Champigny-sur-Yonne Bridge showed that only one of the two abutment bearings was actually loaded.

Longitudinal stresses caused by distortion at the web corners, if one sets:

$$\lambda_s = \sqrt[4]{\frac{3k}{Ec^3 t_w C_9}} \qquad (6.38)$$

become (figure 6.4):

$$\sigma_3 = \frac{3R_h}{(1+C_1)\lambda_s t_w c^2} \qquad (6.39)$$

$$\sigma_4 = C_1 \sigma_3 \qquad (6.40)$$

and should be added to the longitudinal stresses produced by restrained warping.

Finally, the transverse frame stresses, already defined for a couple of unit forces (6.22), are obtained by linear proportion with the actual hemisymmetrical component.

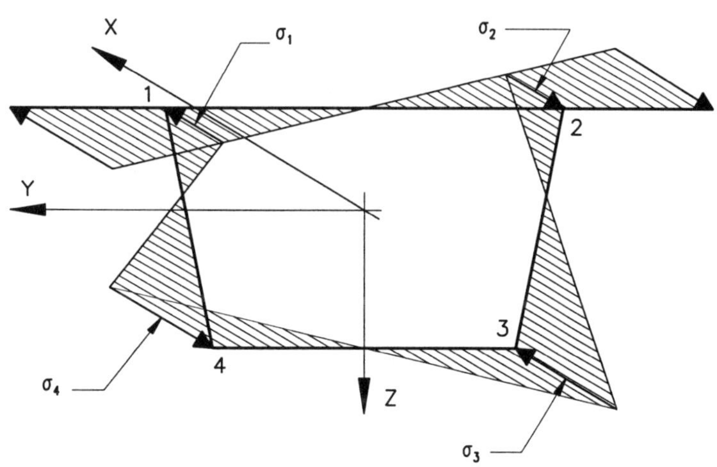

Figure 6.4 - Longitudinal orthogonal stresses due to distortion.

Reliability of the Model

The reliability of this model has been tested[167] by comparing its results with the finite-element analysis in four sections sized according to the statistical criteria of Chapter 5. Figure 6.5 shows the evolution of the in-plane deformation of the cross section under a couple of hemisymmetrical forces, $R_h = \pm 1\,MN$, applied to the section of figure 6.6.

With web inclinations up to *20°* – that is, statistically, in more than *70%* of launched bridges, the model approximates closely the results of finite-element analysis. Higher slopes produce a progressive deterioration of the result, since restrained warping generates shear gradients whose effect is a local increase of C_d. Finite-element analysis confirms also the damping distance of $\eta_d(x)$ and the distortion of the support sections.

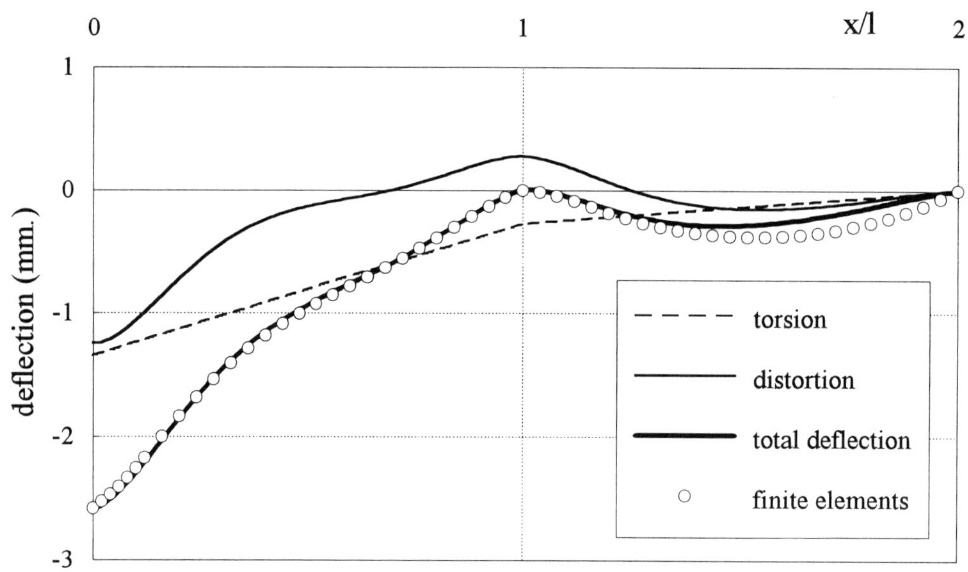

Figure 6.5 - Deflection of node 3 of the cross section of figure 6.6 under a couple of hemisymmetrical forces, $R_h = \pm 1\,MN$.

Only about one half of the local launching surface deformation was absorbed by torsion and therefore could mobilize the high torsional stiffness of the box girder. The remaining share was absorbed by distortion, and produced important longitudinal and transverse stresses.

A hemisymmetrical component of the support misplacement, W_h, (6.8) of only *5 mm* modified the reactions calculated with aligned bearings with hemisymmetrical components by about *40%*. A W_h of *12 mm* decompressed one launching bearing, doubling the load on the other one.

elastic modulus E	MPa	35000
Poisson's ratio υ		0.15
constant span l	m	50.00
external depth H	m	3.20
deck width $(a+2d)$	m	13.00
cantilever net length	m	3.20
top slab thickness t_t	m	0.27
bottom slab overall width	m	5.60
bottom slab thickness t_b	m	0.24
web thickness t_w	m	0.55
a	m	6.09
b	m	5.08
c	m	2.99
d	m	3.46
h	m	2.95
web inclination	°	9.7
z_u	m	1.21
C_d	m^4	3.371
k	kNm^{-2}	44650
λ	m^{-1}	0.0986
$x(\eta_d = 0)/l$ dimensionless damping distance		0.48
$\eta_d(0)$	mm	**1.246**
$\eta_t(0)$	mm	**1.342**
W_h	mm	**2.588**
finite elements	mm	**2.582**
approximation	%	**0.2**
$\eta_t(0)/W_h$	%	**51.9**

Figure 6.6 - Parameters of a cross section designed statistically for a *50 m* long span. (These data can be used to check the correct running of a spread-sheet program).

These overloads cannot be ignored[108], and can be prevented by imposing strict tolerances in the formwork and bearing positioning, but in the most delicate cases require adoption of specific compensation hydraulic devices (figure 4.26).

TIME-DEPENDENT PHENOMENA

In conventional structures, concrete is considered linearly elastic, although in reality it has an unelastic and time-dependent behavior.

This model permits one to adopt simple design methods, but leads one to underestimate or ignore phenomena which can produce serviceability limit states and rapid structural deterioration, and therefore to a wrong evaluation of the stress state and real safety factors of the structure.

Time-dependent phenomena became significant with the diffusion of prestressed concrete, particularly in bridges built by free cantilevering, which require application of prestressing with high eccentricity and after a short curing period.

Although it uses the same materials, the incremental launching construction method produces lower time-dependent effects, analyzed by superimposing the effects of shrinkage and creep of the concrete and relaxation of the prestressing steel.

Shrinkage

Shrinkage is a low, load-independent concrete deformation due to chemical and physical reactions related to the cement setting. Its course is rather long, though in the first fifteen curing days about one third of the total deformation takes place, and in the first year it reaches *80%*.

Shrinkage depends on the mechanical characteristics of the concrete, on the quantity and quality of cement and additives, on the water to cement, W/C, and aggregate to cement, A/C, ratio, on the aggregate rigidity, and on the humidity and temperature of the curing environment.

	W/C			
A/C	0.4	0.5	0.6	0.7
3	800	1200		
4	550	850	1050	
5	400	600	750	850
6	300	400	550	650
7	200	300	400	500

Figure 6.7 - Typical shrinkage values ($\mu m/m$ at six months) of air-cured concrete at *50%* of humidity and $21°C$ of temperature as functions of the W/C and A/C ratios.

Typical shrinkage values of air-cured concrete at *50%* humidity and $21°C$ are reported in figure 6.7 as functions of the W/C and A/C ratios[51]. An increase in the W/C ratio increases shrinkage because of the higher quantity of water which can evaporate, and the higher porosity of concrete which facilitates its evaporation. A reduction in the A/C ratio increases shrinkage because of the higher volume of cement, which is responsible for this phenomenon, and the lower volume of aggregates, which oppose the contraction of volume.

In a generic instant, t, the shrinkage strain measured from a reference instant, t_0, can be expressed with[134]:

$$\varepsilon_s(t,t_0) = \varepsilon_{s\infty}[K_s(t) - K_s(t_0)] \qquad (6.41)$$

in which the final value, $\varepsilon_{s\infty}$, and the function $K_s(t) - K_s(t_0)$, which tends to a unit value for $t \rightarrow \infty$, more rapidly for a thinner element, are generally established by the applicable standards or international recommendations.

Shrinkage is higher on surfaces exposed to the air and lower inside concrete, also because of reinforcing steel which opposes these deformations. The conservation of plane sections requires the presence of strain gradients – that is, of self-balanced stresses. Although creep rapidly reduces these stresses, in the first curing phases they can exceed the tensile strength of concrete and produce distributed superficial cracks, which in turn cause deterioration. Similar stresses arise in the corners of the cross section, and in the case of webs and slabs of different thickness, they are made worse by differential shrinkage.

In incremental launching, the deck is built on the ground, and most of the parameters which govern shrinkage can be easily controlled:

- the W/C ratio can be reduced by using additives, mixing concrete in-situ, and placing it by means of buckets or conveyor-belts instead of pumping;
- further superficial reductions of the W/C ratio can be obtained by vibro-compacting the top slab to expel bleeding water mechanically, and by using formwork with draining felts which reduce concrete porosity, increase superficial strength, protect the surface during the first curing phases, and avoid the use of oil;
- finally, a higher superficial strength can be obtained by means of form vibrators together with the ordinary needle vibrators.

Creep

While shrinkage is a load-independent phenomenon, creep is the spontaneous growth over time of the deformation produced by a constant load.

If a concrete specimen is subjected to a constant compression, σ_0, in an instant, t_0, measured from its casting, its shrinkage-free deformation in a generic instant, t, is[134]:

$$\varepsilon(t,t_0) = \varepsilon_{el} + \varepsilon_{def}(t,t_0) + \varepsilon_{cr}(t,t_0) \qquad (6.42)$$

The instantaneous elastic deformation of concrete, fully recovered when relieving the specimen, is:

$$\varepsilon_{el} = \frac{\sigma_0}{E(t_0)} \qquad (6.43)$$

The deferred elastic deformation, also reversible, is equal to about *40%* of the instantaneous elastic deformation at *28* days of curing, reduced by a coefficient, K_{def}, which depends on the load duration, varies between *0.75* and *1.00*, and can be assumed equal to *1.00* for periods longer than three months:

$$\varepsilon_{def}\left(t,t_0\right) \approx 0.4 \frac{\sigma_0}{E(28)} K_{def}\left(t-t_0\right) \qquad (6.44)$$

Finally, the irreversible deformation due to creep is:

$$\varepsilon_{cr}\left(t,t_0\right) = \frac{\sigma_0}{E(28)} K_{cr1}\left[K_{cr2}(t) - K_{cr2}(t_0)\right] \qquad (6.45)$$

This last component is higher the younger concrete is when loaded, and depends on the hygrometric conditions, the thickness of the element, the quality of concrete, and, above all, its compactness, since the cause of creep is grain sliding caused by the viscous behavior of the cement paste.

The coefficients necessary to evaluate the evolution of creep over time are generally established by the applicable standards or international recommendations, and by introducing in (6.42) these three terms one obtains an expression like:

$$\varepsilon\left(t,t_0\right) = \frac{\sigma_0}{E^*\left(t,t_0\right)} \qquad (6.46)$$

Therefore, the analysis of the total deformation becomes an elastic calculation with a conventional modulus $E^*\left(t,t_0\right)$.

Prestressing Steel Relaxation

In prestressed concrete, time-dependent phenomena in concrete are complicated by viscous phenomena in prestressing steel known as steel relaxation, which occur in a tensioned tendon of constant length, and thus cause loss of prestress.

As steel relaxation develops over a short period of time, tendon re-tensioning a few days after the initial tensioning has been adopted many times, but these operations are expensive, and impossible with coupled tendons or tendons anchored in construction joints. Therefore, tendon grouting after the initial tensioning has become a standard practice, since it guarantees a quick and reliable protection of strands.

The law of evolution of steel relaxation is[45]:

$$\log \frac{\Delta\sigma_p(t)}{\sigma_{p0}} = C_1 + C_2 \log(t) \qquad (6.47)$$

in which $\Delta\sigma_p(t)$ is the loss of prestress at time t, σ_{p0} is the initial tension, and the coefficients C_1 and C_2 depend on the steel properties and are defined experimentally. In stabilized steel, widely used, relaxation is lower and more rapid.

As in reality the length of the tendons is not constant because of the time-dependent phenomena of concrete, the loss of prestress due to steel relaxation is lower than (6.47), and some design specifications provide for reduction coefficients.

Effects of Time-Dependent Phenomena on Launched Bridges

In a continuous beam supported on sliding bearings, the unelastic deformations of the concrete shorten the beam, but neglecting bearing friction and loss of prestress, this does not modify the stress state of the beam both during launch and in the service stages.

As launch tendons generally extend for two or three deck segments, this prestressing is introduced rather gradually. In turn, parabolic tendons are almost always tensioned upon completion of launch – that is, after a long curing period, and with lowest creep deformations. As a consequence, the irreversible component of the longitudinal shortening of the deck is much lower than the one of a deck entirely prestressed after a short curing, and the losses of prestress are also lower.

In spite of this, in long launched bridges, the front segments arrive in the final position after many months of curing while the back segments are much younger, and both creep and shrinkage vary along the deck. Therefore, the length of the segments should be accurately corrected over time, to avoid irremediable misalignment of the deck support sections with respect to the piers.

Once the deck has reached its final position, concrete creep does not modify the stress distribution. In fact, it produces effects similar to variations in the elastic modulus (6.46), and if these variations are similar along the deck (the secondary effects of the different curing periods of the segments are generally negligible) stresses are not affected.

However, during launch, the support scheme of the deck changes continuously, and the concrete strength varies with the different curing stages of the segments. Consequently, the need to forecast the effects of creep is not limited to service stages but also includes the construction stages, since at each stop between two subsequent launches the deck assumes permanent deformations. In particular, when each launch step is as long as the bridge span (segments as long as the span, shorter segments launched as a whole, stayed front schemes etc.), at each stop the deck is subjected again to the same stress diagrams of the previous stop (although shifted one span forward), and particularly the same diagram of rotations.

As time goes by, concrete creep makes some of these constant rotations irreversible, and the deck undergoes a permanent deformation similar to the one of a prestressed concrete continuous beam built directly in its final position, but opposite in sign, since launch prestressing is straight and cannot produce an equivalent vertical load.

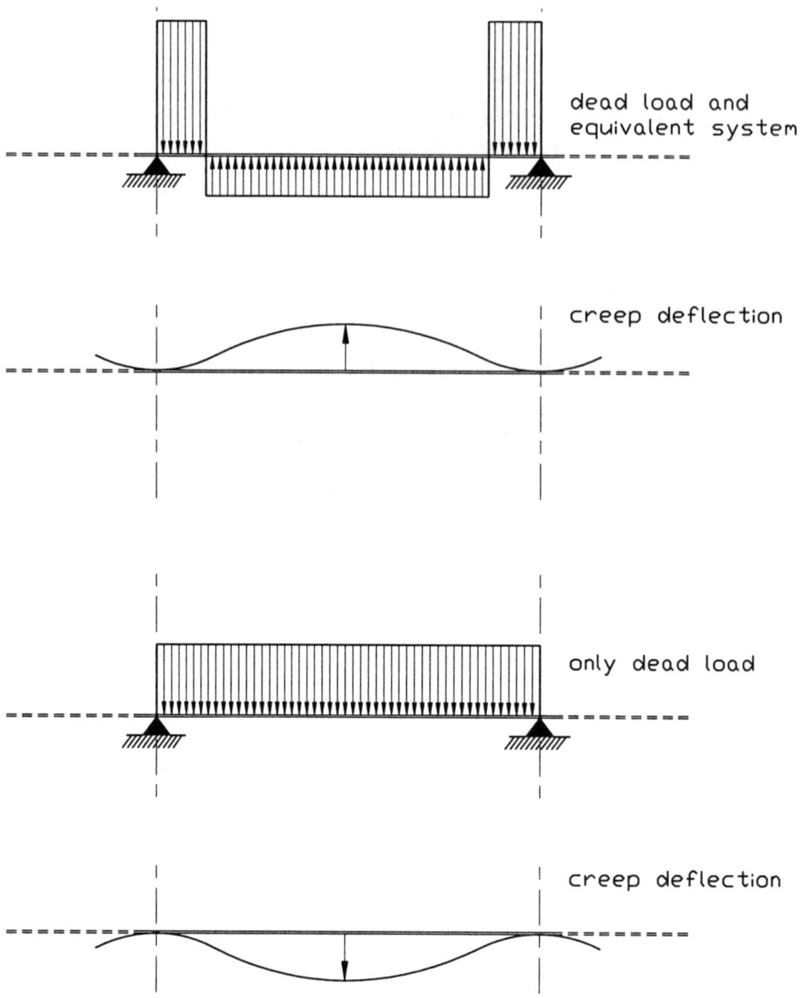

Figure 6.8 - Axial launch prestressing avoids the cambers due to the excessive equivalent load in decks built span-by-span and completely prestressed, but cannot control rotations due to dead load, and a portion of them is made irreversible by creep.

In other terms, if, on the one hand, launch prestressing avoids the cambers produced by the excessive equivalent load in decks completely prestressed span by span, on the other hand, it cannot prevent rotations due to dead load, and a share of them is made irreversible by creep until parabolic prestressing (introduced only after launch is completed) definitively interrupts the phenomenon (figure 6.8).

On the contrary (and this is the most frequent case), when the length of the launch step is half the span, the diagrams of rotations superimpose only every two stops, and at each stop most of the deck is subjected to rotations opposite in sign to the ones at the previous stop. Permanent curvatures tend to cancel each other, their envelope becomes more chaotic, and accumulation reduces.

In every case, as launch goes on, these permanent deformations disturb launch stresses, and this phenomenon can be neglected only in the shortest decks, for which the construction duration is too short to mobilize important permanent deformations, and with segments of different length than the bridge span.

The strain accumulation can be analyzed by assuming that any stress increment, $\Delta\sigma$, produces an irreversible deformation, $\Delta\varepsilon_{cr}$ (6.46), which does not depend on other stress increments, $\Delta\sigma$, applied previously or subsequently[134].

The analysis is repeated at each deck stop. The permanent deformation developed at the stop is added to that accumulated at the previous stops, and the translation of this undulated continuous beam on the unyielding launching supports permits one to simulate the subsequent launch stage. The calculation is less complex if one assumes the continuous beam is undeformed and applies to its supports the settlements which correspond to the deformation of the undulated beam at these points.

Upon completion of launch, the final irreversible deformation is calculated by superimposing the deformation due to parabolic prestressing on the end-of-launch permanent deformation. This correction is generally modest because of the long curing of concrete.

The stress redistribution due to creep requires laborious calculations, which during preliminary design can be avoided by maintaining a margin of about *0.5 MPa* from allowable tensile stress in the worst load condition. Calculations will be refined during construction with experimental data on the evolution of strength, elastic modulus, and shrinkage strain of concrete.

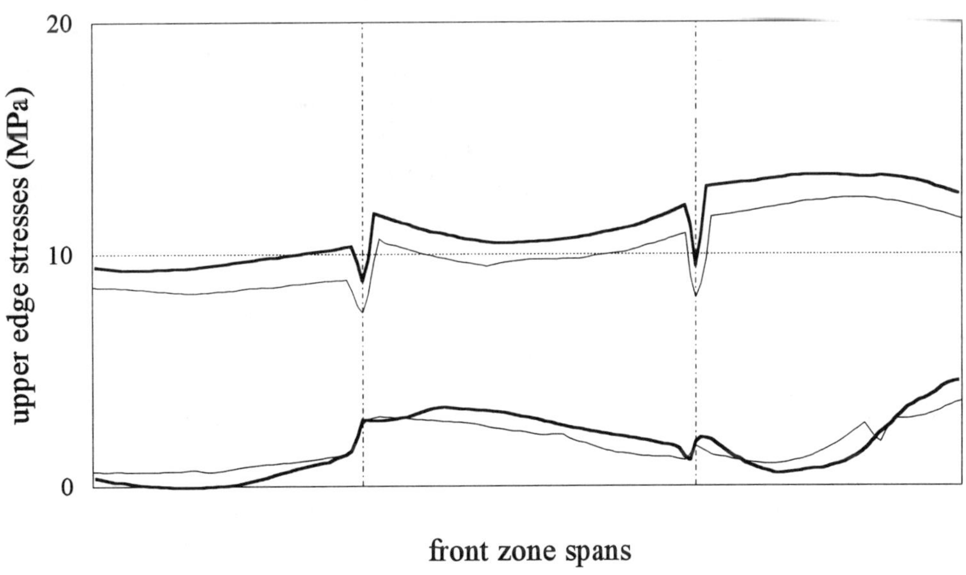

Figure 6.9 - Upper edge stresses in the three front spans of the Amiens Viaduct, calculated by taking into account creep of the concrete and the instant prestress (thicker lines) instead of only the loss-free final prestress (thinner lines, figure 5.61).

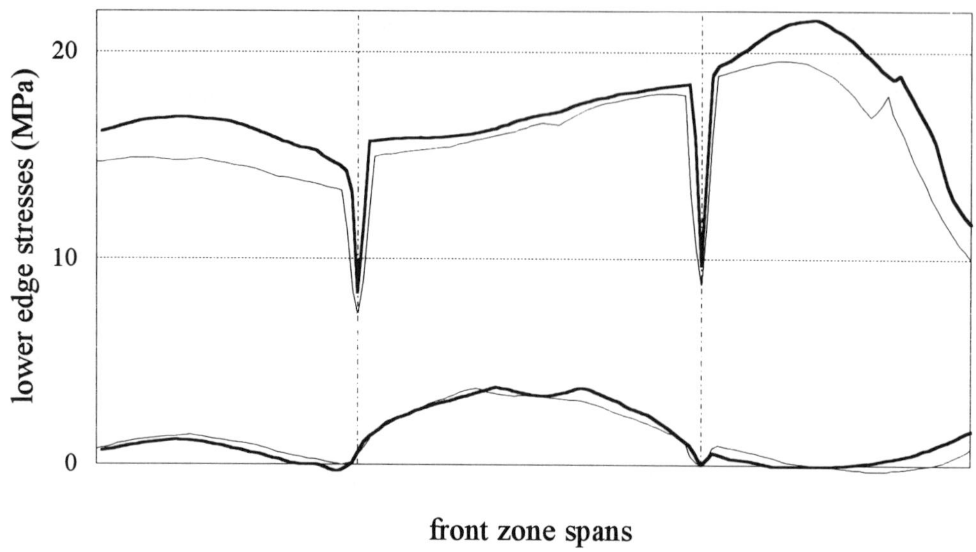

Figure 6.10 - Lower edge stresses in the three front spans of the Amiens Viaduct, calculated by taking into account creep of the concrete and the instant prestress (thicker lines) instead of only the loss-free final prestress (thinner lines, figure 5.62).

Figure 6.11 - A twenty-one month stop in the construction of the Serio River Bridge produced such large creep deflections as to require shimming the nose bottom flanges in order to compensate them (author).

Figures 6.9 and 6.10 show the extreme fiber stresses in the front zone of the Amiens Viaduct[83] calculated with two computer programs, one of which takes into account creep and the instant prestress instead of the loss-free final prestress. The different curves confirm the importance of this phenomenon. In fact, if it is generally assumed that creep reaches *85* to *90%* of its final value in ten years, in the first month *30%* is developed, in three months *50%*, and in one year *75* to *80%*. Although these data are approximate, the construction duration of a medium length bridge is already such as to mobilize most of the time-dependent phenomena, and large deformations can occur in case of stoppage of work.

The creep deformation produced by a twenty-one month stop in the construction of the Serio River Bridge[171] required corrective devices such as shimming of the nose bottom flanges (figure 6.11), and local external prestressing (figure 6.12), in order to complete launch safely. The evolution of the creep-to-elastic-strain ratio according to FIP-CEB Recommendations during this stoppage[171] is shown in figure 6.13 along with some measured data.

Figure 6.12 - Serio River Bridge. At the restarting of launch, the creep deformation caused secondary stresses in the deck front zone which required an integrative external prestressing, anchored to service manholes (author).

These phenomena are even more important in open cross sections, characterized by a higher position of the centroid and therefore by higher differences between edge stresses.

Lastly, long launched bridges present characteristic problems and require dimensional correction of segments during their construction and particular attention in the support zones of the deck:

- the length of segments must be measured spatially and not in plan, since length depends on the final position of the segment in the deck profile;

- time-dependent phenomena are continuous, and at the end of launch the front segments of the deck have developed a higher share of shortening than the back ones. The evolution of these phenomena should be verified during construction by means of measurements of the deck length referred to an isothermal condition (figure 6.13), in order to reach a correct final length of the deck;
- in spite of these controls, it is advisable to increase reinforcement in the support diaphragms of the deck in order to absorb the secondary stresses due to the final eccentricities between diaphragms and bearings.

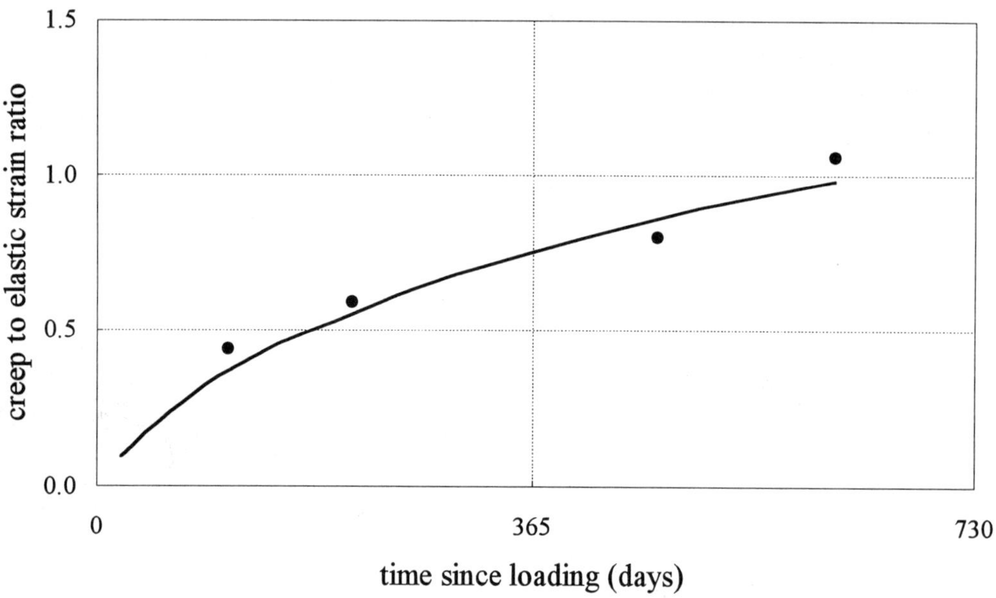

Figure 6.13 - Evolution of the creep to elastic strain ratio according to FIP-CEB recommendations during the stoppage of work of the Serio River Bridge, along with data obtained by measuring the deck length.

In case of high thermal excursions and decks longer than *900* to *1000 m*, all these facts suggest one adopts static schemes with intermediate expansion joints.

LOCAL STRESSES ABOVE LAUNCHING BEARINGS

When the deck slides on the launching saddles, its lower edge is subjected to a rather complex stress state composed of:

- orthogonal and tangential longitudinal stresses produced by the beam behavior of the deck under dead load, the action of equipment necessary to its construction (launching nose, stayed tower, mobile equipment), the launch prestressing, the positional average error (6.7) of the bearings, thermal gradients, and creep deformations progressively accumulated by the deck;

- orthogonal and tangential stresses, both longitudinal and transverse, produced by torsion and distortion in the case of transverse bearing misplacement (6.8);
- orthogonal and tangential transverse stresses produced by thermal gradients in the individual plates of the cross section;
- local vertical compressive stresses produced by the distribution of support reactions toward the webs, in turn increased by the hemisymmetrical components (6.24) produced by positional tolerances of bearings;
- local longitudinal orthogonal stresses due to friction and the longitudinal slope of the launch surface: these being local effects, they compromise the conservation of plane sections.

As most of these stresses are rather high, the local stress state should be checked with adequate strength criteria.

Vertical Compression

During the deck launch, the support reactions transferred by the launching saddles are similar to the vertical loads on the final bearings, which govern sizing of support sections of the deck, generally strengthened and stiffened by diaphragms.

But these loads are applied to current sections of the deck devoid of any stiffening, and it is necessary to check their correct distribution in the longitudinal, vertical, and transverse direction.

Crosswise, the neo-flon plates which transfer the support reaction must be introduced under the deck so as to apply the reaction directly to the web axis. Their edge should be centimeters distant from the external edge of the deck (figure 6.14), to avoid breaking the unreinforced concrete of the corner.

To reduce deck drift problems, it is convenient to design the zones of the bottom slab soffit which come in contact with neo-flon plates as horizontal crosswise. For sloped sections, the soffit scheme of figure 6.15 can avoid the use of trapezoidal neo-flon plates, which are rather expensive.

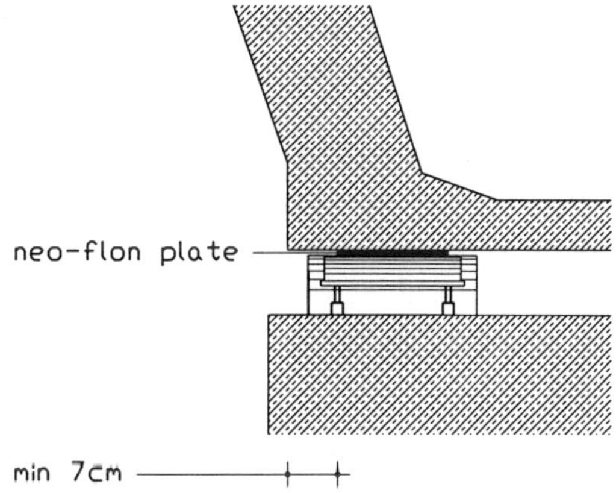

Figure 6.14 - Neo-flon plates should be introduced some centimeters distant from the external edge of the deck so as not to damage the concrete cover in the corner.

Figure 6.15 - By designing the zones of the bottom slab which come in contact with neo-flon plates as transversely horizontal, it is possible to avoid trapezoidal neo-flon plates.

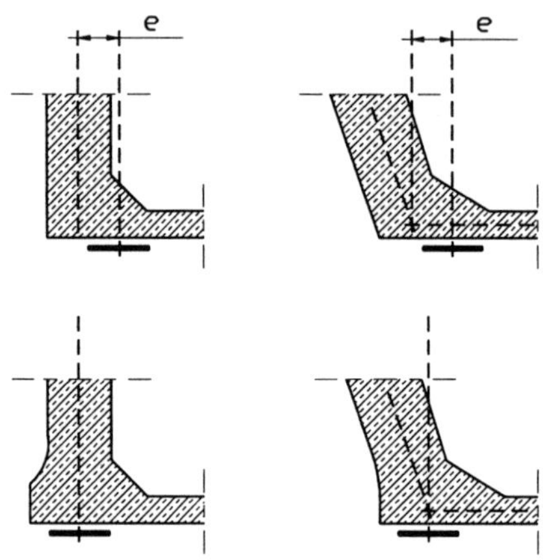

Figure 6.16 - Lower node schemes. When the neo-flon plate width is such as to produce a transverse eccentricity of the support reaction from the web axis, a lateral widening (or, in case of sloped webs, a vertical external edge) can help to center the support reaction.

The problem of centering the support reaction on the webs grows with structural lightening: *0.50 m* thick webs hardly suffer transverse bending, but *0.30* to *0.35 m* concrete webs, or steel webs, can require a lateral widening at the support, or, if sloped, an external vertical edge (figure 6.16).

The contact surface is dimensioned for the highest support reaction, R_{max}, limiting the average compression on Teflon to about *12 MPa*[108], since it is impossible to completely avoid local defects which increase actual compression toward the allowable value, about *17 MPa*:

$$\sigma_{Teflon} = \frac{R_{max}}{b_{nf}\ \Sigma l_{nf}} \approx 12 \text{ MPa} \qquad (6.48)$$

The width, b_{nf}, of neo-flon plates is generally 0.30 to $0.40\,m$ and should be 10 to $20\,mm$ narrower than the plates used on the formwork extraction rails, so as to place the former in the prints of the latter, in contact with a deck surface as regular as possible, especially in case of board-marked finished surfaces.

The length, l_{nf}, of each neo-flon plate is statically uninfluential, and is chosen according to different preferences:

- short plates are light and easy to handle but require frequent introductions, and therefore a higher attention of the workers;
- long plates facilitate accurate positioning, specially when superficial defects of the deck tend to rotate them, and the introduction of shim steel plates, necessary when transverse load differences between launching saddles (6.24) compress neo-flon plates so much that the insertion of the subsequent unloaded plates becomes difficult.

In order to facilitate the insertion and the removal of neo-flon plates, the launching saddles have two rounded corners about $0.15\,m$ long (figure 4.24), and therefore their length is:

$$l_s \geq 0.30 + \Sigma l_{nf} \qquad (6.49)$$

while their width is slightly higher than the plate one.

Finite-element analysis confirms the $45°$ distribution of the support reaction inside the webs (figure 6.17), which implies sharp variations of the vertical compression in proximity of launching saddles, to be taken into account in the local stress verification.

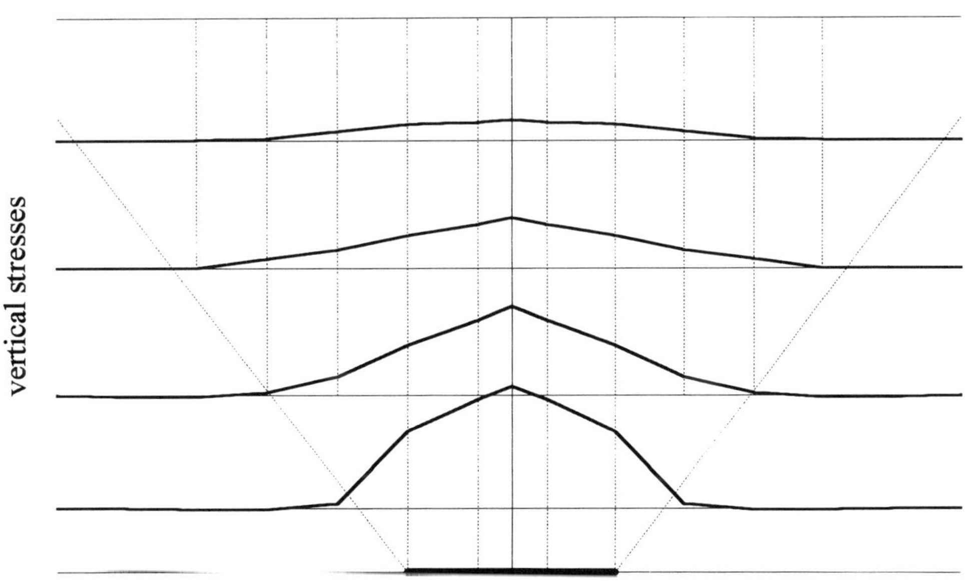

Figure 6.17 - Finite element analysis confirms the $45°$ distribution of the vertical support compressive stresses inside the web.

Longitudinal Compression

The force transferred to the deck by neo-flon plates is not perfectly orthogonal to the contact surface. In fact, the deck is generally launched along a sloped surface, and the vertical component of the contact force – that is, the support reaction of the continuous beam, is not orthogonal to the deck axis. Besides, the contact force is transferred dynamically. Sliding friction along the launching saddles rotates the resultant of forces and produces additional longitudinal components.

The longitudinal force at each saddle is easy to define once the thrust force, the average slope of the deck along the launch direction, the deck slope at each saddle, and the distribution of the support reactions of the continuous beam are known. The diagram of axial force in the deck increases with similar steps proceeding from the front support toward the section in which thrust force is applied.

While this general compression of the deck is often negligible, the local compression produced by these longitudinal forces is rather high before its distribution to the whole cross section, and adds to the longitudinal compression due to the bending moment of the continuous beam and to launch prestressing. Near the lower edge of each support section, the diagram of longitudinal compressive stress differs from the theoretical Navier's behavior, and reaches significantly higher values. Both deck-monitoring[83] and finite-element calculation confirm this phenomenon, which compromises the conservation of plane sections when passing on the launching saddles (figure 6.18). These perturbations of the stress diagram dissipate within short distances, comparable with the deck depth.

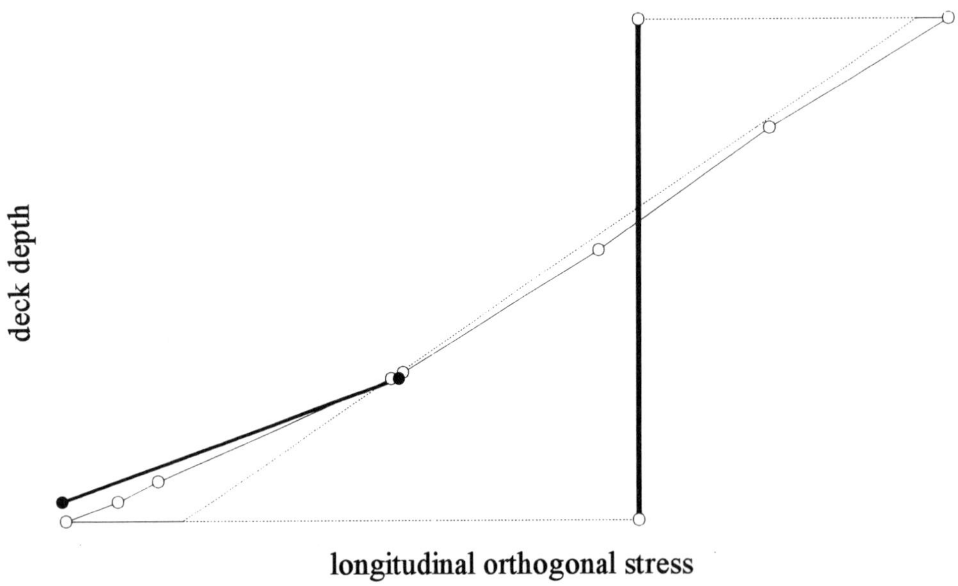

Figure 6.18 - Both deck monitoring (black dots) and finite elements (white dots) confirm the presence of local longitudinal compressive stresses near the lower edge of the support sections. The stress diagram differs from theoretical Navier's behavior, dotted in the figure, and reaches significantly higher values.

Even shear stress distribution cannot be evaluated theoretically. Finite-element analysis confirms high perturbations (figure 6.19), which dissipate within half the deck depth.

These phenomena are important on the current launching saddles, and become even more important when using friction launchers, since the longitudinal force they transfer to the deck is much higher than that mobilized at the pier supports. When designing these units, lifting jacks must be equipped with ball-and-socket caps to improve the adhesion to the deck of the contact plates, whose surface should be adaptable to avoid excessive local pressures (figure 4.23). In addition, the contact surface must be as wide as possible to reduce the average pressure. These measures are even more necessary when using several pairs of friction launchers placed on the piers, since in the case of only one set the distance of the adjacent deck supports is shorter than the final spans of the deck, and the total thrust force opposes smaller support reactions and bending moments.

Stress Verification

When passing over launching saddles, the lower edge of the deck webs is subjected to a triaxial stress state.

With accurate placement of launching saddles and neo-flon plates and low thermal gradients in the deck plates, transverse stresses in the lower nodes are lower than those in the other two main directions, and the state of stress can be assumed as plane. Therefore, principal stresses are derived from Mohr's circle[27]:

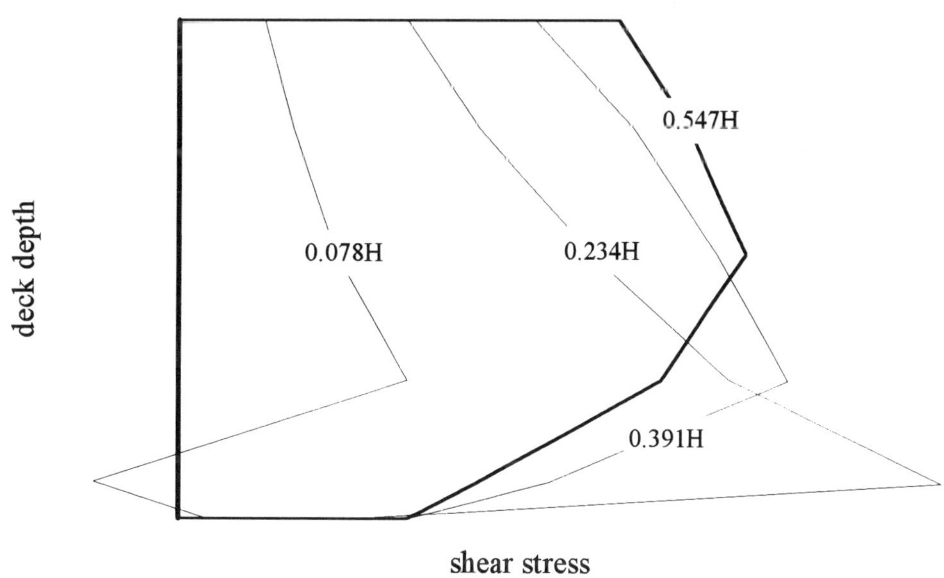

Figure 6.19 - Shear stress distribution near supports. The distance of the analyzed cross sections from the support axis is indicated in relation to the deck depth. Finite elements confirm perturbations which dissipate within half the deck depth.

$$\left.\begin{matrix}\sigma_1\\ \sigma_2\end{matrix}\right\} = \frac{\sigma_x + \sigma_z}{2} \pm \frac{1}{2}\sqrt{(\sigma_x - \sigma_z)^2 + 4\tau^2} \qquad (6.50)$$

and should be compared to the critical stresses deriving from a concrete strength criterion based in biaxial compression.

For instance, by assuming ultimate biaxial compression is *20%* higher than the monoaxial one[14]:

$$\beta = \frac{f'_{bc}}{f'_c} = 1.20 \qquad (6.51)$$

and the ratio of the monoaxial tensile strength to monoaxial compressive strength is according to FIP-CEB MC90 and EC2 (characteristic strength, 28 days):

$$\alpha = \frac{f_c}{f'_c} = 0.21 f_c'^{-1/3} \qquad (6.52)$$

once the octahedral orthogonal stress is defined:

$$\sigma_{oct} = \frac{\sigma_1 + \sigma_2}{3} \qquad (6.53)$$

and also the octahedral tangential stress:

$$\tau_{oct} = 0.471\sqrt{\sigma_1^2 + \sigma_2^2 - \sigma_1\sigma_2} \qquad (6.54)$$

Nadai's equation for biaxial compression ($\sigma_1 \geq 0$ and $\sigma_2 \geq 0$) is:

$$\tau_{oct} - 1.414\frac{\beta - 1}{2\beta - 1}\sigma_{oct} - 0.471\frac{\beta}{2\beta - 1}f'_c = 0 \qquad (6.55)$$

and for tension-tension or tension-compression is:

$$\tau_{oct} - 1.414\frac{1 - \alpha}{1 + \alpha}\sigma_{oct} - 0.943\frac{\alpha}{1 + \alpha}f'_c = 0 \qquad (6.56)$$

As an alternative, Chen's equation for biaxial compression is[14]:

$$\sigma_1^2 + \sigma_2^2 - \sigma_1\sigma_2 - 0.315 f'_c(\sigma_1 + \sigma_2) - 0.685 f_c'^2 = 0 \qquad (6.57)$$

Assuming load lines with the same shape as the ultimate one and, for instance, an elastic limit of $0.6 f'_c$ in axial compression, it is possible to define the biaxial elastic dominion of concrete in the $\sigma_1 - \sigma_2$ plane. The state of stress is easy to check by drawing in this diagram the point corresponding to the couple of principal stresses (figure 6.20).

A last consideration: these stress states, applied many times and for long periods during launch, could cause irreversible local strains due to creep of the concrete, but in most cases, the concrete curing period is so long as to avert this risk.

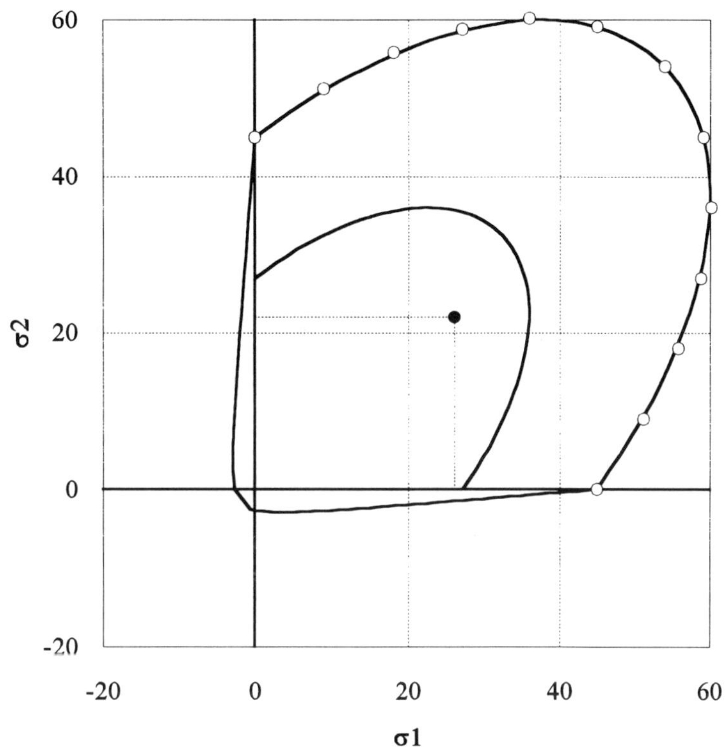

Figure 6.20 - Once the biaxial elastic dominion in the principal stress plane is delimited (white dots are the Chen ultimate values for a B45 grade concrete, the thick continuous line is Nadai equation, and the interior line the Nadai elastic dominion), the state of stress is easy to check by drawing the dots which correspond to the couples of principal stresses.

LIVE LOADS

Upon completion of the launch, a launched bridge does not differ from a deck built directly in its final position, apart from the secondary stresses produced by creep.

Before going into service, the carrying capacity of the deck has to be increased to guarantee the absorption of the additional stresses deriving from service loads. While permanent service loads (curbs, sidewalks, paving, safety fences etc.) are uniformly distributed along the deck and produce the same effects as a dead load

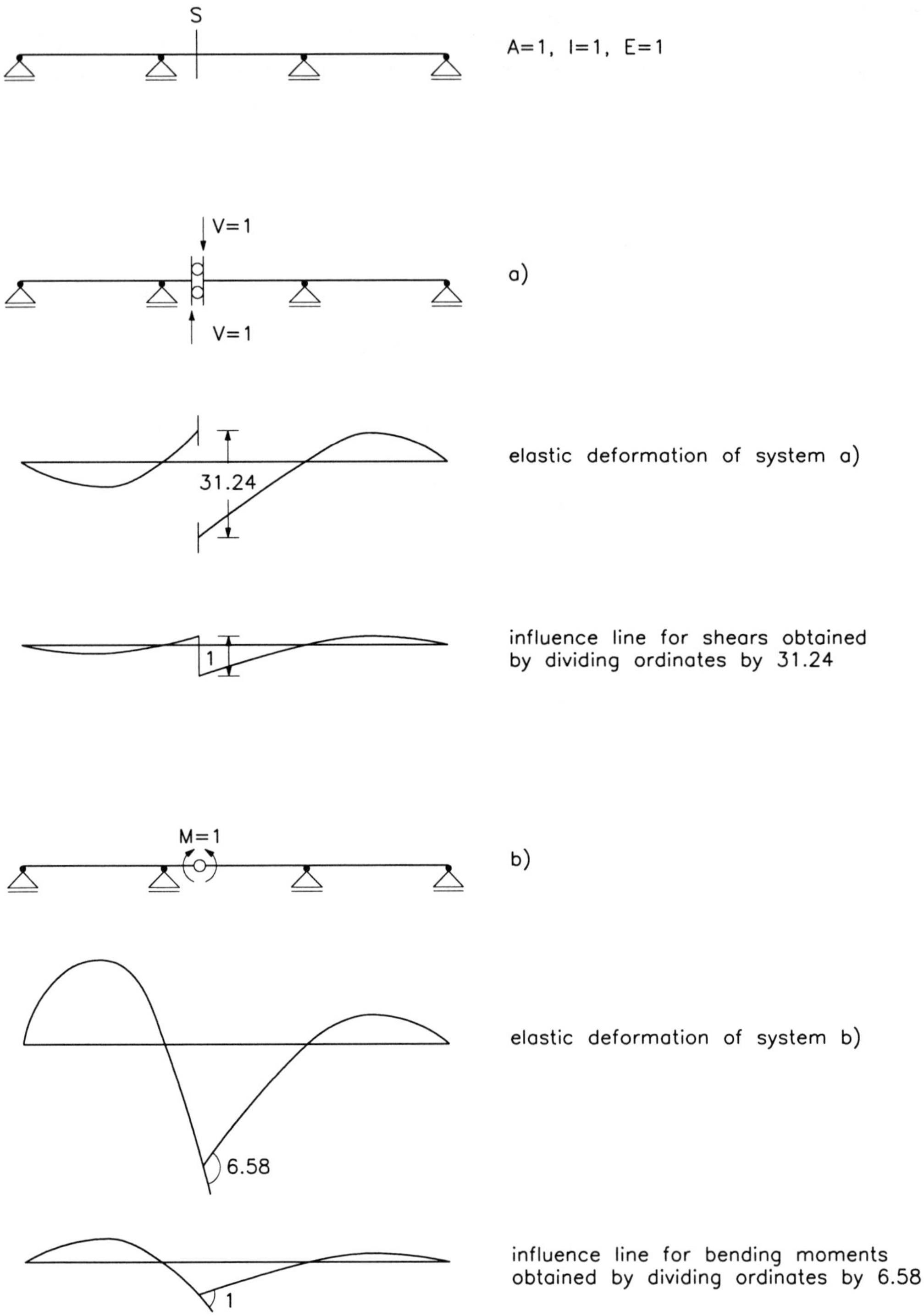

Figure 6.21 - Influence lines for moment and shear force.

increase, live loads are generally mobile, and it is necessary to identify the arrangement which maximizes stresses and strains in each section.

The determination of the worst arrangement of live loads is carried out by means of influence lines, which can be drawn at points, although in a continuous beam even with few spans this procedure involves a great deal of work.

Therefore, an indirect method is generally used, which consists of using particular elastic lines of the beam obtained by applying systems of fictitious forces. The elastic line is intuitive, permits one to define immediately the zones to load, and respects the external restraints of the beam (apart from the ones whose reactions are in question) permitting a fast graphic check of computer outputs.

Their determination is carried out by means of the Colonnetti-Land theorem, which states that in a generic section of the beam the influence lines for moment and shear are proportional to the beam elastic lines obtained by cutting the section and applying equal and opposite forces to the cut edges.

For the influence line for moment, the forces should be such as to produce only a relative rotation at the cut edges, and for shear, only a relative displacement (figure 6.21).

Once influence lines are drawn (with (5.65) when adopting the RTM method), the maximum stresses are found by arranging live loads appropriately. For flexural stresses, which in bridges represent the main problem, this search is limited to a few characteristic schemes:

- for a concentrated load, the highest positive moment is obtained placing the load in the positive peak of the influence line, and the highest negative one with the load in its maximum negative ordinate;
- for a train of concentrated loads (more loads equally spaced) the worst position for positive moment is searched by trial and error, limited to the positions with one of the loads in the peak of the influence lines; it is demonstrable that the position in question lies between these;
- for a load distributed over a fixed length, the worst position occurs when the ordinates of the influence line at the two ends of the loaded zone are equal;
- finally, in the case of a "segmentable" distributed load, of fixed intensity and any length, the worst arrangement is with the load distributed in all the zones of the influence line with the same sign.

Similar considerations characterize the use of influence lines for shear force.

VERIFICATION OF DESIGN ASSUMPTIONS BY DECK MONITORING

In launched bridges, the stresses which arise during launch are so high as to influence the bridge dimensioning, and therefore its cost. Consequently, each hypothesis regarding bearing tolerances, thermal gradients, time-dependent phenomena, and such must be carefully evaluated, since they influence directly the competitiveness of the project.

In the most delicate cases, the correctness of design assumptions may be checked by monitoring the deck and recording some launch parameters. These controls give useful information about the elastic modulus of the concrete, the coefficient of thermal expansion, the influence of topographic and geometrical defects, the effects of shrinkage and creep, and the stress distribution when passing over the launching saddles.

It is convenient to choose the section to instrument among the most stressed ones, and therefore in the front zone of the deck: this also permits early retrieval of the first data. The areas monitored by the strain gauges should overlap (figure 6.22), to eliminate incongruous data. Strain gauges at the lower edge of the webs can be positioned in more directions to measure simultaneously longitudinal and vertical compression, and tangential stresses when passing onto the launching saddles[83]. Strain meters immersed in the deck concrete can be complemented with external extensometers which measure the distance between two points of reference on the deck surface. This allows monitoring a section even after its casting.

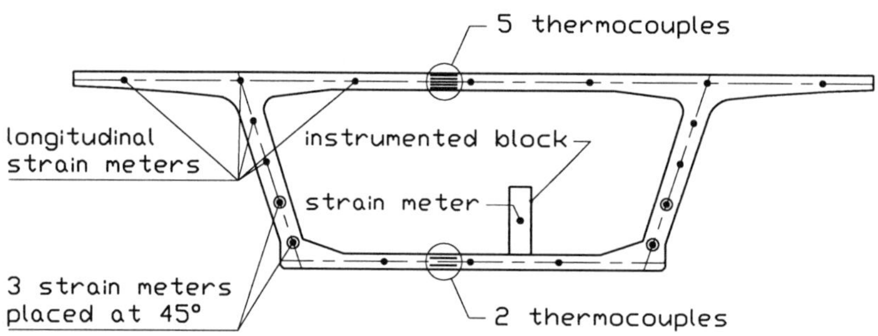

Figure 6.22 - Sample of an instrumented section. The areas of reading of the strain gauges overlap to eliminate incongruous data. Strain gauges at the lower web edges are positioned so as to read longitudinal, vertical, and *45°* inclined strains when passing on the launching saddles. Thermocouples permit one to identify thermal strains and to subtract them from recorded data. The instrumented block placed in the box cell is necessary to evaluate shrinkage.

Besides the investigation of launch stresses, monitoring can be aimed at analyzing time-dependent phenomena. In this case, the section to instrument is chosen between the most stressed ones in the final static scheme.

To obtain stresses from the recorded strains it is necessary to know the elastic modulus and the coefficient of thermal expansion of the concrete. Therefore, immersed strain gauges are generally equipped with thermocouples. Moreover, it is convenient to apply several methods for measuring the same parameters, so as to compare the results and eliminate most of the uncertainties. Lastly, since important local stresses are foreseen when the instrumented section passes onto the launching bearings, a preliminary finite-element calculation is convenient to define the theoretical stress distribution and to permit a better comprehension of the measured data.

Elastic Modulus of the Concrete

Knowledge of the actual elastic modulus of the concrete is of fundamental importance for the calculation of stresses.

The statistical formulae which relate the elastic modulus to the concrete's compressive strength in the different design specifications are generally too approximate to analyze the behavior of instrumented sections, and direct determination is necessary. Because of the great influence of this parameter on the results, it is convenient to use several methods at the same time.

A first method consists of measuring the strains of cylindrical specimens subjected to monoaxial compression. The small dimensions of specimens and the different curing conditions do not lead to reliable results, but only rough data.

A second method is based on tendon stressing, and correlates the average shortening recorded by many extensometers distributed along the section with the total force applied by tendons. Strains must be read immediately before and after each tensioning, in order not to superimpose thermal strains and the first creep strains on the instantaneous elastic ones. The mean of read values, weighted according to the surface of influence of each extensometer, permits one to define the average deformation plane of the section. Generally, the individual strains deviate from this average plane. These differences can be high when recording the strains due to tensioning of launch tendons, since the concrete is relatively young, and are lower when tensioning the parabolic final tendons upon finishing of the launch.

Although these deviations do not modify the elastic modulus, they indicate that this measuring method tends to become imprecise as the number of tendons increases, since they add very small individual strains affected by measurement errors, and the hypothesis itself of conservation of plane sections is questionable in these tensioning sequences. In turn, by operating simply with the difference between the final average strain and the initial one, one could introduce strain components due to creep, underestimating the elastic modulus. Further uncertainty comes from the support friction.

A third method is based on measuring the support reaction, R_A, in two subsequent launch positions (figure 6.23), and leads to a reliable flexural stiffness, EI, but is affected by the manometric tolerances of the weighting jacks (ordinary jacks topped with steel launching bearings)[14]. Once the reactions, $R_{A(1)}$ and $R_{A(2)}$, in the two positions are read, it is[83]:

$$\begin{cases} M_1 = R_{A(1)}x_1 - Pd \\ M_2 = R_{A(2)}x_2 - Pd \\ \Delta M = R_{A(2)}x_2 - R_{A(1)}x_1 \end{cases} \qquad (6.58)$$

and in a fiber of the instrumented section at distance, z, from the centroidal plane, the difference in curvature is proportional to the strains measured in the two support positions:

$$\Delta\varphi = \frac{\varepsilon_2 - \varepsilon_1}{z} \qquad (6.59)$$

from which:

$$EI = \frac{\Delta M}{\Delta\varphi} \qquad (6.60)$$

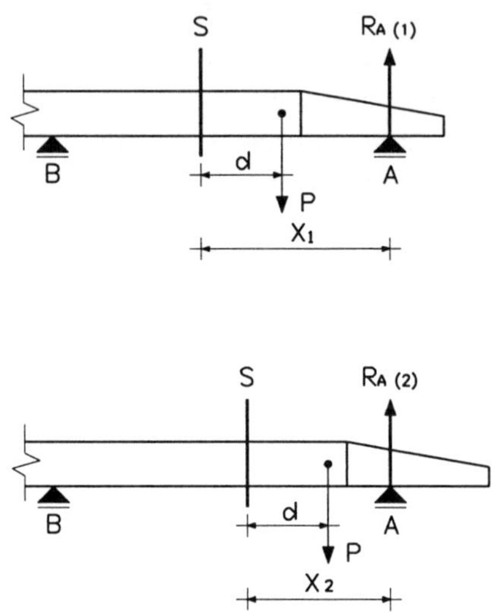

Figure 6.23 - Evaluation of the elastic modulus of the concrete by measuring the support reaction, R_A, in two subsequent launch positions.

By operating quickly – that is, with the hypothesis of constant thermal gradients, and applying the same logic, the elastic deflection of the launching nose tip can be surveyed by means of a laser beam collimated on a graduated plate.

When using friction launchers, the deck raising can be correlated with the vertical force transferred by the lifting jacks, and the recording of operational parameters of the launchers permits one to obtain a great deal of data without instrumenting any section. In figure 6.24, the evolution of the elastic modulus of the Serio River Bridge[171] is shown.

Only specimens, though less representative, permit measuring the elastic modulus directly. In all the other methods, its indirect determination suffers the uncertainty of the actual cross section of the deck.

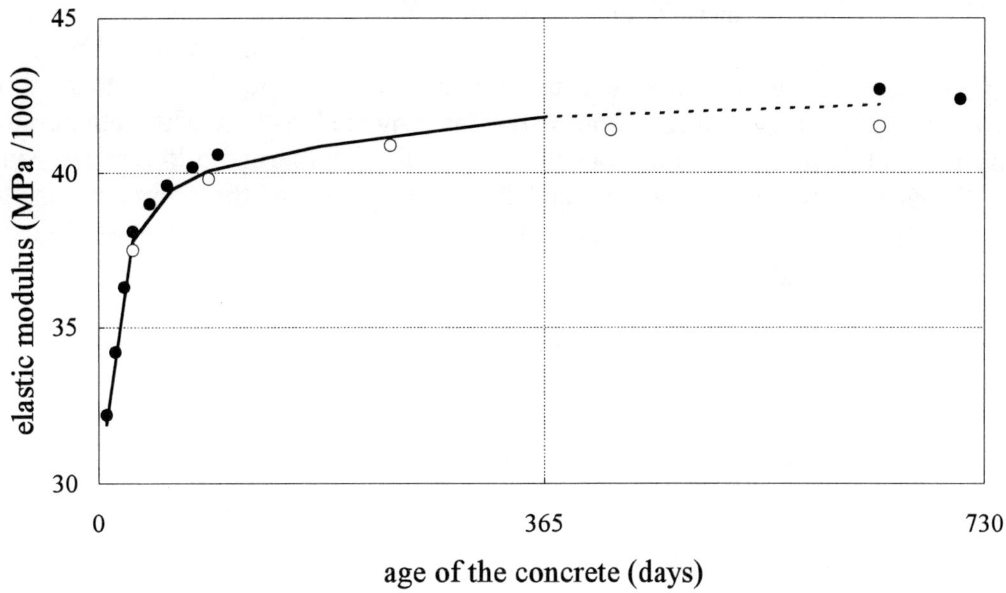

Figure 6.24 - Evolution of the elastic modulus of the concrete in the Serio River Bridge. Continuous line was calculated according to the Italian Norm and the development of compressive strength suggested by BS 5400. Black dots were obtained indirectly (nose deflection and deck raising at the launcher). White dots were read in specimens.

Thermal Correction of Results

Thermal correction of results is a problem neither simple nor to be underestimated, since the evolution of temperature in the deck is sharp and can affect the measured strains. It can be approached in two ways:

- by measuring strains frequently during operations which take a long time (for instance, tendon tensioning) and by interpolating the data so obtained. Once the elastic deformation is estimated, it is possible to cancel the effects of all time-dependent phenomena;
- by measuring temperatures in the instrumented section and by calculating the expansion, ε_0, and the rotation, φ, with (6.5), to define from these the self-balanced thermal stresses, and then to subtract the strains. This method assumes a constant distribution of temperature along the deck, and also assumes that the coefficient of thermal expansion and the elastic modulus of concrete are known.

Coefficient of Thermal Expansion

Knowledge of the coefficient of thermal expansion is fundamental when taking readings over an extended period of time, and needed to define the share of deformation due to creep and shrinkage when the instantaneous deformation is known.

The coefficient of thermal expansion must be defined case by case, since it depends on many parameters hard to schematize, such as the mineralogical nature of aggregates, and the humidity and temperature of concrete. It is evaluated by recording the average deformations of the instrumented section when temperature changes, and therefore it requires a sunny day with high thermal differentials. Since it governs both the expansion, ε_0, and the rotation, φ, of the cross section, the average deformation of the centroidal fibers is used, to reduce as much as possible the influence of gradients.

Shrinkage

Time-dependent strains recorded by extensometers combine the effects of both shrinkage and creep, and the share of strain due only to shrinkage can be evaluated only by means of specific measuring techniques.

A first method consists in building a sample segment of the deck, adequately instrumented and placed near the bridge in order to guarantee similar hygrometric conditions.

A second method consists in instrumenting an unreinforced concrete block placed inside the box cell (figure 6.22), by means of an immersed strain gauge. As in the first curing stages shrinkage is proportional to the ratio of the drying perimeter to the cross sectional area, it is necessary to size the block with the same proportions as the deck and to seal its ends to reproduce the actual hygrometric conditions. In spite of this, the block is protected on each surface while the deck has protected internal surfaces but external sides exposed to wind and drying, and therefore the measured data are approximated.

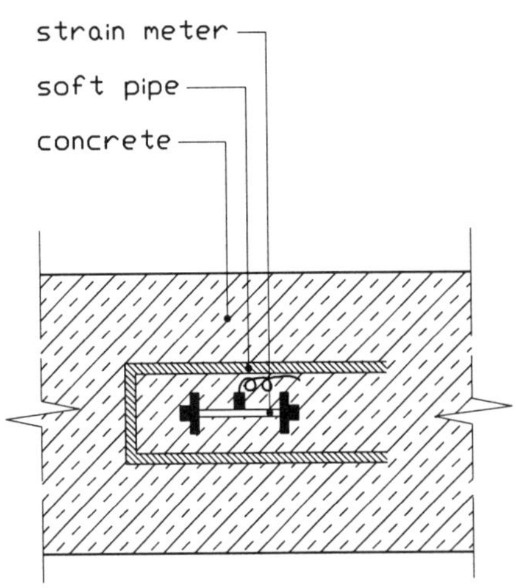

Figure 6.25 - A strain gauge immersed in the deck concrete and mechanically insulated by means of a soft pipe permits one to obtain readings independent from the creep strains.

A third method consists of immersing a strain gauge mechanically insulated from the rest of the section in the deck concrete (figure 6.25), to obtain readings independent from the creep deformation.

Each measuring method confirms that shrinkage strains have a markedly seasonal evolution, with shortening in the dry periods and lengthening in the humid ones[131]. This contradicts laboratory data and produces a total shrinkage substantially lower than the expected one.

Chapter 7

PRESTRESSING

LAUNCH PRESTRESSING

During launch, absorption of edge tensile stresses by means of reinforcement is convenient only in short spans, and in most cases it is better to reduce or cancel them by means of an appropriate launch prestressing. The cyclic changes in sign of both bending moment and shear force do not allow use of the parabolic final tendons for this purpose, since their action would be totally inappropriate in the temporary support configurations which differ from the final one. Therefore, launch prestressing cannot but be centroidal, and its contribution is limited to a uniform compression which reduces tensile stresses in the edges decompressed by bending moment, or, in the case of total prestressing, which guarantees adequate margins from their decompression.

The first thought goes to pre-tensioning, simple to obtain with straight layouts. It avoids the cost of anchorages, couplers, sheaths, and grouting, and guarantees better compaction of concrete and better corrosion protection of strands, compared with their confinement in grouted sheaths.

But, in reality, even allowing edge tensile stresses, the average compression to apply to the cross section is rather high, about 5 to $7\ MPa$, and a surface of many square meters requires total prestressing forces of tens of MN. The application of such prestressing would result in very expensive tensioning devices and braking systems for the deck, which is supported on low friction bearings and movable with much lower forces. Therefore, launch centroidal prestressing is always obtained with post-tensioned tendons, arranged inside or outside the concrete.

This uniform compression is particularly ineffective at the end of launch, since it overloads the deck edges already compressed by the bending moment diagram. Moreover, the straight layout of tendons avoids reducing the shear force near the supports.

As the cost of launch prestressing increases with the span (5.7), for the longest spans incremental launching can compete with free cantilevering only by reducing the influence of temporary stresses – that is, introducing temporary piers which halve the average span of the launch static scheme. In all the other cases, it is better to regularize launching stresses by means of a launching nose or a stayed front scheme and to optimize the use of axial prestressing.

As an adequate quantity of prestressing is available from the beginning of launch, the first approach is to shift tendons at the end of launch from the centroidal position to an eccentric one, so as to use the same strands both for the launch prestressing and for the final one.

The first launched bridge, over the Rio Caroni, was actually prestressed in this way, but savings in material were canceled by the cost of labor necessary to shift the tensioned tendons, and this prestressing scheme was never adopted again.

More specialized schemes were studied, composed of permanent launch tendons integrated at the end of launch with parabolic cables designed to modify the final layout of the prestressing force according to the continuous beam requirements and the need to reduce shear stresses. Thus, launch prestressing became a permanent component of the deck prestressing, and not only because of the cost of its removal. In fact, it increases the safety factor from decompression in case of unforeseen events, such as excessive support settlement, thermal gradients, and seismic phenomena, especially considering that seismic forces can come from totally arbitrary directions. It also improves deck behavior near the inversion points of the bending moment diagram, covering the wide movements of the null points produced by the presence or the absence of live loads.

As this scheme tends to cause higher consumption of prestressing compared with some alternative construction methods, it has been constantly improved to make the centroidal component of the final prestressing even more effective and less expensive.

In spite of many alternatives and various trials, in medium-span bridges the superimposition of centroidal launch tendons and parabolic integrative cables was revealed to be the simplest and least expensive solution. It avoids complex handling of prestressing and permits one to tension each cable once and for all in its final position, and to grout it immediately, avoiding risks of corrosion.

Finally, in the last few years, an innovating prestressing scheme appeared. It consists of permanent parabolic tendons and temporary external tendons arranged to produce an antagonist action which makes the resulting prestressing force centroidal. This scheme requires the superimposition of several tendon groups in subsequent construction stages and is competitive only in particular cases. The high average compression reduces the principal tensile stresses in the webs, resulting in particularly slender and efficient cross sections.

Types of Tendons

In the deck segment supported on the formwork extraction rails, bending stresses are very low. The introduction of centroidal prestressing before launch is mainly aimed at avoiding shrinkage cracks (for this reason it should be applied as soon as possible after dismantling the external formwork), at absorbing the low tensile stresses due to extraction, and at preparing the segment to be launched over the curing supports, and then over the first span.

Concrete curing is very important in the repetitive production sequences which characterize incremental launching, since a delay in launch also delays the construction of the subsequent segment, and curing during working days instead of during the week-end makes the rotation of labor difficult.

Time-dependent phenomena in young concrete suggest introducing launch prestressing progressively by means of long tendons joining more segments, so that the tendons anchored in the back segment produce a low compression in this, and progressively higher stresses in the previous segments.

In order to tension cables directly to their final stresses it is necessary that concrete behind their anchorages not collapse, and this can require:

- the use of precast concrete blocks containing tendon anchorages (figure 3.6). Since defects or voids in the deck concrete behind these blocks can break them during tensioning, the slope of the joint surfaces and the joint reinforcement must be studied so as to guarantee an easy vibration of the deck concrete;
- local heating by means of thermal insulation, warm air, or electric resistances immersed in the concrete. Compared with precast blocks, local heating permits a better continuity of reinforcement;
- the use of higher-strength concrete in the zones around the anchorages.

The advances in incremental launching saw the creation and the optimization of many different schemes for centroidal prestressing:

- permanent straight tendons in the slabs joined to each other by couplers in the construction joints between the subsequent deck segments;
- temporary or permanent straight tendons in the slabs anchored in accessible blisters and joined to each other by longitudinal overlapping;
- temporary or permanent external straight tendons together with one of the previous types or alone;
- temporary or permanent undulated tendons, either internal or external, along with temporary antagonist tendons which make the eccentricity of the resulting force constant, and straight external tendons which raise prestressing force toward the centroidal level to reduce the overall eccentricity.

Coupled Straight Tendons

After the first experiences with shifting tendons, launch prestressing was obtained by straight tendons only as long as the deck segments, joined by couplers in each construction joint of the deck.

In effect, coupled joints present some advantages compared with tendon overlapping (lower secondary stresses, lower losses of prestress due to friction, simpler formwork and reinforcement) but many limitations as well:

- the higher cost of prestressing, because of the shorter average length of tendons, the additional cost of couplers, and the higher consumption of skilled labor;
- the impossibility of removing a portion of the centroidal prestressing at the end of launch;
- the frequent need to increase the slab thickness near the joint in order to contain the couplers (figures 7.1 and 7.7), and the tendency to extend this thickening to the whole deck, resulting in a greater dead load;
- the need to check the joint section accurately, since couplers reduce the moment of inertia considerably (although couplers became smaller and smaller over time);
- the sharp introduction of the whole launch prestressing in young concrete, even though it is centroidal and therefore not able to produce irreversible rotations.

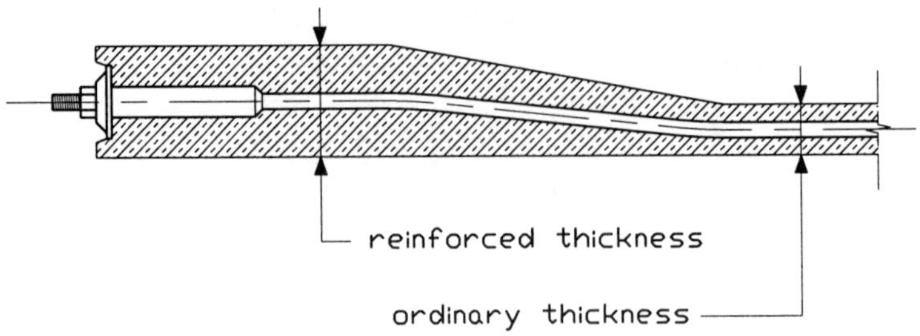

Figure 7.1 - Slab thickening is often necessary to contain tendon couplers.

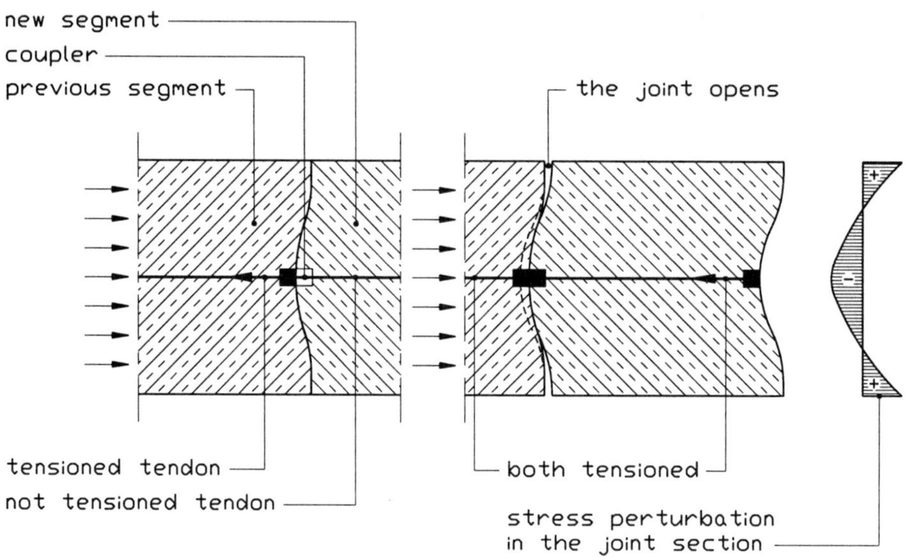

Figure 7.2 - Tensioning of the new tendon segment annuls the pressure transferred by the previous anchorage, and the elastic strain recovery decompresses the distant zones of the joint until tensile cracks are produced.

This last aspect has often been the cause of distributed cracks in the joint section because of the elastic strains on concrete near the anchorages.

The force transferred by each anchorage deforms the concrete locally (figure 7.2), and with it the construction joint against which the concrete of the next segment is match-cast and then compressed by its own launch prestressing.

Tensioning of the newly added tendons annuls the pressure transferred by the anchorages in the previous segment, and the recovery of the elastic deformation of the concrete compresses locally the joint and, for equilibrium, decompresses the zones more distant from the anchorages, until producing tensile cracks.

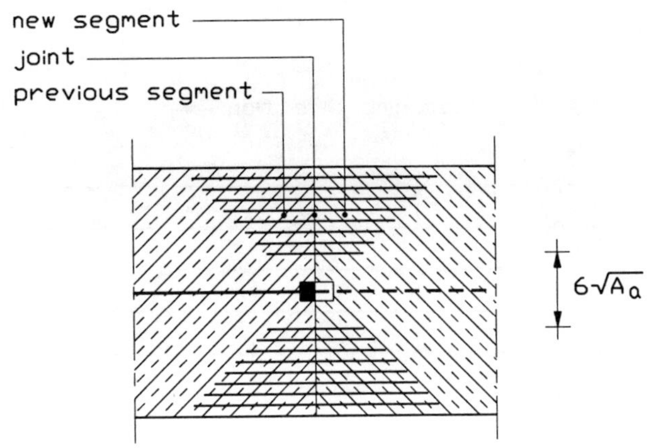

Figure 7.3 - Additional joint reinforcement.

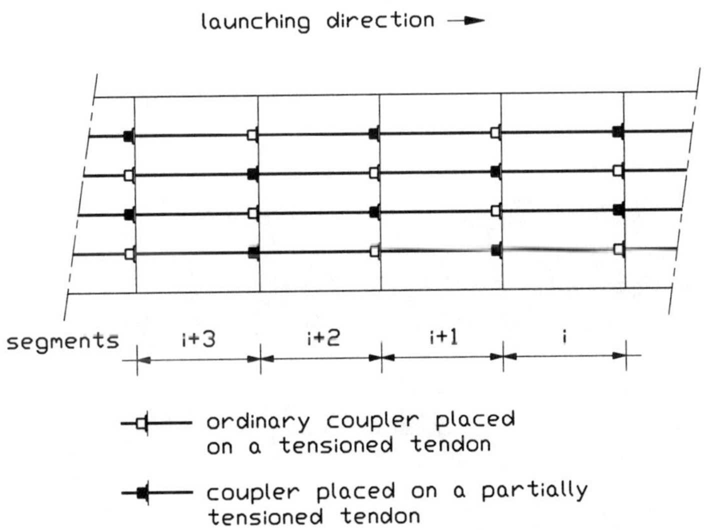

Figure 7.4 - Tendons coupled at each joint but tensioned alternately every two joints.

Therefore, unless launch prestressing is not higher than about $4\ MPa$, the zones of the joint section distant more than $3\sqrt{A_a}$ from the axes of couplers (where A_a is the anchorage area of the coupler) should have a reinforcement ratio of about *0.5* to *0.7%* in the webs and the top slab, and *0.8* to *1.0%* in the thinner bottom slab[112] (figure 7.3). Besides, it is convenient to use many tendons to guarantee a good distribution of the prestressing force in the cross section, and this complicates even more the joint geometry.

Because of these problems, the evolution of this scheme of prestressing was fast, and followed two different directions:

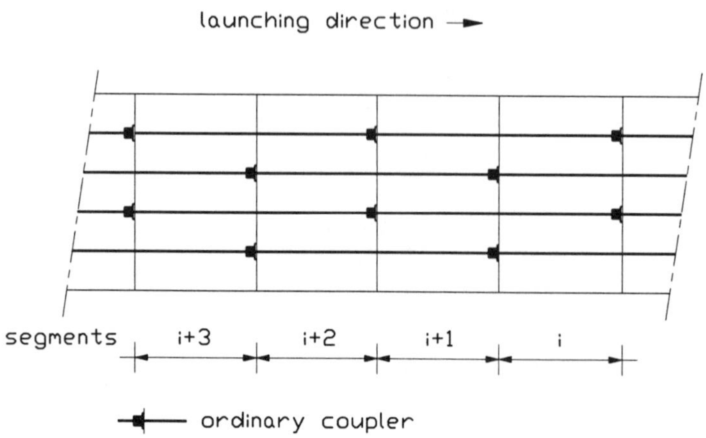

Figure 7.5 - Staggered couplers permit one to prestress two segments with each tendon.

Figure 7.6 - A niche in front of the coupler is necessary to anchor strands installed in empty sheaths after casting. Handling of tendons is simpler and reinforcement is not disturbed by the niches, but the coupling operations are slow and expensive (author).

Figure 7.7 - Coupled prefabricated tendons emerge from the back joint and make the cage assembly in the formwork more complex, and its complete preassembly impossible (author).

Figure 7.8 - Couplers are practically unavoidable in slender decks where access to the box cell is difficult or impossible (courtesy Dywidag Systems International[5]).

- tendons coupled in each joint but tensioned alternately every two joints, so as to make the introduction of prestressing more progressive (figure 7.4);
- staggered couplers so as to prestress two subsequent segments with each tendon and halve the number of couplers (figure 7.5). With this scheme, if the strands are installed in empty sheaths after casting, niches in front of the couplers are necessary to anchor the strands (figure 7.6). On the other hand, if using prefabricated tendons, these emerge from the back joint and make the cage assembly in the formwork more complex, and its complete preassembly impossible (figure 7.7). This must be limited to the webs, and individual panels for the slabs.

In spite of these limits, the schemes with couplers are very competitive in slender decks where access to the box cell is difficult (figure 7.8). Tendon overlapping may be more advantageous in the deeper sections.

Overlapped Straight Tendons

The recent advances in the techniques of strand insertion in empty sheaths through accessible anchorages have made joining by overlapping an often-adopted and competitive solution. The advantages are:

- the handling of short segments of empty sheaths instead of long and heavy prefabricated tendons speeds up the cage assembly, saves labor, and guarantees a better final result;
- concrete curing time can be utilized for fabricating the tendons. This reduces the criticality of the steel cage assembly when it occurs directly in the formwork, and permits a better rotation of labor;
- because of the absence of tendons emerging from the back joint, integral cage preassembly is possible (figure 1.12);
- the average length of tendons increases, and this reduces the cost incidence of anchorages and the losses of prestress due to their re-entry;
- friction losses in the longer tendons can be reduced by symmetrical tensioning from both ends;
- the cost and the implications of couplers are avoided.

Launch overlapped tendons are arranged in the slabs, similarly to the mechanically coupled ones. While their front (launchward) anchorages always emerge inside the box cell, the back anchorages can emerge as well, to be always accessible, or can be placed within the joint cross section. In this latter case, the match-casting of the subsequent segment (figure 1.12) covers them, and the tendons cannot be relieved.

By placing the back anchorages in the construction joint, the cost of blisters halves, but special reinforcing schemes are necessary because of creep strains due to local compression at the anchorages. Unlike coupled tendons, the anchorage forces do not reduce with time, and the creep deformation modifies the stress distribution along the joint, decompressing the zones in front of the anchorages so much that even an average initial compression, σ_m, of about *3 MPa* may be unable to prevent tensile stresses (figure 7.9). Joint cracking may be avoided by additional reinforcement sized for a tensile force of:

$$F \cong 0.3F_p - 1.33A_a\sigma_m \qquad (7.1)$$

in which F_p is the prestressing force in the tendon and A_a the anchorage surface[112]. To limit crack width to less than *0.2 mm*, small diameter bars sized for *0.85%* of the area $4A_a$ can be distributed in it.

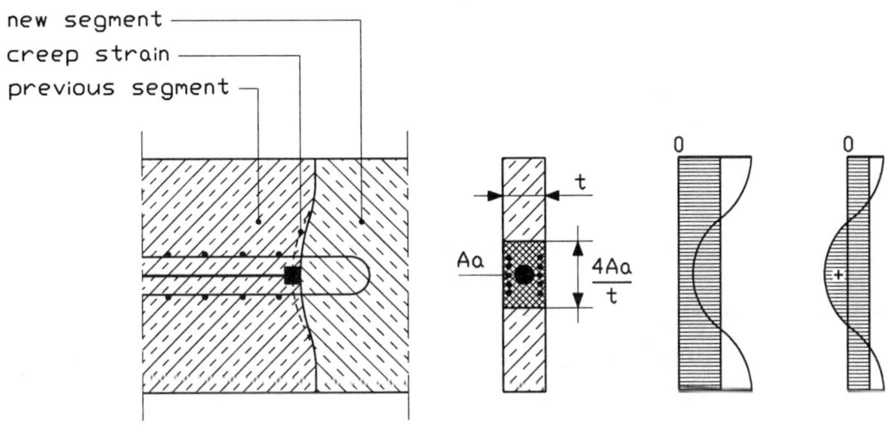

Figure 7.9 - The force transferred by a tendon anchored in the joint cross section does not reduce with time, and creep deformation of the concrete decompresses the zone in front of the anchorage.

On the contrary, back anchorages in blisters are always accessible and permit one to relieve and recover some of the centroidal prestress tendons after launch is completed (figure 7.10). This operation, although it reduces the permanent prestressing in the structure, is generally not very competitive and is adopted only to correct the undesired effects of launch prestressing, as:

- recovered strands are hard to reuse, since they are too short for the parabolic tendons and, if cut longer than the distance between launch anchorages, they become damaged by the jack clamps and by the anchor keys;
- anchorages are lost as well as sheaths, which must be grouted to cover holes in the deck;
- the impossibility of grouting temporary tendons exposes strands to the risk of corrosion under load. Since this risk is acceptable only for short periods of time, the possibility of recovering non-galvanized or PVC-coated strands is limited to short bridges.

Therefore, tendons are generally grouted within one month of their stressing, the period in which the emulsified oils applied to strands in the shop are generally effective.

By inserting strands in empty sheaths, tendons can be very long – that is, they can join several deck segments in order to introduce launch prestressing progressively. For instance, in case of tendons joining three segments, the segment in the formwork is stressed by only one third of the launch prestress, the previous segment by two thirds, and only the third one is fully prestressed.

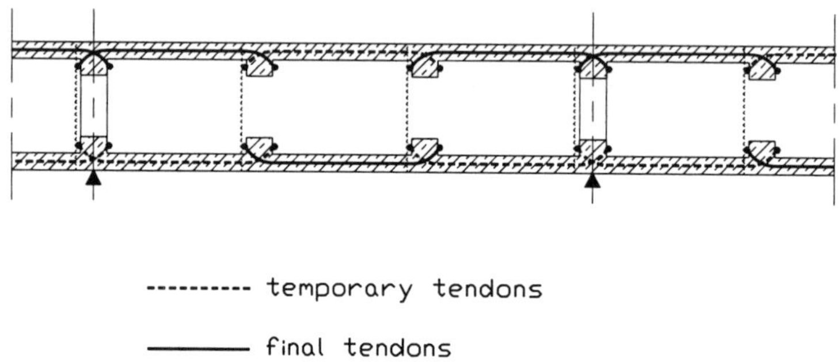

Figure 7.10 - Anchorages in blisters at both ends permit one to relieve and recover some of the centroidal prestress tendons after launch is completed. With this scheme, the optimum length of the deck segments is one third of the current span.

The length of the tendons and the connection scheme of deck segments characterize the progress of launch prestress in the deck, and the allowable bending moment in the deck according to the prestress and curing reached defines the spacing between the curing supports.

External Straight Tendons

The first launched bridge over the Rio Caroni[20] did not use any permanent launch prestressing, since service prestress was obtained by varying the position of launch tendons. For this purpose, they necessarily had to be external.

Later, the scheme composed of launch centroidal internal tendons corrected at the end of launch by parabolic and short, straight tendons characterized most launched bridges. In some bridges, an intermediate solution has been adopted, composed of:

- internal straight final tendons distributed in the slabs of the cross section so as to produce a centroidal prestress force (figure 7.11);
- external straight temporary tendons, also centroidal, to be relieved at the end of launch to reduce deck compression without introducing secondary moments. In the case of total prestressing, these additional launch tendons can be sized to

cover bending moments produced by thermal gradients and bearing settlement, and rarely exceed *25%* of the total final prestressing;

– parabolic, span, and cap tendons prepared and tensioned at the end of launch (figure 7.12).

The use of external tendons for launch prestressing complicates the internal formwork and the steel cage, but offers many advantages, such as savings in sheaths and grouting, the simplicity of tendon removal, and the improvement of the mechanical properties of the cross section due to the lower number of holes.

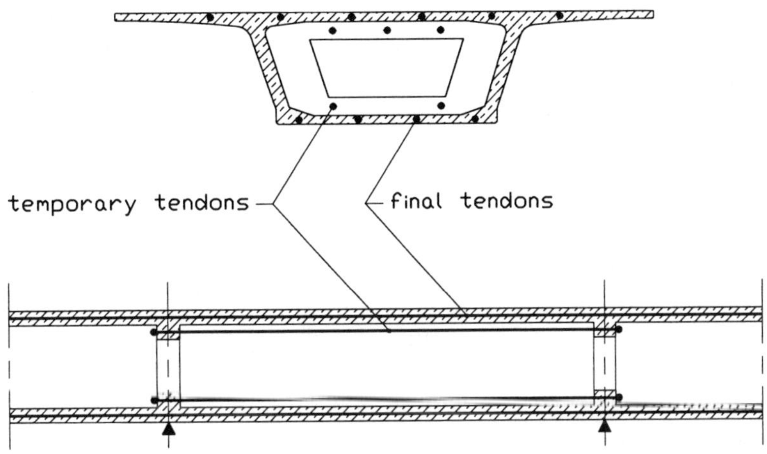

Figure 7.11 - Launch prestressing obtained by integrating internal permanent tendons with external temporary tendons. Both tendon families produce a centroidal prestress force, so as to avoid secondary effects when relieving external tendons.

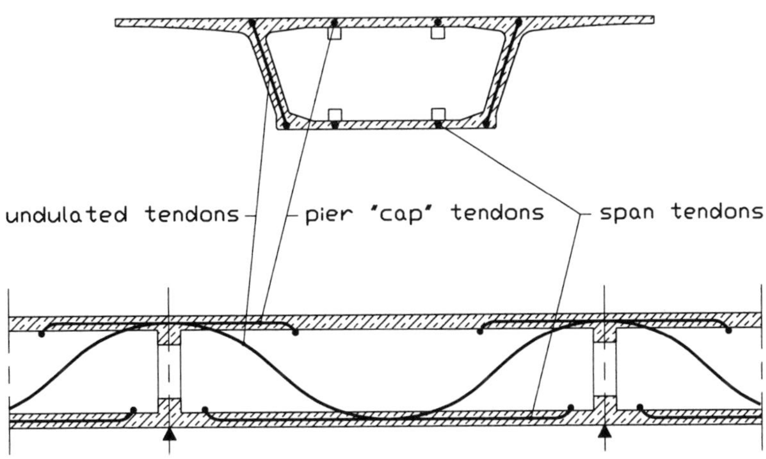

Figure 7.12 - Upon completion of the launch, centroidal prestressing is integrated with parabolic, span, and cap tendons, prepared by inserting strands within empty sheaths.

In order to improve the competitiveness of this scheme, external tendons can be anchored directly to the support diaphragms, permitting reuse of their anchorages if additional external prestressing is necessary in the future. In this case, diaphragms must be built before launch, and the undulation of the bending moment envelope (figure 5.56) increases the cost of launch prestressing. Moreover, diaphragms should be adequately cured before stressing the tendons, and this increases the length of the curing zone. Lastly, in order to guarantee the safety of workers, temporary tendons should be contained in heavy polyethylene pipes anchored to the deck internal surface, even if the tendons are composed of galvanized strands.

The adoption of a large number of temporary tendons is rarely competitive, except when the number of decks to build subsequently is such as to amortize the recovered material, or when it can be reused as parabolic or polygonal final prestressing. On the contrary, external tendons can be useful for absorbing the tensile stress peaks in the front zone of the deck, since they permit sizing of internal prestressing for more uniform stresses (figure 6.12).

Many applications of these combinations of prestressing have been carried out in France. In the highway Aiguilly Bridge, the final internal centroidal prestressing was of 4 to $5\ MPa$, while the external one, sized for the tensile stress peaks, temporarily added another $2\ MPa$.

Antagonist Tendons

Antagonist prestressing met a certain success in recent years. Its efficiency encouraged engineers to improve this technique, and several applications showed such advantages that it is easy to imagine that incremental launching will evolve in this direction in the future.

The principle is to separate the final prestressing from the temporary one. The final prestressing is the typical one for a cast-in-situ deck, with parabolic internal tendons or polygonal external ones, sometimes complemented by cap and span tendons. The temporary prestressing consists of temporary tendons which compensate the eccentricity of the final tendons during launch, until a centroidal prestress force is obtained. Temporary tendons can be, in very variable proportions:

- short and rectilinear, placed inside or close to the bottom slab in the support sections and to the top slab in mid-span, and coupled by overlapping in transverse beams that, in the case of external final prestressing, can be its deviation diaphragms (figure 7.10). In this case, the optimum length of the deck segments is one third of the current span;
- long and undulated, with polygonal layout symmetrical to the final tendon layout with respect to the gravity axis of the deck. In this case, deviation saddles are necessary along the span.

The high axial compression produced during launch by final and antagonist tendons permits one to obtain extremely efficient cross sections, since the reduction of principal tensile stresses limits the web thickness. In the Amiens Viaduct[83], two $0.30\ m$ webs were used for spans longer than $50\ m$.

The need not to compromise this result with anomalies of the shear force diagram near the deviation points (especially in the case of internal final tendons (figure 7.13), which are parabolic and not polygonal as are the external antagonist tendons) makes the positioning of deviation points difficult, and complicates internal formworks.

In spite of these difficulties, these schemes permit one to optimize prestress both in the construction stages and in the service ones. The final prestress is sized only

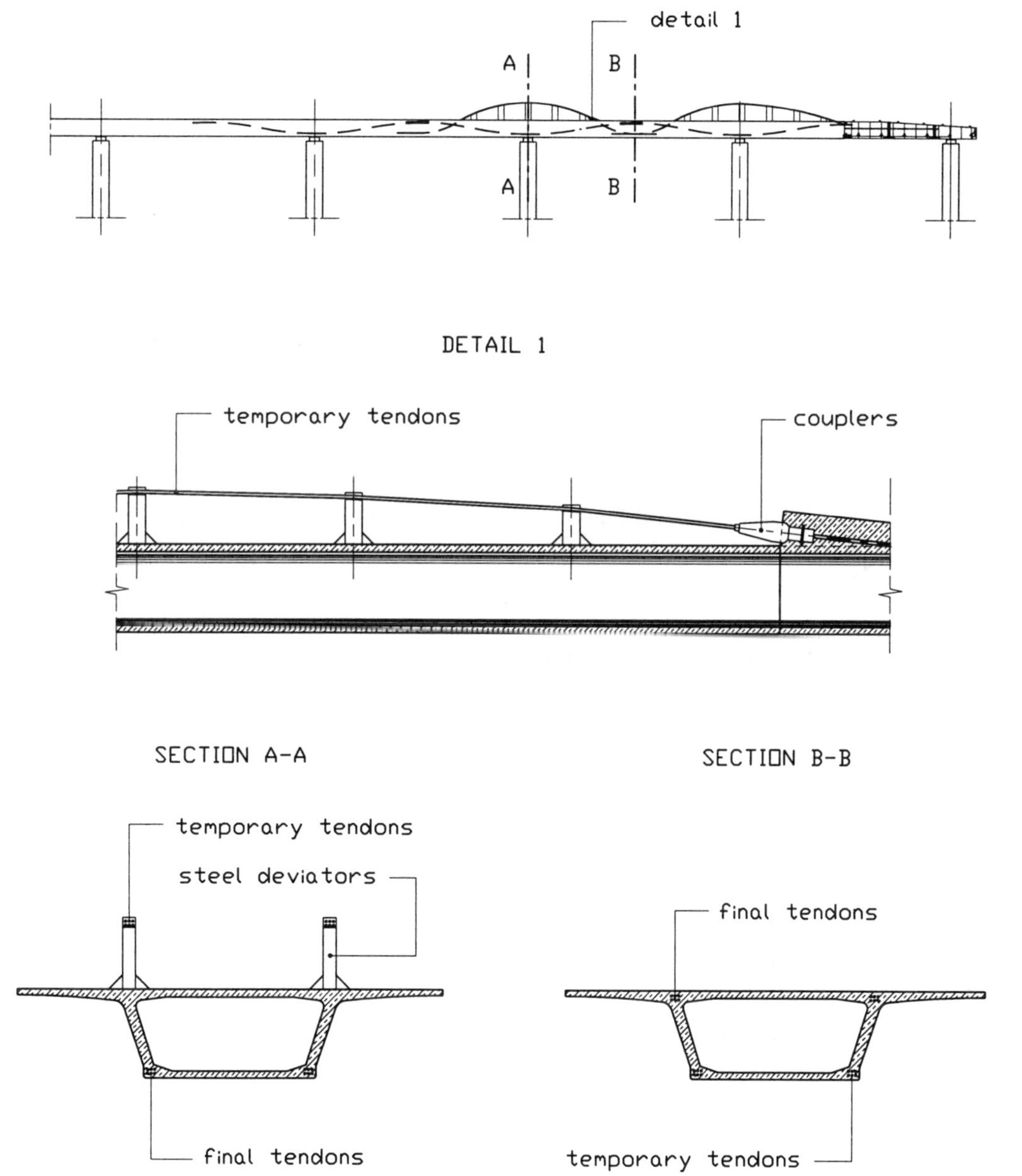

Figure 7.13 - Prestressing scheme for the Sathorn Bridge. Since internal final tendons are parabolic, external antagonist tendons are deviated rather uniformly in order to reduce the shear force anomalies.

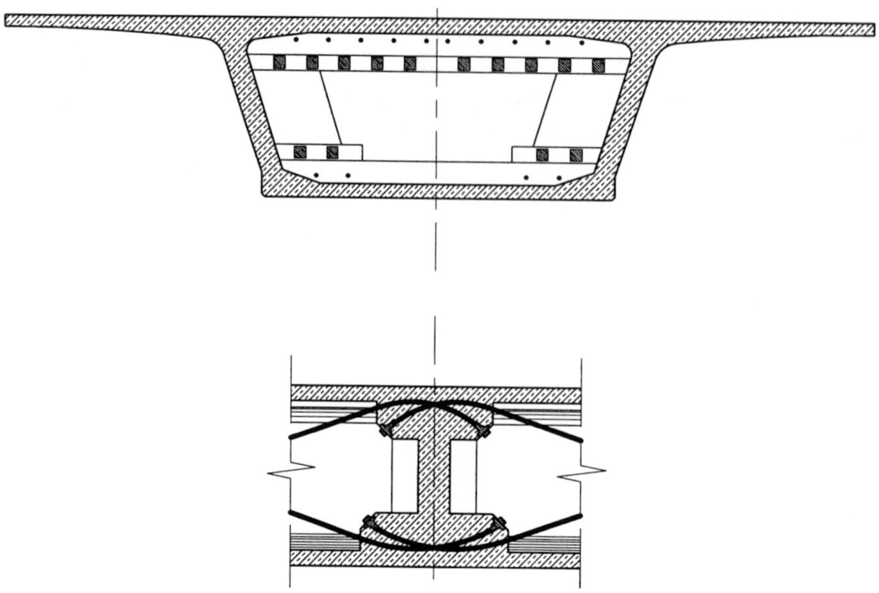

Figure 7.14 - Widened bottom anchor beams increase the length of antagonist tendons, so that their strands can be reused in the final tendons once the portions notched by the anchor keys have been eliminated.

for service loads, and is introduced partly during the deck construction and partly upon the completion of launch. Temporary prestress is only aimed at compensating the effects of the final tendons during launch. Even construction times are shorter, since the end-of-launch prestressing operations involve fewer tendons, although the recovery of the antagonist ones is rather difficult.

Tendons are generally chosen for strength, in order to limit their number and avoid excessive congestion of the box cell. For a safe reuse of their strands, especially for the final prestressing, some precautions should be taken during their temporary use:

- temporary launch prestress should be less than *85* to *90%* of the allowable tensile stress[28], to reduce risks of local plasticization of the strands;
- strand portions notched by the anchor keys must be eliminated. Since, in this case, temporary tendons must be longer than the final ones, the bottom transverse anchor beams can be widened (figure 7.14), or the anchorages can be arranged in steel pipes filled with concrete, which can be reused more times (figure 7.15);
- as an alternative, by tensioning tendons from only one side instead of from both, it is possible to use permanent pressed strand couplers instead of anchor keys in the dead anchorage, or the notched zones (painted for a safer control) can be placed immediately outside the dead anchorage, since, in both cases, the higher elongation at final tensioning permits the opposite notched zones to emerge outside the pulling anchorage;
- the strands should be galvanized and placed in perforated polyethylene pipes which permit corrosion control in case of damage to galvanization during the

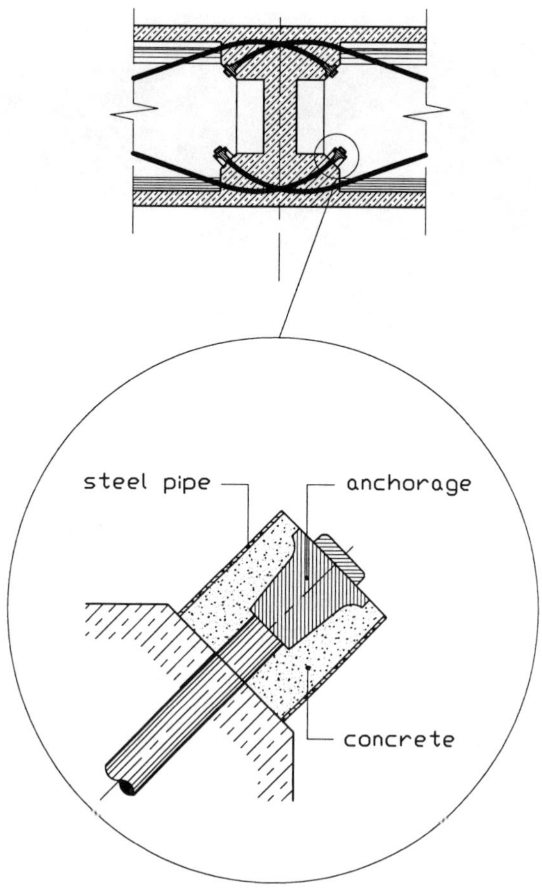

Figure 7.15 - As an alternative to widened anchor beams, the anchorages may be placed in steel pipes filled with concrete. In this case, they can be recovered and reused several times.

- strand insertion. To guarantee the safety of workers in case of breakage of a strand or of the whole tendon, these pipes are anchored to the deck concrete by means of collars spaced *3* to *4 m*;
- only one temporary tendon should be dismantled at a time by backward pulling inside plastic sheaths, which facilitate extraction and protect the tendon from dirt, and immediately reused as a final tendon in the same span, and near the same web in case of curved decks to avoid differences in length;when temporary tendons are used several times, each strand should be inspected and greased before its reuse, and eliminated if corroded.

Antagonist launch tendons were used in the Sathorn Viaducts along with internal final tendons (figure 7.13). Since the cross-sectional centroid was so high as to penalize the mid-span efficiency of antagonist tendons contained in the box cell, they emerged above the top slab (figure 7.16), and were deviated by steel saddles rested on the deck (figure 7.17). In the support zones, they were contained inside the

Figure 7.16 - In the Sathorn Bridge, the antagonist tendons emerged above the top slab (courtesy Virlogeux[217]).

Figure 7.17 - Detail of the steel deviation saddle for the Sathorn Bridge (courtesy Virlogeux[217]).

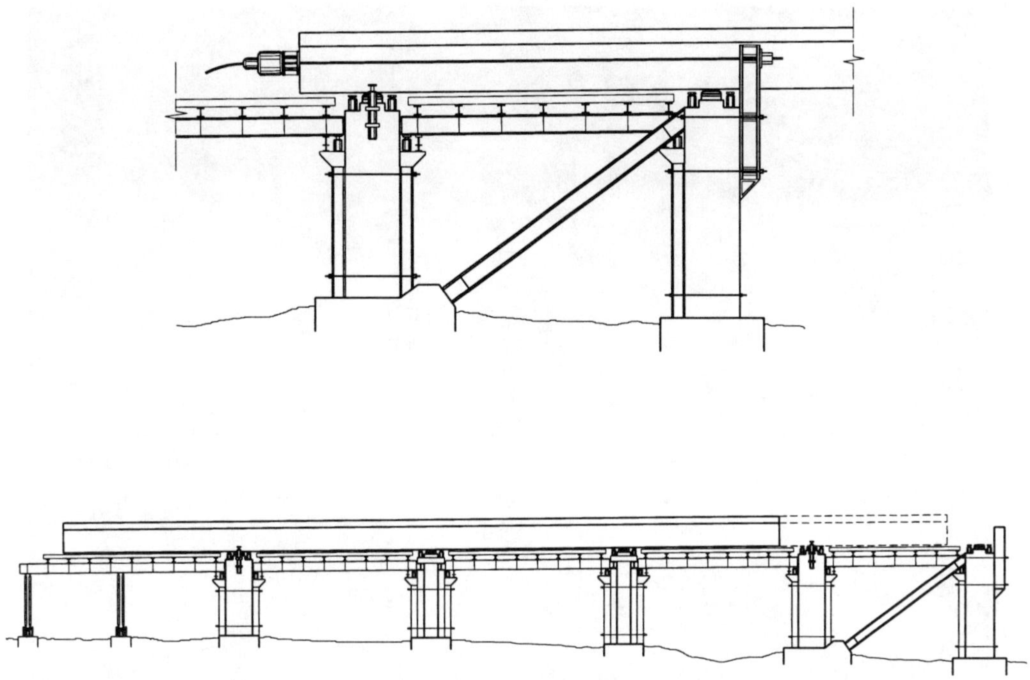

Figure 7.18 - Detail of the formwork for the Sathorn Bridge, rested on temporary piers and equipped with sloped struts aimed at resisting the thrust force. In most cases, both final parabolic tendons and antagonist tendons are anchored at the support zones of the deck. Therefore, each launch is as long as the whole span, and the length of the formwork increases.

deck webs. Both final parabolic tendons and antagonist tendons were anchored in the support zones of the deck, and therefore each launch was as long as the whole span. As a casting yard long enough was not available, the deck was built in a long formwork supported on temporary piers (figure 7.18).

An even more radical application of antagonist prestressing was tested in the Amiens Viaduct. The launch tendons, fully external, were located as follows: two with a final trapezoidal layout deviated at span quarters, two with a final triangular layout deviated in mid-span, four in antagonist position with a rigorously symmetrical layout, and six straight and close to the top slab to raise the prestress force to the centroid of the current section.

Upon completion of the launch, span by span, the four antagonist tendons were relieved and repositioned parallel to the four permanent polygonal tendons stressed before launch (figure 7.19).

Figure 7.19 - Internal view of the Amiens Viaduct upon completion of the launch. Four empty polyethylene pipes are ready to receive the strands of the four antagonist tendons, still anchored at the bottom transverse beam (courtesy Virlogeux[224]).

Prestressing of Temporary Joints

Although most launched bridges keep the continuous beam static scheme used during launch permanently, in some bridges it is necessary to introduce intermediate construction joints. In this case, temporary joints are designed so as to guarantee at the same time the transfer of launch stresses and their easy separation once the deck has reached its final position.

Once the maximum fluctuation of the launch bending moment is defined, the joint prestressing is sized with the same criteria as the deck current section, and distributed so as to produce a centroidal resulting force.

The use of external joint tendons is encouraged by the presence of the deck end diaphragms, which can easily contain their anchorages. A totally external prestressing also reduces the congestion of internal tendons at the end support sections of the decks, but the dead load of the transverse anchor beams increases the undulation of the launch moment envelope.

As an alternative, joint tendons can be contained in the slabs (figure 7.20).

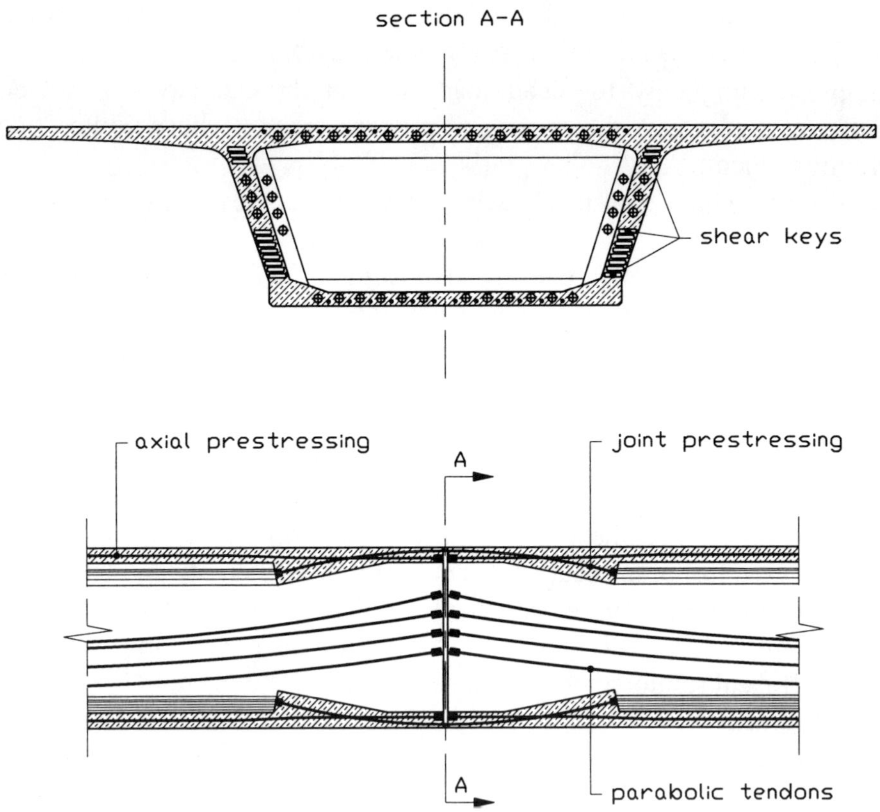

Figure 7.20 - Temporary joint prestressing with internal tendons. Their anchor blisters are lighter than the transverse anchor beams for external tendons, and the undulation of the launch moment envelope is smaller.

The average compression of the joint must guarantee an adequate safety factor at joint sliding (5.44), and, if needed, the joint can be equipped with shear keys.

Dimensioning of Axial Prestressing

In a deck composed of many equal spans of constant dead load, with a launching nose adequately stiff and $l_n / l = 0.60$, the bending moment envelope is approximately the one of figure 5.55.

As already mentioned, the deck can be divided in two zones: a back current zone, characterized by a repetitive course of launch stresses and therefore by constant envelopes, and a front zone characterized by peaks of both bending moment (positive and negative) and shear force. Usually, launch prestressing is sized for the current zone, and then adapted to the particular requirements of the front zone.

Applicable standards and design preferences can guide in the choice between three different sizing criteria for launch prestressing:

- total prestressing for all load conditions, so that neither thermal gradients nor bearing settlement can decompress the extreme fibers;
- total prestressing only for dead load, so that thermal stresses and accidental events can decompress the edges and produce cracks, whose width is controlled by reinforcement;
- partial prestressing – that is, allowing edge tensile stresses independent from the load combinations.

In a bridge built directly in its final position, total prestressing produces high differences in the edge stresses, and since the presence of live loads is a short-duration event while the excess of prestress with respect to the dead load stress is a long-duration one, the differential creep deforms the deck with upward cambers in mid-span.

For many years these cambers, troublesome to drivers and damaging to the bridge aesthetic, have been accepted because of the possibility of taking the most advantage of prestressing. At the same time, the use of reinforcing steel, less efficient than prestressing steel, has been reduced because of the absence of edge tensile stresses.

But structural deterioration in many bridges showed that the presence of unforeseen tensile stresses is possible even with total prestressing, and that in this case an insufficient reinforcement reaches yield soon and cannot effectively limit the width of the cracks[112].

In turn, tendons slide inside the sheaths because of cracking of the cement grout, which has a shear strength of about *10* to *15%* the strength of reinforced concrete, and this prevents from actually reaching the theoretical ultimate loads, which assume complete adherence of tendons.

These defects can be avoided by sizing reinforcement so as to limit crack width, and this increases steel consumption so much that it can effectively contribute to the carrying capacity of the cross section. This leads to partial prestressing.

Generally, the most competitive solution consists in allowing low edge tensile stresses during launch[108] (*1* to *2 MPa*) to be absorbed by longitudinal reinforcement. Upon completion of the launch, the dead load and *10* to *30%* of the live load is carried without edge decompression, while the remaining portion of the live load is carried by partialized sections. Edge tensile stress is not limited, and crack width is controlled according to environmental aggressiveness.

On the contrary, in the case of total launch prestressing, allowing for the stresses produced by bearing settlement and thermal gradients increases considerably the cost of prestressing.

Back Current Zone of the Deck

In the back zone of the deck, each cross section is subject to bending moments which vary from $+M_{max}$ to $-M_{min}$:

$$\Delta M = M_{max} + M_{min} \qquad (7.2)$$

To annul edge tensile stresses, the minimum axial prestressing, F_{min}, should be such that the maximum upward eccentricity of the resulting force:

$$e_u = \frac{M_{max}}{F_{min}} \qquad (7.3)$$

and the maximum downward eccentricity:

$$e_d = \frac{M_{min}}{F_{min}} \qquad (7.4)$$

maintain the resulting force inside the central core of the cross section – that is (5.5):

$$e_u + e_d = \rho_s H \qquad (7.5)$$

from which:

$$F_{min} = \frac{\Delta M}{\rho_s H} \qquad (7.6)$$

In the case of launch partial prestressing, similar considerations will apply, so that defined edge tensile stresses are not exceeded.

In reality, no cross section can have a central core perfectly dimensioned for the maximum fluctuation of bending moment. One edge of the central core will be more critical than the other one, and launch prestressing will be higher than this minimum value (7.6).

By imposing limitation on the allowable edge stresses:

$$\begin{cases} -\sigma_t \leq \dfrac{F}{A} + \dfrac{M}{I} z_u \leq \sigma_c \\ -\sigma_t \leq \dfrac{F}{A} - \dfrac{M}{I} z_l \leq \sigma_c \end{cases} \qquad (7.7)$$

where σ_c is the allowable compressive stress and $-\sigma_t$ is the allowable tensile one, the worst conditions are:

$$\begin{cases} -\sigma_t + \dfrac{M_{min}}{I} z_u \leq \dfrac{F}{A} \leq \sigma_c - \dfrac{M_{max}}{I} z_u \\ -\sigma_t + \dfrac{M_{max}}{I} z_l \leq \dfrac{F}{A} \leq \sigma_c - \dfrac{M_{min}}{I} z_l \end{cases} \qquad (7.8)$$

which define the lowest moment of inertia of the cross section:

$$\begin{cases} \dfrac{I}{H} \geq \dfrac{M_{min}}{\sigma_c + \sigma_t} \\ \dfrac{I}{H} \geq \dfrac{M_{max}}{\sigma_c + \sigma_t} \end{cases} \qquad (7.9)$$

For the allowable tensile stress to be reached both in the upper edge of the support section and in the lower edge of the mid-span section, it must be:

$$-\sigma_t = \frac{F}{A} - \frac{M_{min}}{I} z_u = \frac{F}{A} - \frac{M_{max}}{I} z_l \qquad (7.10)$$

from which[188]:

$$\frac{z_l}{z_u} = \frac{M_{min}}{M_{max}} \qquad (7.11)$$

For the bending moment produced by constant dead load, the optimum section has the centroid at a distance from its upper edge equal to one third of its depth, M_{min} being double M_{max} (figure 5.55). In reality, the other launch stresses affect M_{max} more than M_{min}, and the centroid should be lower:

$$0.33 \leq \frac{z_u}{H} \leq 0.42 \qquad (7.12)$$

This condition is rarely satisfied, since the wide top slab draws the centroid upward, and the minimum launch prestressing will be such as to not exceed the allowable edge tensile stress in the worst condition:

$$\begin{cases} F_1 = \left(\dfrac{M_{min}}{I} z_u - \sigma_t \right) A \\ F_2 = \left(\dfrac{M_{max}}{I} z_l - \sigma_t \right) A \\ F_{min} = max(F_1, F_2) \end{cases} \qquad (7.13)$$

It frequently coincides with F_2 – that is, it is governed by the mid-span launch stresses.

Launch prestressing is generally arranged inside the two slabs, whose axes are respectively distance d_t and d_b from the centroid, and distributed so that the resulting force is centered on the gravity axis, to avoid secondary moments in the deck:

$$\begin{cases} F_t = \dfrac{d_b}{d_t + d_b} F \\ F_b = F - F_t \end{cases} \tag{7.14}$$

The case of launch external prestressing is treated in the same way.

Front Zone of the Deck

In the front zone of the deck, the maximum negative moment and, in a lesser degree, the maximum positive moment, depend on the type and the characteristics of the cantilever support device. Even using an optimized launching nose, the maximum negative moment at the front support of the deck is much higher than at the supports of the back zone, and the back launch centroidal prestress must be increased with a centroidal or eccentric integration. Even mid-span positive moment may require a lower additional prestressing.

The additional prestress, F_a, with eccentricity, e_a, positive if upward from the cross sectional centroid of the deck front zone, is designed to respect the allowable edge stresses in the worse launch condition. Neglecting the secondary flexural effects of additional prestressing, the maximum negative moment is given by (5.32) in the case of a long or optimized launching nose, and by (5.18) in the case of a short nose. At both edges of the front support section it should be:

$$\begin{cases} \dfrac{F + F_a}{A} + \dfrac{F_a e_a - M_{min}}{I} z_u \geq -\sigma_t \\ \dfrac{F + F_a}{A} - \dfrac{F_a e_a - M_{min}}{I} z_l \leq \sigma_c \end{cases} \tag{7.15}$$

The maximum positive moment is given by (5.34), and in the most loaded span section it should be:

$$\begin{cases} \dfrac{F + F_a}{A} + \dfrac{F_a e_a + M_{max}}{I} z_u \leq \sigma_c \\ \dfrac{F + F_a}{A} - \dfrac{F_a e_a + M_{max}}{I} z_l \geq -\sigma_t \end{cases} \tag{7.16}$$

Once the eccentricity, e_a, is defined according to the position of the additional tendons, the calculation of F_a and the verification of the edge stresses are immediate.

The same problems can affect the back end of the deck, especially when using friction launchers. In fact, these devices need a certain distance between the last curing support and the launcher itself in order to produce a high support reaction at the launcher, and this distance can become critical in the cantilever configuration at the end of launch (figure 7.21).

Since the last deck segment is never very cured, instead of locally increasing launch prestress it is preferable to introduce a temporary support between the last curing support and the launcher. This support is kept disengaged during the previous launch stages by not inserting neo-flon plates or the launching saddles themselves.

Finally, the maximum positive moment in the back end span is lower than the moment reached in the front zone of the deck, since the end span is generally shorter than the interior ones. However, the principles for calculating the additional prestressing of the back end span are the same.

Figure 7.21 - Friction launchers need a certain distance between the last curing support and the launcher itself in order to produce a high support reaction. This distance can become critical in the cantilever configuration at the end of launch (author).

Stress Verification

Once the current launch prestressing, F, and the additional prestressing, F_a, in the front zone are defined, extreme fiber stresses have to be checked everywhere, and, for instance, even in the stiffened support sections, whose larger surface reduces the average compression and shifts the centroid downward.

These verifications are easy in the deck current zone, in which the centroidal prestressing does not introduce indeterminate effects, and the bending moment envelope is almost constant. In the first and in the last spans, the cantilever moments and the eccentricity of additional prestressing complicate calculation.

Once the bending moment envelope is drawn, and from this the envelope of the extreme fiber tensile stresses, its superimposition on the compression diagram (figure 5.63), allows one to locate the excessive tensile stresses. They can be absorbed without secondary effects by increasing locally the centroidal prestressing with additional tendons, or with a higher number of strands in the same sheaths.

Since total launch prestressing could produce excessive compressive stresses in service, temporary additional tendons can be used, but when the specifications allow partial launch prestressing, it is preferable to control the crack width with reinforcement.

SERVICE PRESTRESSING

Upon completion of the launch, the location of lightened and stiffened cross sections finally matches the stress diagrams due to dead load. Launch prestressing tends to be excessive, since the better distribution of dead load reduces moment and shear diagrams with respect to their launch envelopes, and at the same time the correct distribution of the mechanical characteristics of the cross sections reduces the edge stresses even more.

In this context, the structure has to be modified to resist service loads.

For this purpose, additional centroidal tendons would not be very advantageous, due to both their low efficiency and the cross-sectional stiffening required to resist this additional compressive stress where it adds to the permanent compression due to the static scheme of the continuous beam. Therefore, it is preferable to shape the additional tendons so as to obtain the highest eccentricity of the resulting force, and, if needed, to relieve some tendons of the launch prestress.

The importance and the effectiveness of this correction depends on the quantity of centroidal prestressing which remains in the deck, and therefore on the launching technique. Using temporary piers, launch prestressing can be relatively low, and the correction to introduce after launch is important and effective. In case of full span launch, on the contrary, the launch prestressing is generally so high that only a few tendons can be added, which cannot improve substantially the efficiency of the final prestress unless some launch tendons are relieved.

To evaluate the efficiency of undulated prestressing in resisting permanent loads, the flexural effect it produces can be assimilated to a distributed upward load, p_{und}, equal to[33]:

$$p_{und} = \frac{8F_{und}\Delta e}{l^2} \qquad (7.17)$$

where F_{und} is the undulated prestressing force, Δe is the amplitude of undulation, and l the length of the span (figure 7.22). This value becomes more precise the closer the resulting tendon layout is to a parabola.

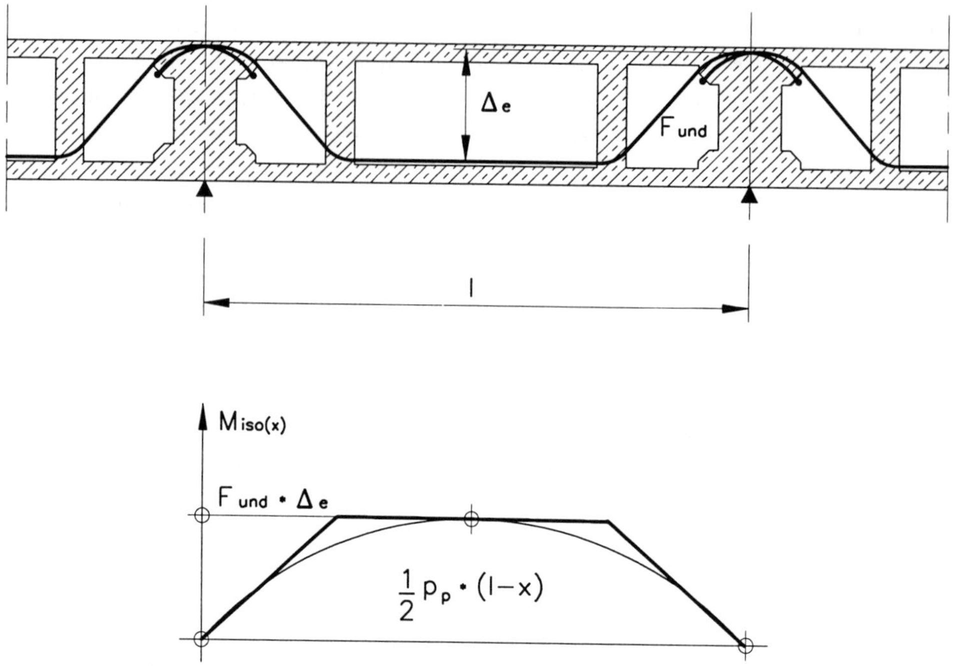

Figure 7.22 - Negative bending moment produced by undulated prestressing in a simply supported beam.

The lifting action of undulated tendons balances a part of permanent loads (dead load, q, and superimposed dead loads, p_{perm}), and the ratio of the prestressing equivalent load to the permanent loads is a good indicator of the distribution of longitudinal stresses in the deck. It is convenient to reach at least[33]:

$$\frac{p_{und}}{q + p_{perm}} > 0.60 \qquad (7.18)$$

but values of about 0.70 to 0.80 are preferable, limiting the creep deformations and forcing the stress state in the unloaded structure to approach a centroidal compression, ensuring the longest durability.

The additional tendons can be arranged in the cross section in several ways, according to the total force needed: in the conventional parabolic layout inside the webs, in a polygonal external layout deviated by diaphragms or saddles, or in an external parabolic layout in contact with the internal surface of the webs[108], convenient only in case of large tendons, 6 to $8\ MN$.

Internal Tendons

The additional internal tendons (figure 7.23), can be divided into parabolic tendons arranged in the webs along the principal tensile stresses, cap support

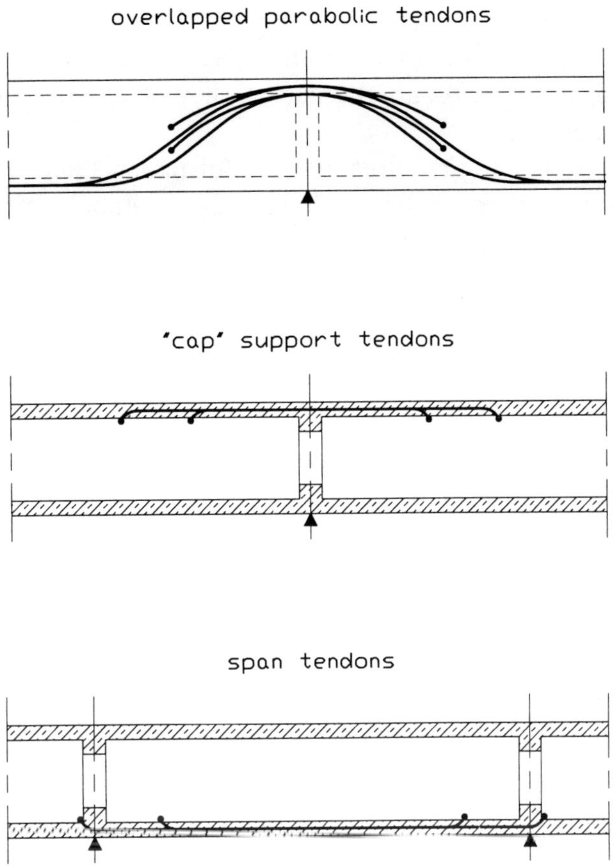

Figure 7.23 - Internal additional tendons can be divided in parabolic tendons arranged in the webs, straight support tendons, and straight span tendons.

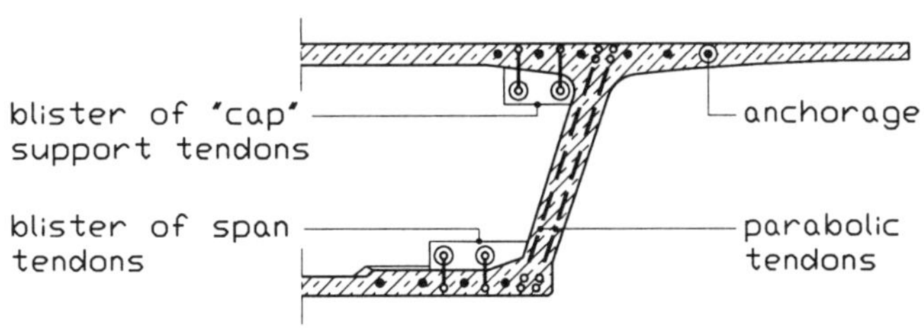

Figure 7.24 - Distribution of the internal additional tendons in the cross section.

tendons placed in the top slab, and span tendons placed in the bottom slab (figure 7.24). The last two types of tendons are generally used only in the longest spans.

Parabolic tendons can extend over several spans, but in this case they suffer high friction losses. When limited to only one span, they can be anchored at the zero-moment zones of adjacent spans, so that they overlap in the support sections. This layout doubles their effect in terms of both negative moment and, if anchored near the bottom slab, shear force. As an alternative, they can be overlapped at any point of the structure (figure 7.25).

In both cases, overlapping requires crossing of tendons and anchoring beams or blisters, and this makes parabolic tendons rather cumbersome. Once the shear stresses are reduced by the inclined portion of the tendons, additional prestressing requirements at the supports and at mid-span can be fulfilled by using local straight tendons.

Cap tendons are difficult to tension. Unless the section is so deep as to allow the transit of electric forklifts or the use of specific servo-supports (figure 7.37), the stressing jacks must be suspended manually from the top slab by means of holes crossing the slab. However, cap tendons are rarely indispensable, since the parabolic tendons are added to the centroidal launch prestressing, and this is usually enough to cover the negative moment.

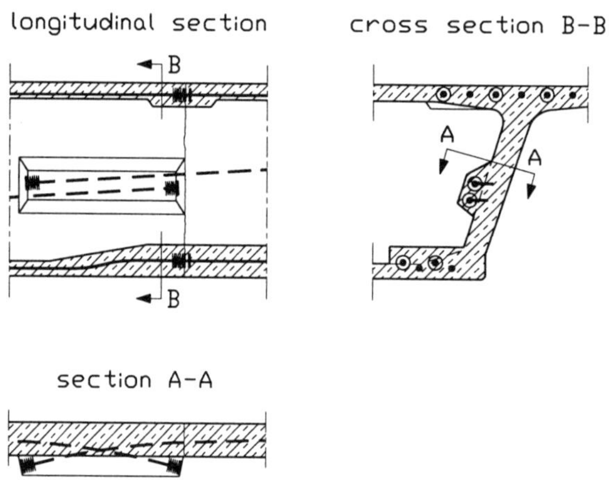

Figure 7.25 - The use of special blisters allows overlapping of parabolic tendons at any point of the structure.

On the contrary, span tendons are convenient, and they are frequently used in the end spans of the continuous beam, especially when it is not possible to reduce the length of these spans with respect to the internal ones (2.3). Their secondary effects can help in reducing the need for cap tendons.

The additional internal tendons are stressed only upon completion of the launch, and, to avoid risk of corrosion, their strands are inserted immediately before tensioning. Since their sheaths are embedded in the deck during construction, they must be watertight and stiff, since burrs of cement grout or buckling under casting or vibration would prevent strand insertion. For this purpose, stiffening polyethylene pipes can be inserted inside the sheaths before casting, and removed at its end to be reused in the next segment.

External Tendons

External prestressing is adopted more and more frequently in several countries, because of both the immediate savings deriving from the shorter duration of construction, and the long-term savings deriving from higher quality.

These two aspects produce different results in terms of short-term competitiveness of the project, and represent two different design approaches[33]. If the primary objective is the reduction of construction costs, external prestressing limits the dead load of the deck, and with it, the need of prestressing, and simplifies construction. If jobs are bid through restricted tenders between highly qualified contractors, external prestressing can be required or encouraged to improve the quality of the structure with simple and sure tendon layouts, lower friction losses, better casting and injection conditions, and replaceable tendons. Lastly, even when the design specifications prescribe a minimum web thickness, and therefore limit the structural lightening, the higher quality reduces maintenance costs, and this balances the higher construction costs.

Generally, the use of external prestressing offers many advantages:

- the web thickness can be limited to that strictly necessary to resist the principal compression, and this reduces the dead load of the deck (especially in the longest spans, where the web thickness is often overdesigned to contain tendons) and increases the cross sectional efficiency;
- launch internal prestressing can be produced by a few powerful tendons placed in the nodes of the cross section, no longer affected by the parabolic tendons. This reduces the thickness of slabs and allows one to couple tendons directly in the nodes, avoiding anchor beams and blisters;
- the concrete of monolithic webs is stronger than a concrete containing many sheaths[112], which reduce the resisting section, disturb stress distribution, and complicate vibration;
- by eliminating the internal tendons and their supports, the density of the reinforcement is reduced, overlapping and cover can be optimized, the duration of steel cage assembly is shorter, the general economy of construction improves, and the negative influence of poorly skilled workers is limited;
- the visual control of the sheaves of strands permits one to guarantee their optimum layout both in the straight zones and inside deviators, and this allows a good control of tensioning and elongation. Because of the low losses (friction is null in the straight zones and can be reduced inside the deviators[33]), this leads to higher average tensile stresses in the strands, which have no consequences in terms of either relaxation (nowadays, steel is almost always stabilized) or fatigue, because of low stress fluctuations;
- low friction allows extension of the same tendons for several spans, and this reduces the number of anchorages. In asymmetrical structures this can lead to a

rough covering of the bending moment envelope, which can be improved by a higher distribution of anchorages, although this intervention is rarely convenient;

- external tendons are not affected by concrete cracking (unlike internal tendons, which come in contact with the atmosphere near cracks) and represent a more valid solution in case of partial prestressing, because of the particular sensitiveness to corrosion of tensioned prestressing steel;
- it is possible to use powerful tendons composed of sheaves of strands anchored to the webs (figure 1.6), grouped so as to allow the use of ordinary stressing jacks and anchorages. Strands are embedded in a cement mortar poured inside left-in-place sheet metal forms after tensioning, with the best likelihood of control. Their use is generally advantageous only for high prestressing forces, and therefore mainly in the case of launching over temporary piers;
- maintenance costs are lower, since tendons can be inspected easily (they are the elements most subject to wearing, along with expansion joints) and this allows easy repair or replacement while maintaining the structure in service.

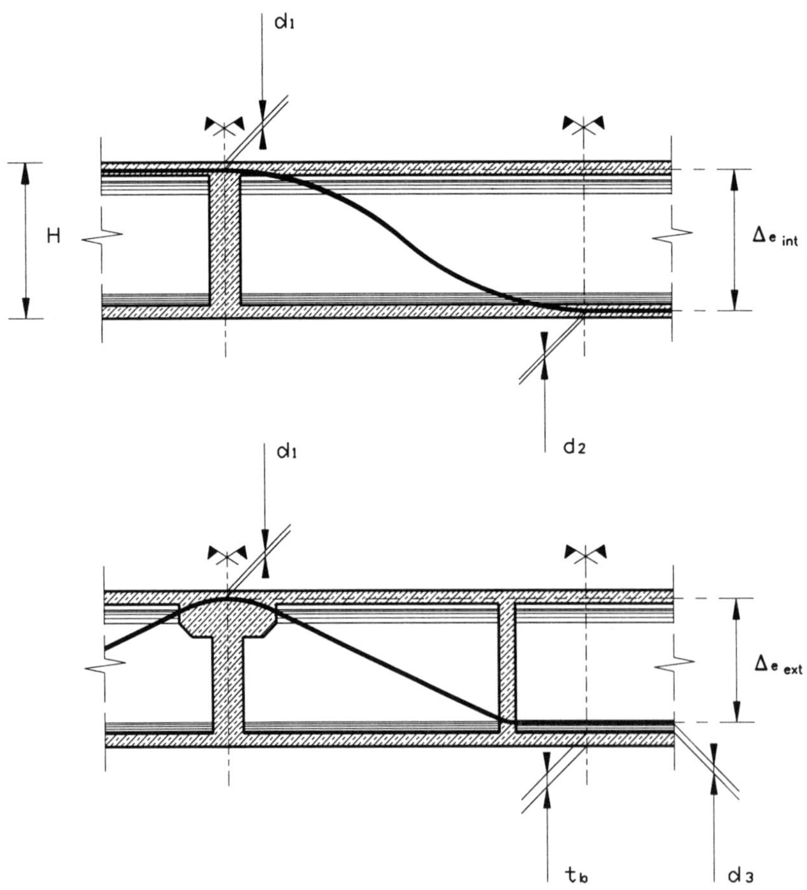

Figure 7.26 - The maximum eccentricity of external tendons is lower than that reached with internal prestressing.

Appropriate theoretical models permit one to solve the problems observed in the first applications of this prestressing technique, due to the lack of participation of tendons in crack control, and the strain differences between the deck and the tendons which produce sliding in the deviation saddles, and therefore corrosion[33].

Finally, the geometrical efficiency of tendons – that is, their maximum eccentricity, is lower than that reached with internal prestressing (figure 7.26):

$$\rho_t = \frac{H - (d_1 + t_b + d_3 + \phi)}{H - (d_1 + d_2 + \phi)} \qquad (7.19)$$

The technological aspects of external prestressing are treated in this chapter only in regards to incrementally launched bridges; in subsequent chapters they will be expanded to prestressed composite sections and bridges built by balanced casting.

External tendons pose the problem of resisting local forces and moments near anchorages, which is more delicate than with internal prestressing due to both the absence of the distribution of forces due to friction and the convenience of using powerful tendons to avoid excessive congestion of the box cell.

The simplest solution is to anchor tendons to massive elements, and therefore to the support diaphragms. To obtain a tendon layout suitable for both moment and shear, it is then necessary to introduce some deviations, with saddles, ribs, or diaphragms.

Support diaphragms must satisfy many requirements.

First of all, the anchorage of many powerful tendons requires the presence of a transverse beam in the top slab (at least *3 m* wide lengthwise) which guarantees at the same time an adequate distribution of the prestressing force toward the webs, the stiffening of the slab, the crossing of the tendons anchored on its opposite sides, and the balance of their forces. To facilitate the handling of stressing jacks, the anchorages can be placed under the tendons which emerge from the transverse beam. In turn, by placing tendons as high as possible, prestressing moment and shear reduction can be increased. An advantageous layout is the one of figure 7.27, obtained by moving away the centers of curvature of the tendons. Tendons come out from the concrete as high as possible with the desired slope, and the anchorages are lower, accessible, and not enclosed in uncomfortable niches.

Vertical beams are necessary along the webs to transfer the deviation forces toward the supports and to stiffen the webs to facilitate stress distribution inside the box girder. Finally, the support diaphragm is completed by a bottom transverse beam which stiffens the slab, closes the cross frame to resist the stresses due to prestressing and torsion, and, in the case of antagonist launch prestressing, permits one to anchor the lower tendons.

Support diaphragms are often prestressed with sheathed or polyethylene-coated strands which descend in the webs and the bottom slab, since large diameter, closely spaced reinforcing bars cannot prevent cracks.

As regards deviators (figure 7.28), they can be divided into two groups.

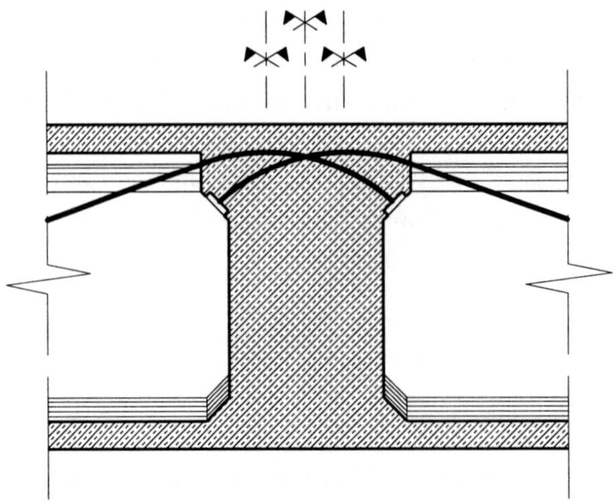

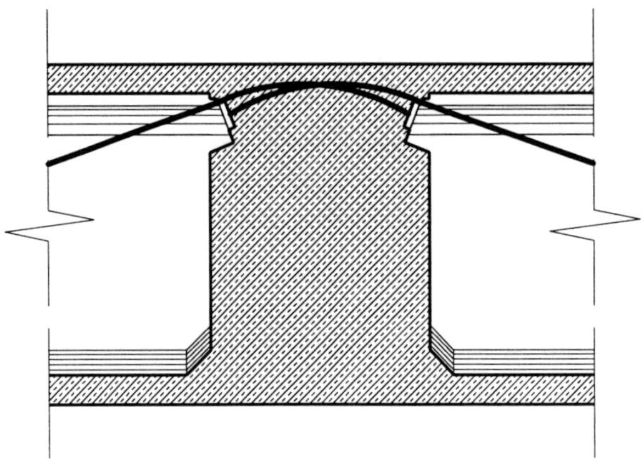

Figure 7.27 - By moving away the centers of curvature of the tendons, these come out from the concrete as high as possible. At the same time, the anchorages are lower, accessible, and not enclosed in uncomfortable niches.

A first group is based on the compressed strut vertical action which transfers the resultant of the tendon deviation forces directly to the webs. It is composed of the diaphragms, which cover the whole width of the cross section, and of ribs, which are lighter but less efficient in resisting the transverse bending produced by the distance between the tendon deviation points and the webs. They permit one to maintain the tendons in almost vertical planes, simplifying their layout and reducing angular

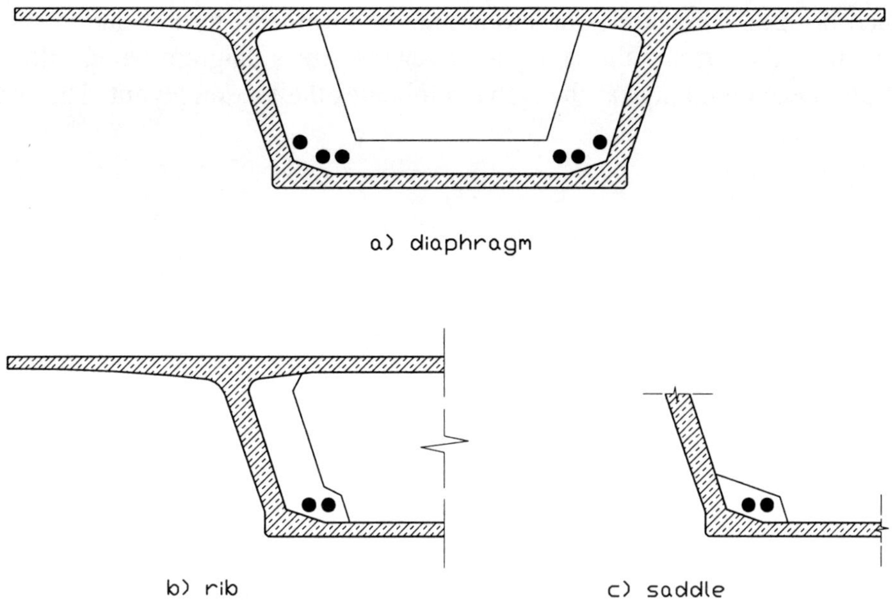

Figure 7.28 - Different types of deviators.

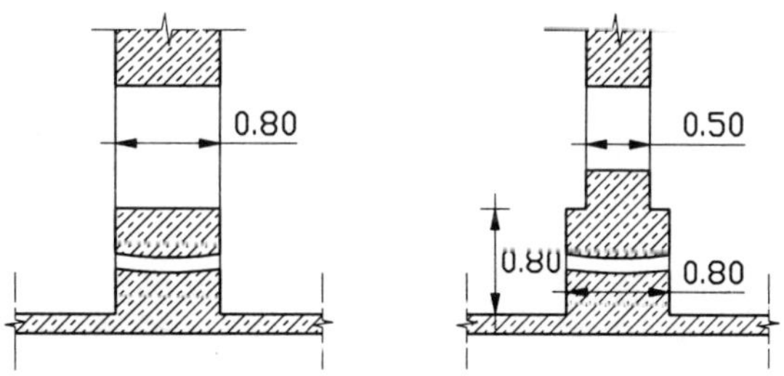

Figure 7.29 - Deviation ribs and diaphragms can be lightened by reducing their thickness in the zones not affected by tendons.

deviations, and to deviate more tendons simultaneously. Moreover, they resist distortion, although with different efficiency. Their dead load generally suggests use of only two deviators in each span. Further lightening can be obtained by reducing their thickness, usually about *0.80 m*, to *0.50 m* in the zones not affected by tendons (figure 7.29), or by increasing their number and deviating all the tendons, to reduce the angular deviations and therefore their thickness. Finally, diaphragms and ribs make the passage of non-deviated tendons difficult, especially in curved decks, where they subtend the chords between the deviation points while the webs follow the arches.

A second group of deviators is composed of saddles, small concrete blocks placed in the nodes between the webs and the slabs. The saddles are very useful in launched bridges. They are light and simple to form, and permit one to deviate the tendons individually, and therefore, to cover the stress diagram better, although the deviation of each tendon near the web complicates the tendon layout. The saddles do

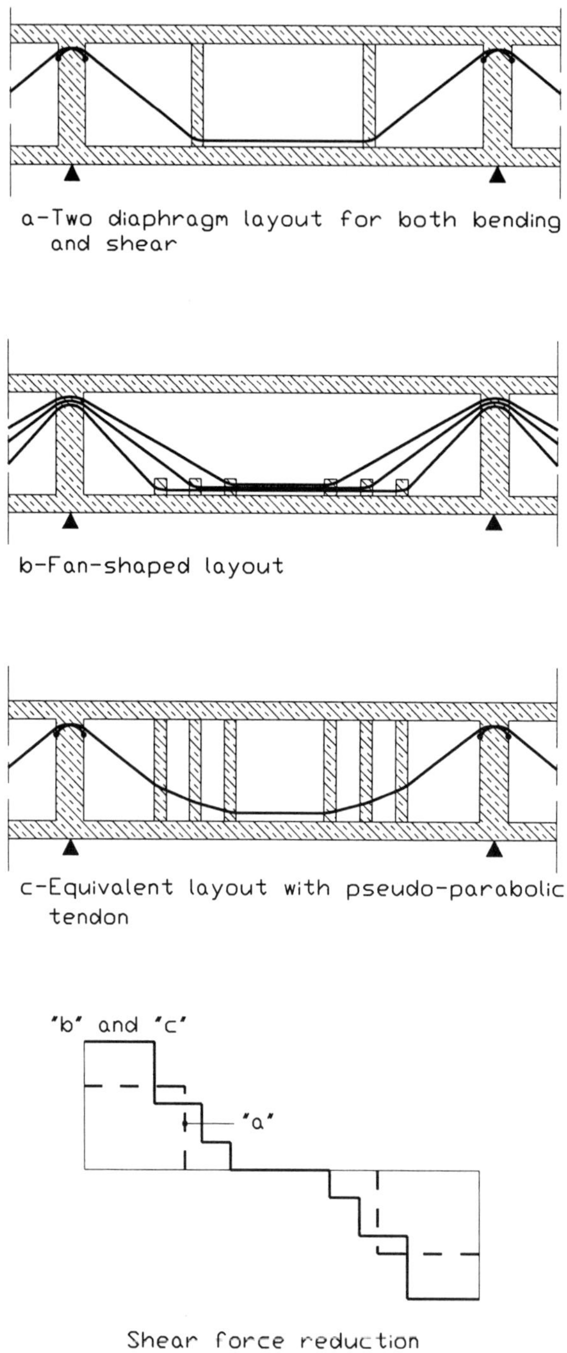

Figure 7.30 - Shear force reduction in the deck support zones depends on the longitudinal distribution of deviators.

not act as compressed struts, and require that the whole deviation force be resisted by reinforcing steel. Its elongation cracks the concrete and dictates the use of galvanized or epoxy-coated bars, whose performance in cracked conditions is not well known[33].

Diaphragms, ribs, and saddles contain mechanical devices aimed at reducing the effects of the geometric imperfections of the tendon layout: steel pipes sheathed with soft products near the external surfaces of the deviators, and pipes over-bent or with flared ends[33].

The web thickness depends on the shear force reduction near the support sections of the deck, and therefore on the longitudinal distribution of the deviators (figure 7.30):

- by fan-shaping the tendons from the support section, it is possible to obtain a shear force reduction diagram close to the triangle, and at the same time a resulting tendon layout close to the funicular of permanent loads. The use of four or six deviators in each span permits one to attain extremely thin webs, and since in this case the launch stresses require a high centroidal prestressing, the presence of many diaphragms can be used to deviate or to anchor antagonist tendons[33]. On the contrary, the double-fold layout is not convenient, since the diagonal of the parallelogram produces the same effects with lower angular deviations and friction losses;
- as an alternative, all the tendons can have the same undulated layout instead of being fan-shaped, to reduce the angular deviation in the saddles closer to the supports without penalizing the average angular deviation, and to avoid interference and geometric errors;
- since the dead load of deviators suggests reduction in their number, the tendon layout is often trapezoidal, with only two intermediate deviation points placed at about one third of the span for flexural reasons. In this case it is not possible to obtain the optimum layout for shear, which requires deviation points at the quarters of the span, and the webs are thicker.

Very powerful tendons have been used in long span bridges[20] (figure 1.6). They were obtained by deviating the strands within steel saddles composed of curved plates anchored to the internal surface of the webs, so as to obtain external tendons that were parabolic instead of polygonal. After tensioning, sheet metal forms were hung on the strands, filled with fine crushed aggregate and then injected with cement grout. The shear connection to the roughened surface of the webs was obtained with distributed stirrups bent during the web casting and then straightened, or with high strength bolts which guarantee a light prestressing of these curbs against the webs, limiting the shrinkage cracks of the grout.

Transverse Prestressing

In decks wider than twelve meters, the transverse prestressing of the top slab begins to be worthwhile.

It is definitely used for widths over *17* to *18 m*, since transverse bending produced by live loads depends on the width by the quadratic law, and slab thickening increases the stresses produced by thermal gradients and distortion, and with them the quantity of reinforcement needed. At the same time, losses of prestress due to anchorage reentry are less critical, and the cost of anchorages less prohibitive.

Above certain span lengths, even the general need for dead-load reduction makes a light transverse prestressing very useful, as it permits one to reduce the thickness of the top slab, its deformations, its cracks, and the cyclic stress fluctuations in its reinforcement. However, with a non-antagonist launch prestressing, the thicker webs can make its application delicate, as slab shortening produces transverse bending and shear stresses in the cross section.

The grade of prestressing can be chosen freely. Total prestressing is superfluous and even difficult to obtain, since live loads are much higher than dead load, and generally a partial prestressing dimensioned according to empirical criteria and experience is enough. For instance, one can absorb the transverse tensile stresses due to permanent loads and *30%* of the live loads with closely spaced (*0.5* to *1 m*) light tendons (*300* to *600 kN*), whose small anchorages do not increase the slab thickness, placed with pulling anchorages alternating with dead ones, as in figure 7.31.

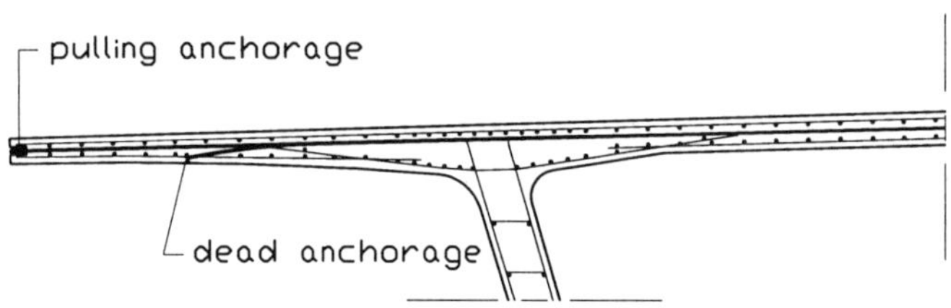

Figure 7.31 - Transverse prestressing. Closely spaced small tendons can be placed with pulling anchorages alternated to dead ones.

In the case of longitudinal partial prestressing, some transverse tendons will fall in cracked zones, and waterproof rigid sheaths should be used, such as polyethylene pipes.

Finally, an important contribution of transverse reinforcement is necessary, as the different layers of bars prevent obtaining an adequate eccentricity of the transverse tendons, especially in the zone distant from the webs.

Vertical Prestressing

The impossibility of undulating the resulting launch prestressing force prevents one from modifying the shear force diagrams with it, and each cross section of the deck has to resist the highest launch shear stresses. Therefore, the web thickness must be such as to limit the highest principal compression within allowable values, and the shear reinforcement must limit the width of cracks produced by tension.

In spite of this, the principal tension can exceed the allowable value. When it occurs in localized zones, rather than increasing the web thickness, it can be convenient to use local vertical prestressing, obtained with bars or U-shaped strands anchored on the top slab (figure 7.32).

A distributed vertical prestressing can change completely the stress state in the webs, reducing or eliminating the principal tension, and avoiding cracking.

For vertical prestressing to be efficient, the webs must be rather deep, and this limits its adoption to long spans. Moreover, some design specifications require one to analyze shear stresses (and consequently the principal tension), neglecting a portion or the whole sheath diameter, and so in these cases the introduction of vertical prestressing may not even be convenient.

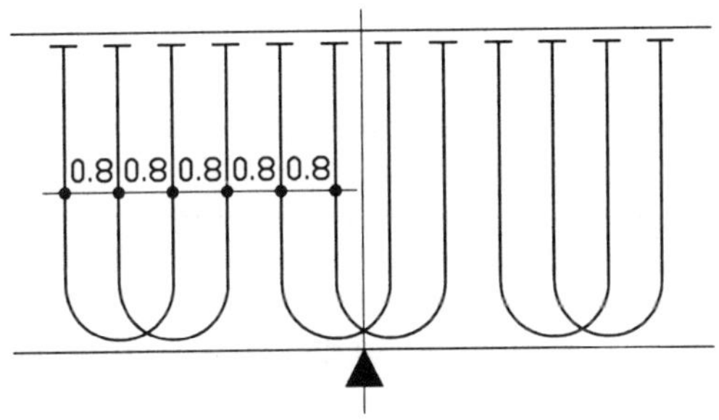

Figure 7.32 - Vertical web prestressing with U-shaped strands anchored at the top slab.

In any case, vertical bars or strands are short, difficult and expensive to install, and their effects are unpredictable because of the anchorage reentry, which makes the actual distribution of vertical compression irregular. For these reasons, it is convenient to use these schemes only to resist a share of the principal tension, and to absorb the remaining share with conventional reinforcement.

As an alternative, the use of precast webs vertically prestressed by adherent strands permits one to reach construction and prestressing savings and an optimum stress state at the same time.

RECENT TRENDS

Although the use of conventional external tendons (strands contained in polyethylene pipes and deviated at certain points) is rather recent in bridges built by incremental launching, the particular requirements of this construction method have already resulted in solutions so innovating as to represent the most recent developments in bridge prestressing.

The first applications developed the idea of mixed prestressing: final straight internal tendons anchored in the nodes of the cross section in the correct position for the final static scheme, temporary straight internal antagonist tendons removed at the end of launch, and polygonal external tendons assembled and tensioned at the end of launch and anchored to the support diaphragms (figure 7.33).

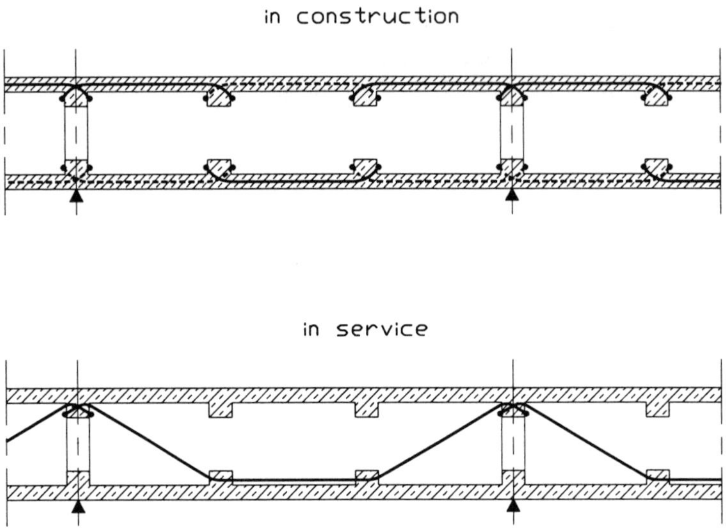

Figure 7.33 - Mixed prestressing composed of permanent straight internal tendons, temporary straight internal antagonist tendons, and polygonal external tendons tensioned at the end of launch.

Compared with the conventional tendon layout (launch prestressing in the slabs and parabolic tendons in the webs) this scheme permits lightening both the slabs and the webs. It was used in many projects, among them the Roquebillere Bridge[186], launched with temporary piers.

Subsequent advances saw a final internal centroidal prestressing integrated with straight external tendons during launch (figure 7.34), and corrected at the end of launch with straight cap and span internal tendons and polygonal external tendons. This scheme was used in the Poncin Viaduct[87], also launched with temporary piers.

In both cases, the launch prestressing was obtained by superimposing several groups of tendons according to the deck segmentation. This made possible, in the Roquebillere Viaduct, relieving and removing some tendons falling in zones naturally compressed by the final dead load bending moment at the end of launch.

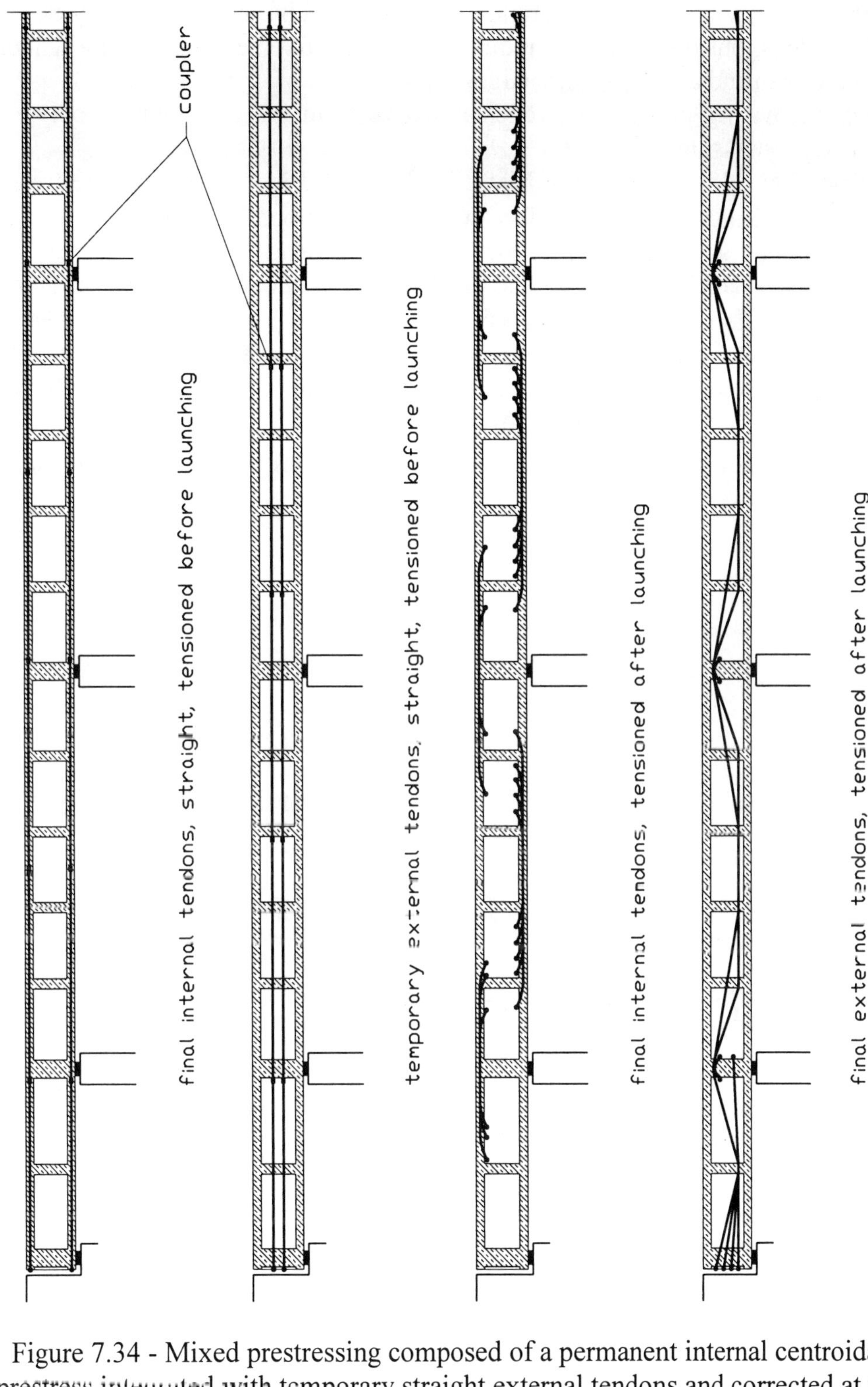

Figure 7.34 - Mixed prestressing composed of a permanent internal centroidal prestress integrated with temporary straight external tendons and corrected at the end of launch with straight internal tendons and polygonal external tendons.

In the Poncin Viaduct these operations were avoided by using temporary external launch tendons, but their removal required addition of some final straight tendons, both in the span and at the supports (figure 7.34).

Other innovating prestressing schemes derive from the evolution of polygonal antagonist prestressing.

In Bridge No. 33 of the A55 at Marseille[78] (figure 7.35), *12.0 kg* of prestressing steel per square meter of deck surface were tensioned before launch as permanent internal tendons sized to resist most of the dead load. Their eccentricity was compensated by antagonist polygonal tendons and reduced by additional straight tendons, all external and temporary, which added a further $17.4\,kg/m^2$ of prestressing steel. At the end of launch, and after relieving the temporary launch tendons, the deck prestressing was completed with some polygonal and some span tendons, all external and sized for service loads, for a further $7.9\,kg/m^2$ (figure 7.36). The parabolic internal tendons avoided excessive obstruction of the shallow box cell.

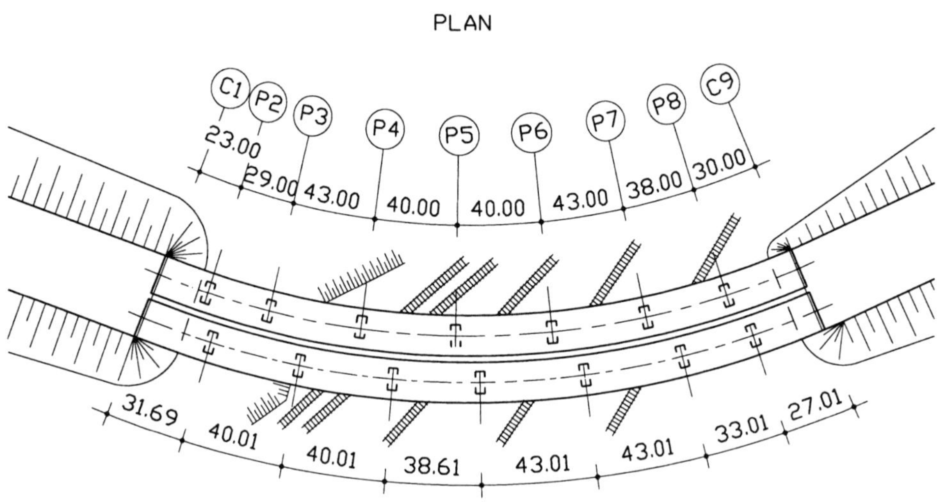

Figure 7.35 - Bridge no. 33 of the A55 at Marseille.

This concept was improved in the Durance River Bridge, built with totally external prestressing. A share of the polygonal tendons was tensioned in its permanent position before launch, compensated by antagonist tendons and integrated by additional straight tendons, all temporary and external, which were relieved and repositioned in the correct final polygonal layout at the end of launch.

This scheme was further improved in the Amiens Viaduct, with some final external straight tendons.

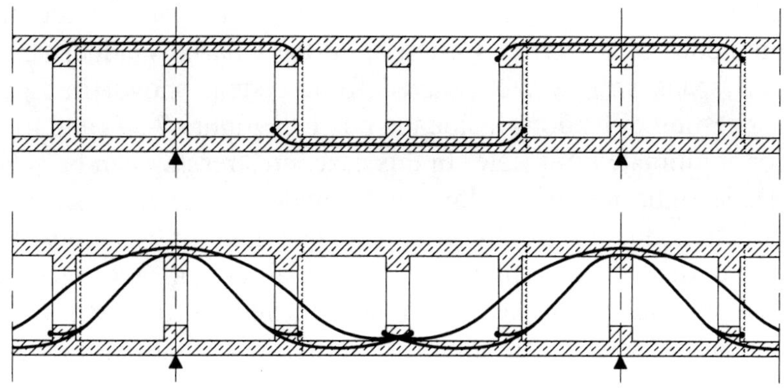

final tendons stressed before launching

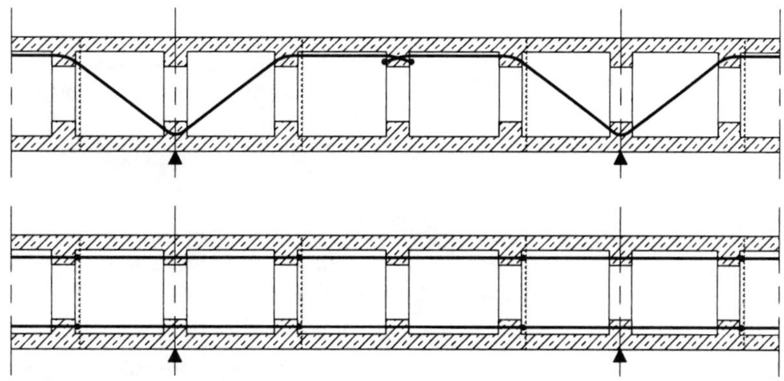

temporary external tendons stressed before launching

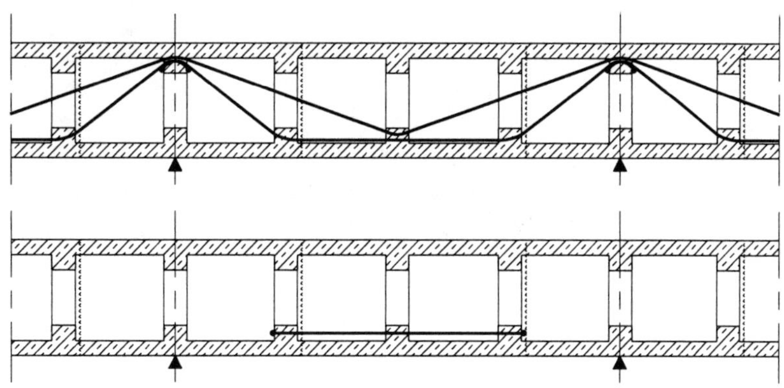

final external tendons stressed after the end of launch

Figure 7.36 - Mixed prestressing composed of permanent conventional internal tendons compensated during launch by antagonist polygonal tendons and additional straight tendons, both external. At the end of launch, prestressing is completed with external polygonal and span tendons.

Some authors[33] prefer to maintain few straight tendons inside the concrete (placed in the nodes of the cross section as in the Charix Viaduct[208], launched on *64 m*-long full spans with a cable-stayed front system shown in figure 5.35) to reduce the congestion of tendons inside the box cell (figure 7.37) and to improve the deck behavior at ultimate limit state. In this case, prestressing can be schematized as in figure 7.38: straight internal tendons in the nodes of the cross section, tensioned during construction and joined by couplers; one half of the polygonal external tendons stressed during construction in the final layout; the second half used temporarily in an antagonist layout to be relieved at the end of launch and repositioned, span by span, in the final layout; temporary straight external tendons which lift the prestressing force to the centroidal level; and, finally, some internal span tendons in the case of long spans.

Figure 7.37 - In the Charix Viaduct, the number of external tendons was reduced by internal straight tendons placed in the cross section nodes. Notice the hydraulic servosupport of the stressing jack (reproduced with permission, SCETAROUTE - Berenguier).

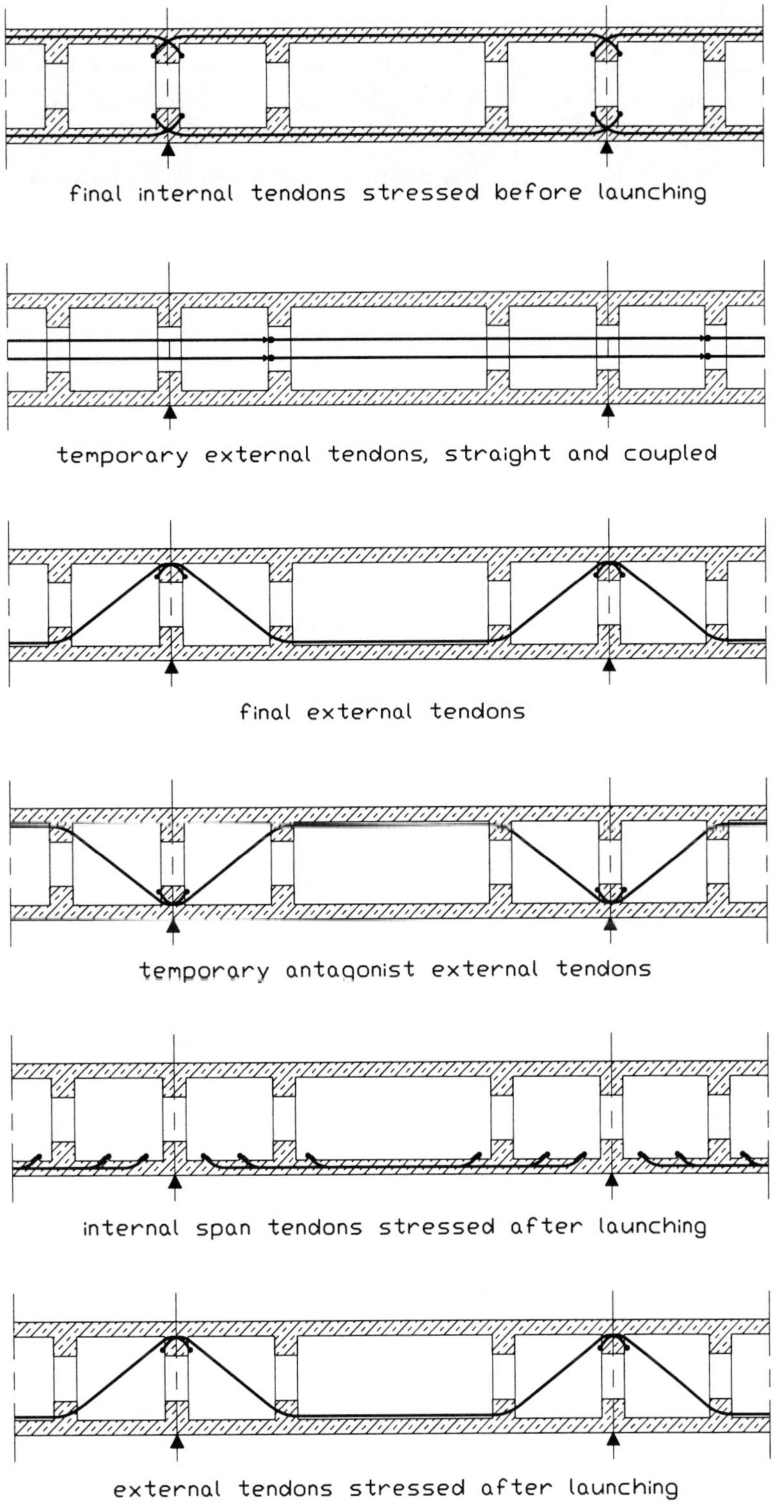

Figure 7.38 - Few internal straight tendons placed in the cross section nodes reduce the congestion of tendons inside the box cell and improve the deck behavior at ultimate limit state.

The low residual centroidal prestressing deriving from the removal of most of the launch tendons permits sizing the polygonal tendons to resist most of the permanent loads (with the purpose of resisting fully their flexural effect) and this leads to an extremely efficient final prestressing. In the Amiens Viaduct, it reached $p_{und}/(q+p_{perm})=0.75$ (7.18), and the additional strength was obtained with straight tendons.

REINFORCEMENT

In a box girder built by incremental launching, reinforcement is generally distributed as follows:

- two orthogonal bar layers placed at the edges of the top slab to resist longitudinal and transverse stresses due to permanent and live loads, secondary stresses, and distortion;
- two similar bar layers in the bottom slab, sized to resist the same stresses and to distribute the launch support reaction toward the webs. It is necessary to check the compressed strut in each cross section of the deck, and the horizontal reinforcement for the tension rod action (figure 7.39);
- two orthogonal bar layers in the webs which resist the principal tension due to launch shear forces, and the edge stresses due to transverse bending, thermal gradients, and differential shrinkage;
- local reinforcement to resist the thrust launch force, which depends on the type of launchers. Using friction systems, it is necessary to resist the superficial stresses produced by the contact plates, while when using back transoms or crossing pins, it is necessary to resist local concentrated forces, and to check the web-slab nodes for the shear forces produced by the thrust (figure 7.40).

Transverse reinforcement is generally composed of straight and bent bars designed according to the casting phases of the cross section. Bent bars are more expensive, but necessary to stiffen the steel cage, acting as spacers of the bar layers.

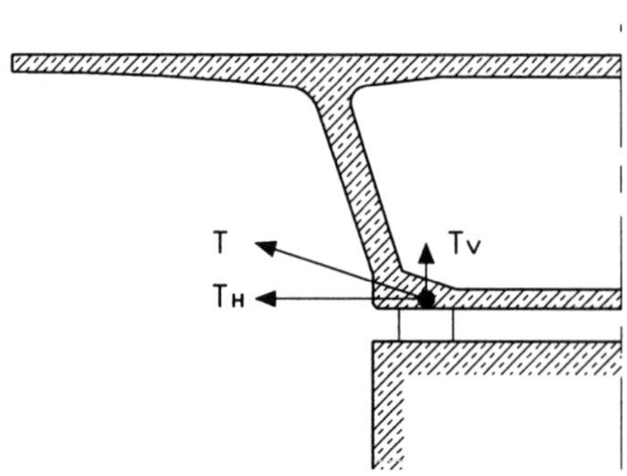

Figure 7.39 - The distribution of the launch support reaction toward the webs produces a tie-strut action in the cross section lower nodes.

Finally, a good practice is to distribute wire mesh on the whole deck surface, to further stiffen the steel cage and to limit the concrete cracks due to shrinkage and thermal gradients. The section of this additional reinforcement should be about *0.1%* of the deck cross section, and about *0.3%* in the bottom slab, subject to the highest stresses both during launch and in service.

An adequate stiffness of the steel cage is necessary not only to guarantee a correct handling of preassembled cages, but even during casting and vibration, to prevent movements which create cavities around the bars, which in turn permit corrosive agents to come rapidly in contact with a larger steel surface and accelerate degradation.

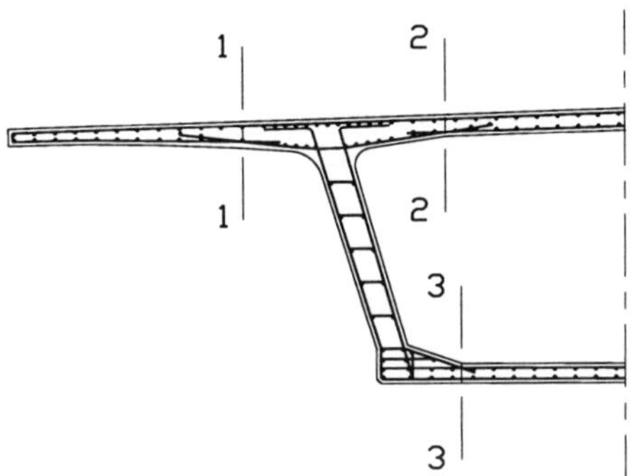

Figure 7.40 - When the launch force is transferred by back transoms or crossing pins, it is necessary to check the web-slab nodes for the shear forces produced by the thrust.

Chapter 8

PIERS AND BEARINGS

PIERS

The application of launching techniques to prestressed concrete bridges was permitted by the commercial availability of Teflon (PTFE), a self-lubricating material which allows reduction of the sliding friction at low cost.

Previously, the higher weight of prestressed concrete structures prevented use of the equipment for steel construction. Long and heavy rollers would have been expensive and hard to amortize. Greased wooden sledges would have transferred horizontal forces to the piers often higher than the design ones, resulting in expensive overdesign.

The development of sliding bearings based on the steel-Teflon contact permitted launch of longer and longer prestressed concrete bridges. The railway Veitshoechheim Bridge[110], *1260 m*-long and weighing *433 MN*, was launched with a total thrust force of *18.2 MN* – that is, *4%* of its weight.

This average friction, confirmed by many horizontal launches, is affected by the lack of simultaneity of the support breakaway (figure 4.32), and, when dimensioning a pier, it is prudent to use the generally higher static friction coefficient (figure 4.33).

Forces on the Pier Cap

The force transferred by the deck to the pier cap during launch may be divided into three components:

- a vertical component, which is the support reaction of the continuous beam, generally calculated as horizontal: it can be eccentric from the pier axis both longitudinally and transversely;
- a horizontal launchward component, proportional to the vertical component and due to both friction and the longitudinal slope of the launch surface over the pier;
- a horizontal transverse component produced by the guide (which prevents drift phenomena in the deck), whose value does not depend on the vertical load but on the driving force and, when launching along a curve, by the vectorial resolution of the thrust force.

Neglecting this last component and the transverse eccentricity, the forces transferred to the pier lie in the vertical surface which contains the deck axis.

Once defined with (4.3):

$$C_f = tg\phi \qquad (8.1)$$

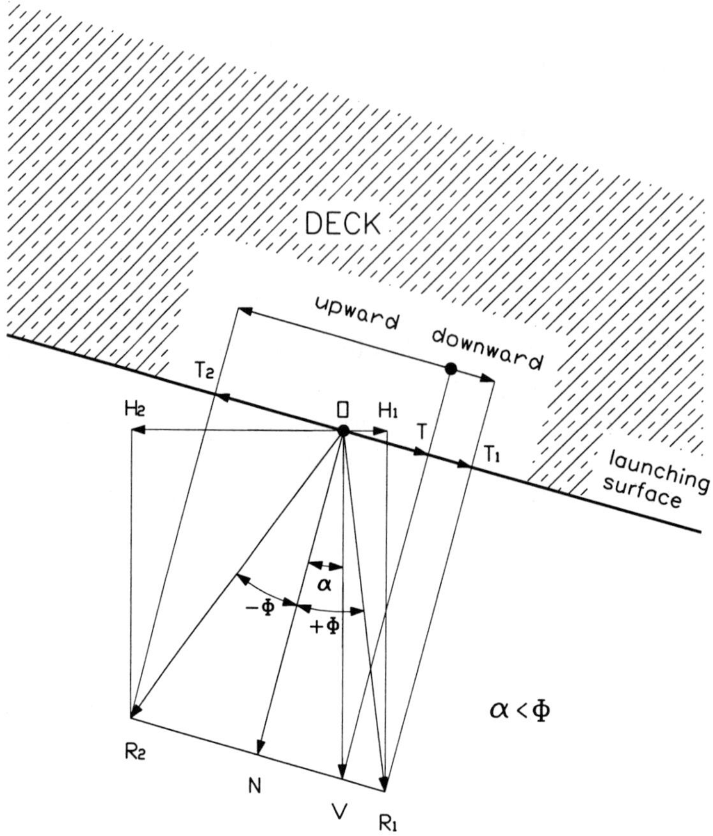

Figure 8.1 - Pier cap forces in the case of slope of the launch surface lower than the friction coefficient ($\alpha < \phi$).

the friction coefficient between the neo-flon plates and the stainless steel sheet which covers the launching saddle, and with α being the longitudinal slope of the launch surface over the pier, figures 8.1 and 8.2 define, respectively for $\alpha < \phi$ and $\alpha > \phi$, all the possible vectorial combinations.

Defining the vertical component of the load transferred by the deck with the vector $\overline{OV}$, its orthogonal and tangential components with respect to the launch surface are:

$$\overline{ON} = \overline{OV}\cos\alpha \qquad \overline{OT} = \overline{OV}\sin\alpha \qquad (8.2)$$

Applying the thrust force, the resultant begins to rotate, and launch starts only when its inclination exceeds the equilibrium angle $R_2OR_1 = 2\phi$. Therefore, the limit equilibrium resultants are:

$$\overline{OR_1} = \overline{OR_2} = \overline{OV}\frac{\cos\alpha}{\cos\phi} \qquad (8.3)$$

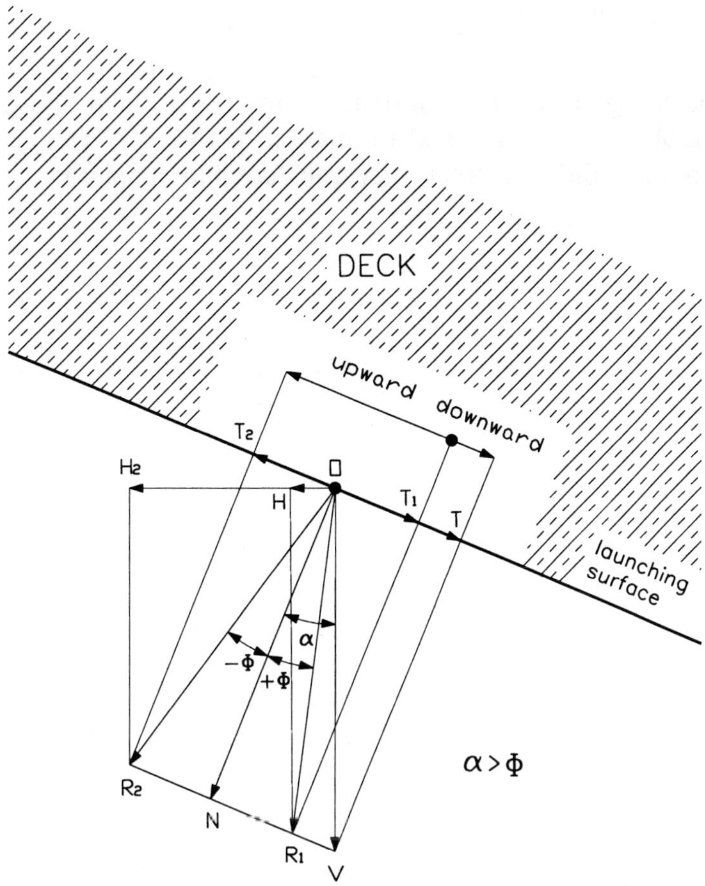

Figure 8.2 - Pier cap forces in the case of slope of the launch surface higher than the friction coefficient ($\alpha > \phi$).

In case of launch uphill, to rotate the resultant from $\overline{OV}$ to $\overline{OR_2}$ it is necessary to apply a force F:

$$F = \overline{TT_2} = \overline{OT_2} - \overline{OT} = -\overline{OV}\left(\cos\alpha \; tg\phi + \sin\alpha\right) \qquad (8.4)$$

and the horizontal force transferred to the pier is:

$$H = OH_2 = -OV \frac{\cos\alpha}{\cos\phi} \sin(\phi + \alpha) \qquad (8.5)$$

On the contrary, in the case of launch downhill, if a thrust force has to be applied (that is, for $\alpha < \phi$, as in figure (8.1)), the resultant rotates from $\overline{OV}$ to OR_1 under a force F:

$$F = \overline{TT_1} = \overline{OT_1} - \overline{OT} = \overline{OV}\left(\cos\alpha \ tg\phi - \sin\alpha\right) \qquad (8.6)$$

If the slope is higher than friction (that is, for $\alpha > \phi$, as in figure (8.2)), the deck slides without any intervention, and to brake it the force F (8.6) must be directed uphill. In both cases, the horizontal force transferred to the pier is:

$$H = \overline{OH_1} = \overline{OV}\frac{\cos\alpha}{\cos\phi}\sin(\phi - \alpha) \qquad (8.7)$$

Therefore, when launching uphill, the slope adds to the friction and the horizontal forces transferred to the piers are higher. When launching downhill, these forces are lower.

If no external constraints impose a particular launch direction, by launching uphill the deck movements are better controlled, but the thrust force and the horizontal forces transferred to the piers are higher. By launching downhill, these forces are lower, but it is always convenient to be able to pull the deck upward in case of need – that is, the same forces are mobilized. Therefore, if possible, it is always convenient to launch uphill.

During launch, the horizontal force transferred to the pier can accidentally exceed the value based on friction and slope (a typical and not infrequent case is the inverted insertion of a neo-flon plate). In the case of tall piers, it is prudent to use real-time monitoring and emergency stops, when not directly supporting the pier with horizontal braces or cable-stays.

Once the system of forces on the pier cap is defined, their eccentricity with respect to the centroidal axis of the pier must be analyzed. In theory, neo-flon plates should assure a uniform distribution of the vertical reaction, R, on the contact surface. In this case, the vertical force is perfectly centered on the launching saddle, which can be shifted backward from the centroidal axis of the pier to reduce the moment produced at its basis by the horizontal component, H, of the force applied to its cap. The optimum eccentricity, set h_p as the pier height, is:

$$e = \frac{H}{R}h_p \qquad (8.8)$$

In reality, the inevitable tolerances and the imperfect parallelism of the saddle and deck surface cause different compression of the neo-flon plates – that is, they make the vertical force, R, eccentric. This eccentricity must be evaluated case by case in relation to the dimensions of the plates. In addition, the pier always has a tolerance in verticality, especially if built with slip forms. Finally, the vertical load transferred to the pier is eccentric even in the transverse direction, because of the geometric imperfections of the support and/or of the deck, which produce different reactions on the two launching saddles, and therefore a moment about the longitudinal axis. This aspect is rarely critical in permanent piers, sized for transverse service loads, but in temporary piers it can require the use of launch saddles controlled by hydraulic devices able to homogenize their reactions – that is, to control the torsion in the deck and the transverse bending in the pier (figure 4.26).

In the longest decks, it can be necessary to place friction launchers above the piers to better adjust the thrust action. In this case the piers should be checked for the whole local thrust force, and their cable-staying (figure 4.20) is practically indispensable.

Finally, in the last few years, new hydraulic devices, schematically similar to friction launchers, have been developed to apply to the deck a thrust force equal and opposite to their own internal friction, so as to cancel the horizontal force transferred to the pier. These devices are rather expensive[108], and reserved for the most critical cases.

Stress Verification

The pier, as a cantilever partially fixed at its basis, is very sensitive to the moments and the horizontal forces applied to its cap, and instability must be checked even during launch.

The slenderness ratio of the pier is defined as:

$$\lambda = \frac{l_0}{r} \qquad (8.9)$$

in which r is the radius of gyration (5.2) of its cross section, and l_0 is the effective length – the double of the height only in case of a perfectly fixed footing, and progressively longer as the degree of constraint decreases. If this ratio is not very high, it is sufficient to calculate stresses neglecting the pier deformations, and to check the critical cross sections for combined compressive and bending stress with the coefficients given by the different specifications, in which Euler's critical load:

$$P_E = \pi^2 \frac{EI}{l_0^2} \qquad (8.10)$$

should be defined with a reduced elastic modulus to take into account the creep deflection of the pier due to the constant application of a horizontal force to its cap. In fact, the pier cannot recover its elastic deflection during launch stops.

On the contrary, if the pier is very slim and deformable, the stresses should be defined by taking into account its deformation, and therefore with a non-linear approach.

Similar verifications should be carried out even in the transverse plane of the pier, in which (except for temporary piers) the low launch forces generally do not require non-linear analysis. At the same time, the high torsional stiffness of the deck generally permits one to neglect the rotation of the pier cap, and therefore to use an effective length which depends on the transverse mobility of the deck, and is:

$$\frac{h}{2} \leq l_0 \leq h \qquad (8.11)$$

that is, from one half to one quarter of the longitudinal one in the best case of a perfectly fixed footing.

In the case of tall temporary piers, this principle may be used even in the longitudinal direction: by placing four launching saddles on each pier instead of two (figure 5.38) it is possible to take advantage of the flexural stiffness of the deck to fix the pier cap.

Further Considerations about Temporary Piers

The static analysis of the temporary piers is characterized by the frequent presence of longitudinal cable-stays, anchored to the footing of the adjacent final piers to reduce the cost of foundations.

In a monolithic pier supporting both the launching saddles (figure 1.8), the problem is the same as for a permanent pier.

In the case of a couple of pillars placed under the deck webs (figure 5.41), the cable-stays are generally contained in the longitudinal plane, since the large distance between the two pillars, the lower transverse forces, and the shorter effective length generally allow one to avoid transverse stayed schemes. Therefore, transverse forces are resisted only by cross bracing connecting the pillars.

In terms of instability, the problem does not depend much on the type of pillar (precast reinforced concrete modules or steel tubing) since in any case the pillars and their bracing form vertical trusses to be checked by taking into account the influence of the shear stresses.

If the X-type braces of this truss are sloped angle α from the horizontal plane, the critical load is[27]:

$$P_{cr} = \frac{P_E}{1 + \dfrac{P_E}{2E_b A_b \sin\alpha \cos^2\alpha}} \qquad (8.12)$$

in which E_b is the elastic modulus of the braces and A_b their cross section.

Since this formula is based on the hypothesis of instability in the elastic field[27], this condition should be checked by calculating:

$$\sigma_{cr} = \frac{P_{cr}}{2A_p} \qquad (8.13)$$

where A_p is the cross section of each pillar.

Then, local instability must be checked. If the pillars are made of thin reinforced concrete precast modules, of width, b, and thickness, t (figure 5.37), they tend to buckle, and by assuming bracing distributed so as to hinge them along their vertical edges, the critical load is[134]:

$$P_{cr} = \frac{\pi^2 E t^3}{3(1-\nu^2)b^2} \qquad (8.14)$$

which is valid when the height of the temporary pier is at least three times the module width, and instability occurs always in the elastic field. Finally, pillars composed of steel tubing and/or composite tubes should be checked according to the usual criteria.

PERMANENT BEARINGS

Generally, the deck is launched over special saddles (although the use of permanent bearings topped by temporary launch steel plates is not unimaginable). Therefore, at the end of launch the temporary saddles must be replaced with permanent bearings.

The large bearing movements of continuous beams many hundred meters long and their high loads generally dictate the use of steel-Teflon bearings. Their installation follows the stages shown in figure 8.3:

- a battery of jacks is placed on the pier cap. The deck is lifted and the neo-flon plates, the launching saddles, and their support seats are removed;
- the pier cap is cleaned, the elevation of the deck soffit is surveyed, and the actual support reaction is calculated by means of the oil pressure in the hydraulic circuit. At the same time, the permanent pads for the bearings are built;
- finally, the permanent bearings are installed under the deck, and over them steel shimming wedges or a rheoplastic mortar casting absorbs the differences in height due to the longitudinal slope of the deck.

Before lifting the deck, the residual kinetic friction in the neo-flon plates and the longitudinal slope of the launch surface bend the pier. Since the static friction of the permanent bearings is often lower than the kinetic friction in the dirty and worn neo-flon plates and the wedged shimming creates a horizontal instead of sloped deck support plane, this residual deflection is quickly recovered when the deck is lowered onto the permanent bearings, and this compromises their adjustment.

Therefore, this deflection must be eliminated before loading the permanent bearings. This is obtained by placing the batteries of jacks on new and greased Teflon plates which permit the pier cap to move horizontally when the launch saddles are unloaded (figure 4.29).

This is necessary also when the bearing replacement occurs simultaneously with prestressing operations, to avoid transferring to the piers the horizontal forces due to the prevented longitudinal deformations of the deck.

Finally, the support reactions of the deck can be corrected by shimming the bearings, or the bearings can be placed on flat jacks to continuously monitor the support reaction.

Using this construction method, the bearings (weak point of the structure in terms of maintenance costs) are not affected by any construction operations (oil and cement grout spilling, form dismantling, creep and shrinkage deformations of the deck, elastic deformations due to prestressing, etc.) and can be installed under the deck in optimum conditions. Moreover, their installation is completed in a few weeks, and therefore practically in isothermal conditions. This facilitates their adjustment, and reduces the risk of reaching limit stops during the thermal movements of the deck.

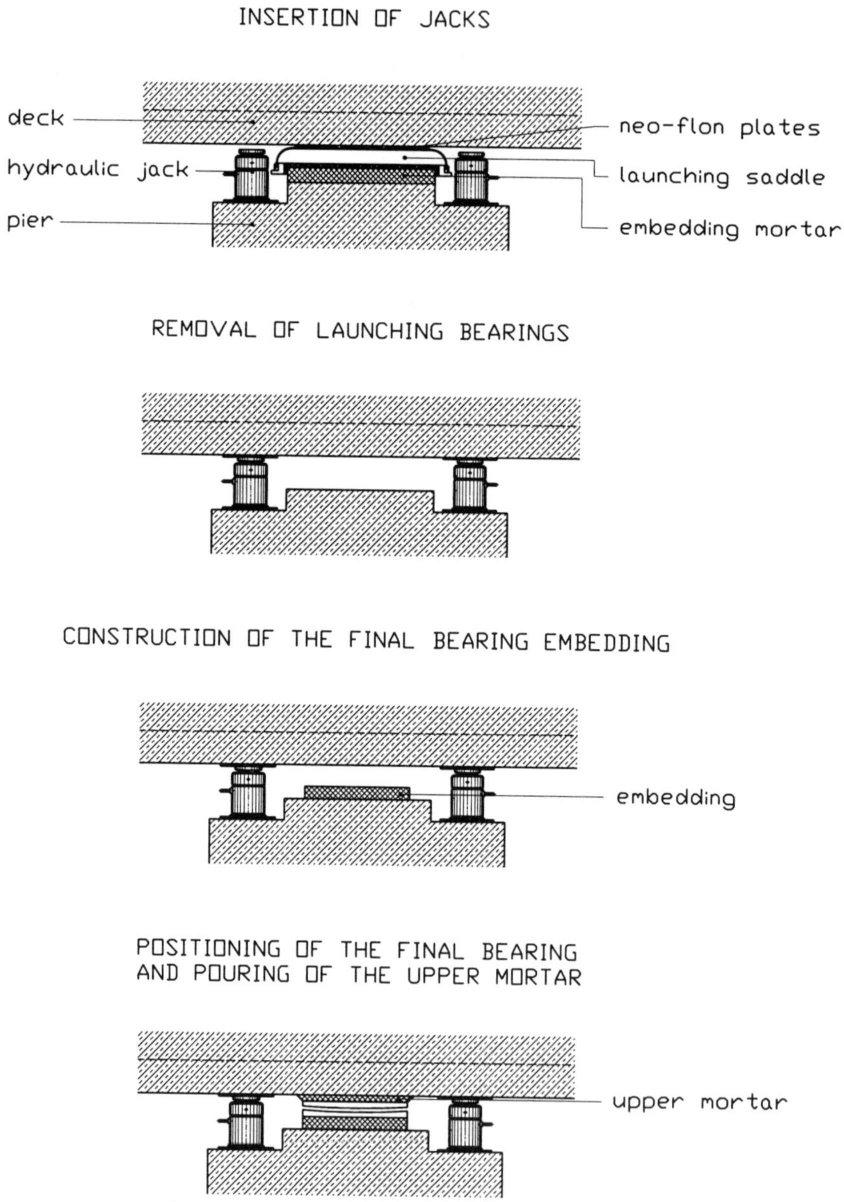

Figure 8.3 - Installation of the permanent bearings.

Chapter 9

EVOLUTION OF INCREMENTAL LAUNCHING

Whichever the construction technique of a bridge may be, the limitation of dead load is fundamental, since it is the most important load the deck has to resist and its reduction creates a reserve available for live loads, and because it governs the design of the construction equipment – that is, its cost.

In incremental launching, the dead load produces temporary stresses which affect the structural sizing more and more as span increases. Since these stresses depend on the unit dead load linearly and on the span with the quadratic law, the most spontaneous intervention is to reduce the launch span with respect to the final one, and the use of temporary piers, in certain conditions, can be definitely advantageous.

Since the deck works mainly in bending, the structural efficiency, ρ_s (5.4), of its cross section affects the total carrying capacity, and therefore the general efficiency, ρ_g (5.1), of the deck. The efficiency of the cross section depends on the distribution of the masses and on the efficiency of the individual materials, whose properties are rather different. Instability penalizes the use of steel in compression, a few millimeter-thick steel plate resists the same shear force as a many decimeter-thick concrete web, although it weighs only the twentieth part, but at the same time the tensile strength of a steel plate is only one fifth of that of prestressing steel.

The efficiency of a material can be evaluated with the ratio of its strength to its specific weight[169]:

$$e_M = \frac{f}{\gamma} \qquad (9.1)$$

A *B45* concrete has a compressive efficiency of $e_M = 45000/25 \approx 1800\ m$, and a low tensile efficiency. As a consequence, shear efficiency is low as well.

The tensile efficiency of a steel plate is higher, $e_M \approx 6400\ m$ for a BS 4360-50 steel, but its compressive efficiency is reduced by instability.

Prestressing steel reaches much higher values. A BS 5896/3 super strand-1770-15.7 has $e_M \approx 22500\ m$, and the cable is generally the least expensive way to use steel in tension. By relating the efficiency of materials in their most suited load conditions to that of prestressing steel, $\rho_{B45} = 0.08$ and $\rho_{St50} = 0.28$.

Because of these different values, steel box girders are definitely more efficient than prestressed concrete ones, as the masses are more distant from the centroid and the material performance is almost four times better. However, instability prevents one from fully exploiting the strength of steel plates, and thus reduces the actual efficiency of this structural scheme.

The high e_M of prestressing steel leads to its frequent adoption in resisting flexural stresses. In the continuous beam, this requires the presence of two slabs to compress without instability – that is, composed of an inexpensive material such as reinforced concrete. Deviation of prestressing tendons reduces shear forces, which can be fully resisted by the webs.

The dead load of a box girder is the sum of the weight of the top slab, the bottom slab, and the webs, and these three components can be treated separately in order to lighten the cross section. The thickness of the top slab depends on the live loads and on the necessity of reaching an adequate punching shear strength. Once sized for longitudinal bending, its transverse behavior can be improved with transverse prestressing and/or ribs. The thickness of the bottom slab depends on the need of reaching an adequate transverse flexural stiffness and of containing internal tendons, and on the convenience of optimizing the action of launch prestressing through an adequate position of the cross sectional centroid (7.12). Not being directly affected by live loads, its thickness is less restrained.

Using internal prestressing, the lightening of the webs is limited by the need of containing and deviating prestressing tendons, and in narrow decks their area can reach up to *30%* of the whole cross section. This means a *40%* overload of the area of the slabs (which could practically generate the flexural strength of the cross section alone) and a low structural efficiency because of their position close to the centroid. In addition, they increase the cost of materials and labor, being the most difficult element to cast of the whole cross section.

External prestressing produces a first improvement of the cross-sectional efficiency and reduces the cost of materials and labor. As an alternative, the deck width of a launched bridge may be increased, since, unlike with other construction methods, the dead load of the deck does not dramatically affect the cost of the equipment. Thus, it can be advantageous to avoid twin box girders and to adopt a double-width single-cell section, since the web thickness necessary to resist the principal compressive stresses may be sufficient to deviate the internal tendons.

The next alternative consists in launching only a part of the cross section, and to complete it once this carrying core has reached its final position. This permits one to reduce dead load during launch, but most of the advantages of repetitive production in an easily accessible formwork are lost.

At the same time, the efficiency of the material (9.1) can be improved by increasing strength with the same density – that is, adopting *HPC*, or by reducing density with the same strength – that is, adopting structural lightweight concrete.

Once even these steps have been taken, the prestressed concrete box girder concept is generally abandoned for the non-prestressed beam composed of a reinforced concrete top slab and two or several steel webs, connected or not by an orthotropic steel bottom plate which closes the box cell. These composite sections are not suitable for front launching because of both the reduced compression area in the negative moment region and the vertical compressive stresses in the webs when passing on the launching bearings. Therefore, they are generally built by launching (or lifting) only the steel girders and then casting in-situ the concrete slab. Since the construction of a concrete bottom slab is rather complex, torsional stresses are generally resisted by steel diaphragms and/or bracing, and more rarely by an orthotropic steel bottom plate.

Finally, a similar overall efficiency may be accomplished with prestressed composite sections which combine the use of external prestressing, reinforced concrete slabs, and solid or trussed light webs. This results in efficient decks (composed of masses far from the centroid and prestressed in an effective way) that make the most out of prestressing steel and are light and easy to build. The considerable shear strength of steel plates suggests their use as webs, but the same general efficiency may be reached with trussed webs composed of either steel or reinforced concrete members, since instability reduces steel's compression efficiency (9.1) to values comparable with a good concrete.

The simultaneous use of external tendons, reinforced concrete slabs, and solid or trussed light webs permits one to:

- reduce the self-weight of the cross section without penalizing its carrying capacity – that is, to increase general efficiency (5.1). On a *40 m* span, a conventional box girder with internal prestressing requires about $0.55\ m^3/m^2$ of concrete, which reduces to $0.45\ m^3/m^2$ in case of external prestressing. A prestressed composite section requires only $0.35\ m^3/m^2$, and lightening is of about *25* to *35%*;
- increase the radius of gyration (concrete is concentrated at the flanges) and the structural efficiency (5.4), with lower costs of prestressing (5.7);
- specialize the contribution of each material. Concrete resists the whole bending moment thanks to prestressing, whose deviation reduces shear forces to values which can be directly absorbed by the webs, and each material works in uniform stress diagrams instead of triangular;
- reduce the costs compared with the conventional composite sections;
- simplify construction activities and equipment;
- reduce construction duration;
- obtain structures with an excellent final aspect.

Prestressed composite sections (PCS) may be grouped into two types, the main difference between them consisting in the transmission of shear stresses. Trussed webs tend to eliminate material not working in the Mörsch lattice scheme. Solid steel webs tend to benefit from the higher shear efficiency (9.1) of steel plates compared with reinforced concrete.

The level of specialization of the contribution of each material depends on the strain distribution within the cross section and requires a careful design, since the different behavior of the materials at prestressing and over time generates secondary stresses.

Further limitations derive from the construction technique. In front launching the migration of support reactions produces local stresses which affect the dimensioning of solid steel webs and discourage the use of trussed schemes. On the contrary, these may be exploited in the construction on falsework – that is, in bridges launched by rotation or symmetrical approach on movable bearings.

SECTIONS LAUNCHED IN PARTIAL CONFIGURATIONS

A simple way to reduce launch stresses is to subdivide the construction of the cross section by launching only a carrying core and completing the cross section after launch by casting the remaining parts in-situ.

This technique is widely adopted for conventional composite sections (figure 9.1), while its use is rather unusual in prestressed concrete bridges.

For the shorter spans, the incremental launch of parallel U girders completed with a cast-in-situ concrete slab has sometimes been adopted, but the stiffened top flanges of the U girders increase the consumption of materials. This construction method can benefit from the use of precast concrete plates as left-in-place formworks, but this solution is not appreciated by many owners because of the questionable static contribution of the precast planks in the longitudinal direction.

Since the ratio of the highest negative launch moment to the positive one (7.11) is rather variable because of thermal gradients and bearing settlement, it is convenient to launch a partial section with a central core as high as possible.

Figure 9.1 - Composite bridges are generally built by launching only the steel girder (Figure 1.1) and casting in-situ the concrete slab. Since the use of left-in-place precast concrete plates is not appreciated by many owners, launching formworks may be necessary (Author).

Sometimes this has been accomplished with a box girder composed of the two webs and the two portions of slab which join them, while the cantilevers of the top slab have been cast-in-situ using mobile formworks sliding along this carrying core. Another attempt in this direction (figure 9.2) saw the construction of the cantilevers on left-in-place precast concrete planks. Their dead load reduces the actual savings of materials, but there are indirect savings in the site layout, the launching nose, the thrust devices, and the self-launching formwork.

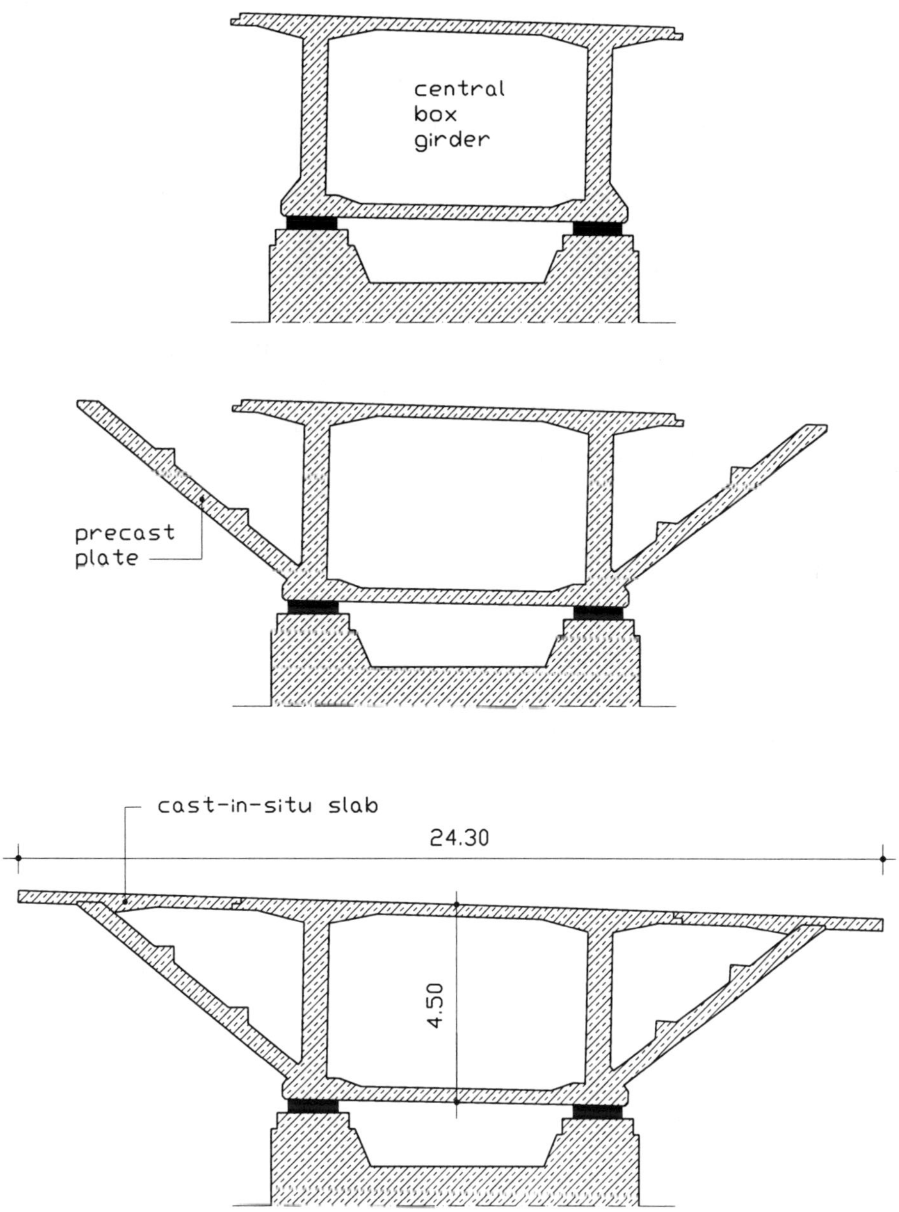

Figure 9.2 - In the Edlitz Valley Bridge, the cross section was completed in-situ after the launch of the carrying core.

HIGH PERFORMANCE CONCRETE

In the last twenty years, the international research in High Performance Concrete has led to a great many new materials, and new abbreviations and acronyms. Among these, the most general one is likely Chemically Bonded Ceramics (CBC), which emphasizes the ceramic character of the material (inorganic and not metallic) and the prevalently chemical aspect of the bond.

The association of CBC materials to the family of ceramics is justified also by their production processes, much more similar to those for conventional neo-ceramics (powder pressing, high shear mixing, lamination, extrusion, calendering etc.) than to those for ordinary concrete.

Only some of these materials are suitable for the construction of large and complex works using conventional techniques (mixing and transport in a concrete mixer, casting in formworks etc.), and the remaining ones are reserved for more sophisticated uses, such as the production of small elements of regular shape.

Generally, CBC materials aim at reaching the highest mechanical strength permitted by the absence or the minimum presence of defects or pores in the matrix microstructure, and can be grouped into two classes:

- Macro Defect Free (MDF) materials, characterized by a gummy plasticity of the mixture which permits its processing by extrusion or calendering between rollers;
- Densified with Small Particles (DSP) materials, characterized by a more conventional rheology and based on the principle of filling the voids among the cement granules (dimensions 1 to $50\mu m$) with sub-micronic particles (0.01 to $0.1\mu m$). They are processed by both humid powder pressing and pouring mixtures of plastic consistency into molds.

Within this last family, the DSP materials can be divided into two subtypes: the current HPC, and the RPC, which represents its peak of technological development.

High Performance Concrete (HPC) is currently produced in many countries. It is characterized by a compressive strength of 60 to $120\ MPa$ and is obtained with:

- high strength cements mixed with mineral additives with high pozzolanic activity, such as silica fumes, rice husk ash, microfine slag, etc.;
- lowest water to cement W/C ratio (considering as cement both the cement itself and the mineral additives) obtained with super-water-reducing additives;
- high quality rough surface crushed aggregates (basalt, granite, diabase etc.), assuring both high mechanical strength and high adhesion to the cement matrix. In fact, one can obtain such a dense and compact matrix as to exceed the mechanical strength of conventional aggregates.

Along with the best mechanical performance, however, HPC has some limitations when strained to the ultimate limit state because of the typical brittle rupture of the ceramic materials, characterized by low breaking energy. The low ductility of the material should be taken into account in the structural design, and compensated by means of an adequate distribution of reinforcement.

The advance of Reactive Powder Concrete (RPC) is aimed at eliminating these defects. Although it is only in the research stage, the results let us imagine that the soft-casting production of a concrete with a compressive strength of $200\ MPa$ and ductility, bending strength, and fracture energy much higher than the present HPC will open new horizons in the use of concrete, and will represent a valid alternative to structural metals even for bridges.

At present, the design specifications of many countries allow the use of HPC, with a maximum strength of *80 MPa* in Japan, Sweden, and in the CEB-FIP International Recommendations, *100 MPa* in Finland, *105 MPa* in Norway and Holland, *115 MPa* in Germany, and without any limits in USA and Canada.

The advantages of using HPC in launched bridges are numerous:

- higher mechanical strength of the concrete along with external prestressing permits one to reduce the thickness of both the slabs and the webs, increasing the structural efficiency (5.4);
- reduced influence of launch stresses due to structural lightening increases the quantity of draped tendons, and with it the efficiency of prestressing (7.18);
- reduction of reinforcement;
- decreased structural deformability and lower prestress losses due to elastic shortening, shrinkage and creep. Many specifications allow significant increases of the elastic modulus, and the Finnish norm allows a *30%* reduction of the creep coefficient compared with conventional concrete;
- lower depth of the cross section, resulting in more slender and aesthetically pleasing structures;
- reduced risk of shrinkage cracks, which compromise the durability of the bridge and its resistance to aggressive agents;
- lower cost of launching equipment, and an increase in the span which can be overcome without temporary piers.

A design strength of *60* to *120 MPa* does not subject the concrete handling to particular restraints, but the production of the cement hydration heat must be controlled, since it can produce micro-cracks. Therefore, it may be necessary to:

- limit the maximum temperature inside the concrete. In the Elorn Bridge, in France, internal temperatures of *+80°C* were measured, but their effects were lower than those expected in a conventional concrete, with a reduction of the *28* day strength of about *10%*;
- limit thermal gradients. In the Great Belt Link, in Denmark, maximum differences of *20°C* were allowed between the external surface and any internal point, and of *12°C* between the mean temperature of the setting segment and the previous one. These conditions were obtained and maintained by combining heating or cooling of the fresh concrete, heating of the adjacent segment, and cooling of the setting young concrete;
- postpone the formwork dismantling to reduce the superficial shrinkage cracks. In the Elorn Bridge, formworks were maintained against the young concrete for *72* hours;
- immediately protect the slab extrados with plastic sheets;
- operate with the lowest possible W/C ratio, checking the number of micro-cracks wider than *0.01 mm* measured in cylindrical cores;

- adhere to quality assurance procedures, with training programs for the staff of both the owner and the constructor;
- carefully program the whole production process, including methods and limit times for transport, pouring and vibration of concrete, as well as the eventual mixing of additives in the yard.

LIGHTWEIGHT CONCRETE

Lightweight concrete is used when dead load constitutes a high percentage of the design loads of the bridge. Reduction in dead load increases ρ_g (5.1), limits the cost of reinforcement, prestressing, and equipment, and lightens piers and abutments.

Soon the savings so obtained exceed the higher cost of concrete. In the Bridge No. 33 of the A55 highway at Marseilles[78], an alternative project with lightweight concrete turned out to be *3%* less expensive than the conventional concrete tender project, due to the lower cost of prestressing.

The cost of lightweight concrete is higher than the cost of conventional concrete, due to the higher production cost of the aggregates and the few sources of supply, which increases the cost of transport.

This concrete can reach a compressive strength of about *60 MPa* with a density lower than $18.5\ kN\ m^3$. Being a very adaptable material, its use can be convenient in many settings, such as for bridges in seismic areas or very high above the ground, or built by rotation or balanced cantilever.

The physical properties of light aggregates are rather particular, and many details of the project must be adapted to the material:

- the lower ductility requires dense reinforcement, to increase the deformation capacity of the cross section in the plastic dominion in order to dissipate energy before collapsing;
- the low thermal conductivity of the light aggregates reduces the loss of heat and accelerates the growth of mechanical strength. The ratio of the *7* day strength to the *28* day one is generally of about *0.8* to *0.9* versus the conventional concrete average value of about *0.7*;
- the axial tensile strength (which affects the adhesion between concrete and reinforcement, and therefore the number and the width of cracks) is *15* to *30%* lower than a conventional concrete, and, for instance, Eurocode provides:

$$f_c = 0.33 f_c'^{2/3}\left(0.3 + 0.7\frac{\gamma}{2400}\right) \qquad (9.2)$$

- in which γ is the concrete density in $Kg\ m^3$ and f'_c the cylindrical compressive strength in MPa;
- the elastic modulus is *25* to *50%* lower, and this increases the losses of prestress but reduces the influence of secondary stresses. For instance, Eurocode provides, at *28* days:

$$E_{28} = 9500\left(f_c' + 8\right)^{1/3}\left(\frac{\gamma}{2400}\right)^2 \qquad (9.3)$$

- shrinkage is about *50%* higher because of the lower rigidity of aggregates, and this further increases the losses of prestress;
- creep is about *20%* higher, and this increases the accumulation of irreversible strains during launch.

PRESTRESSED COMPOSITE SECTIONS (PCS)

In the last decades, many factors increased the competitiveness of conventional composite sections and reduced their lower threshold of utilization to spans of about fifty meters, which previously were the dominion of prestressed concrete box girders.

Progress in iron metallurgy led to rolled steel of high and reliable mechanical properties. The development of design codes based on the strength of materials led to a better knowledge of safety. The progress of assembly techniques replaced riveted joints with friction bolted or welded ones.

In the last years, this evolution has been further accelerated by the cost reduction of steel plates and the parallel general increase in labor cost.

The qualitative advantages of the conventional composite sections are numerous: the high tensile and shear efficiency (9.1) of the steel plates combined with the good compressive efficiency of concrete, the rapidity of construction, the possibility of building most of the structure in a workshop, the long durability permitted by different and renewable protective treatments, the architectural quality of structural elements, whose function is clearly recognizable.

The optimum field of utilization of a conventional composite section is the simply supported beam, since the bottom flange compressive stresses in the continuous beam's negative bending zones require thick plates or closely spaced stiffeners to prevent instability, and in both cases the structural cost increases. For the same reasons, the launch of a conventional composite section generally consists of launching only the steel girder, and the top slab is cast in-situ in a second stage.

Since the presence of a concrete bottom slab in the negative bending zones is competitive only for long spans, the cross section is almost always open, and only aesthetic reasons or particular requirements justify, in straight alignment decks, the higher cost of closing the box girder with an orthotropic steel bottom plate.

On the contrary, the construction of a concrete bottom slab in a formwork supported on the ground is much simpler, and led to the concept of box girders composed of two concrete slabs and two steel webs, and built by incremental launching or precast segments placed with launching gantries.

In turn, the presence of two concrete slabs begs the massive utilization of prestressing, and concrete slabs, light webs, and prestressing tendons are combined in new types of prestressed composite section (PCS).

If the deck is not very wide, the top slab can be supported on the two webs only. A thickness of *0.25* to *0.30 m* allows spans between their axes of *7 m* and two side cantilevers *3 m* wide – that is, an overall width of about *13 m*.

In such a cross section, longitudinal bending moment is mainly resisted by axial forces in the concrete slabs, and the shear forces, reduced by the tendon deviation (figure 7.30), are mainly resisted by the webs. Therefore, the behavior of the structure depends on the shear force distribution between the webs and the polygonal prestressing tendons. The different efficiencies (9.1) suggest shaping the tendons to balance the shear forces produced by the permanent loads and only one half of the live loads, so that the webs resist only the shear fluctuations due to the presence or absence of live loads.

Thus it is possible to obtain particularly slender cross sections, which on the one hand guarantee the highest efficiency in service, but on the other hand make difficult and delicate any operation in which the dead load shear forces are not resisted by shaped tendons.

Not all these structural schemes are equally suitable to full span incremental launching, while all of them give the best results in construction by free cantilever and derived schemes, such as rotation or front approach.

Stiffened Plate Prestressed Composite Sections (PCS)

The most intuitive way to substitute the concrete webs of a box girder with steel beams is to use stiffened plate webs (figure 9.3). The deck has a conventional aspect, since tendons are hidden inside the box cell, the individual behavior of the materials is well known, and so their combined application is amply tested.

In reality, however, this simplicity is only apparent, and several problems (unusual in the field both of steel structures and of prestressed concrete) discourage the adoption of this structural scheme.

First of all, the webs absorb a high portion of prestressing, up to *20* to *25%*, which is uselessly wasted and in turn necessitates use of stiffeners to prevent buckling, since thicker plates, often less expensive, makes things worse.

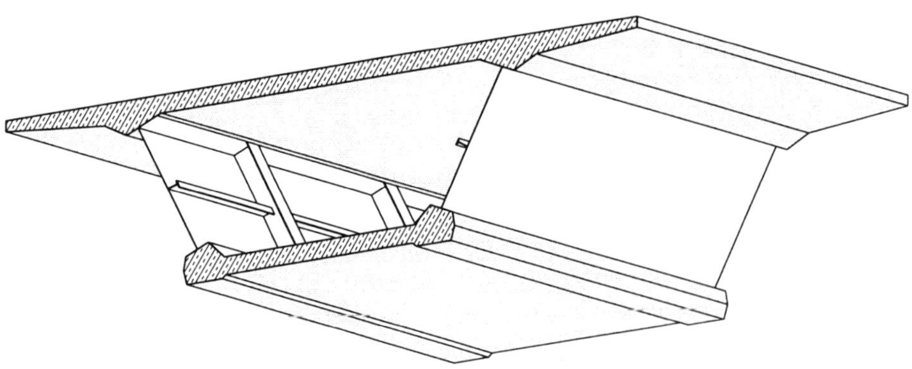

Figure 9.3 - Stiffened plate prestressed composite section.

Further problems derive from the rheological properties of concrete, which modify the state of stress with time. While the effects of shrinkage are similar to those in the conventional composite sections, the effects of creep are more complex, as they tend to transfer further prestressing to the webs. Creep, therefore, decreases progressively. Since time-dependent phenomena can always be evaluated by means of a reduced elastic modulus (6.46), even the homogenizing of materials based on the modular ratio varies with time, modifying the mechanical properties of the cross sections, the position of their centroids, and the eccentricity of prestressing forces, and accentuating the migration of prestressing toward the webs. This makes the static analysis of the webs rather complex.

A cross section composed of only two steel plates and two concrete slabs would be excessively deformable, and vertical stiffeners must be introduced to reach a transverse flexural stiffness of the webs similar to those of concrete slabs, resulting in additional costs. The use of bracing systems (steel diaphragms, light but deformable, or concrete diaphragms, very heavy in such refined structures) is also necessary to complete the action of the tendon deviation diaphragms.

The connections between the webs and the slabs must be studied carefully, being the weakest points of the structure. Corrosion phenomena observed in the upper nodes of many composite sections should discourage placing triple contacts (air, concrete, and steel) near main structural elements. In the box girders the situation is worse, as condensation water inside the cell flows along the steel plates and gathers in the lower nodes of the steel webs. Therefore, while the upper nodes can be designed with conventional schemes (migration of prestressing superimposes longitudinal shear forces lower than in a conventional composite section) the lower nodes present particular aspects.

Web-Slab Lower Node

The structural node between the steel web and the concrete bottom slab can be developed with the schemes of figure 9.4.

The simplest solution, scheme A, consists of casting the slab over the steel flange of the web, and is penalized by the presence of a triple contact on a main structural element. Therefore, it is convenient to modify this scheme by welding a longitudinal stiffener along the extrados of the slab (scheme B) to move the triple contact away from the main element[61].

A second solution is to place the concrete slab under the steel flange (scheme C). This moves the triple contact away from the web and offers a better aesthetic result, but presents some disadvantages. Concrete pouring and vibration under the steel plate is difficult and can require sloped webs (scheme D), with the higher node cost balanced by the smaller width of the bottom slab. Launch of the steel web alone is impossible.

The stresses transferred through the connection between web and slab are not very different from those in conventional composite sections, and consist of distributed shear forces and orthogonal forces concentrated around the shear discontinuities (described in a subsequent chapter). Therefore, conventional connectors can be used in case of slabs cast over the steel flanges[7], while in the suspended slabs they must resist even their dead load, and therefore must be embedded more deeply in the concrete.

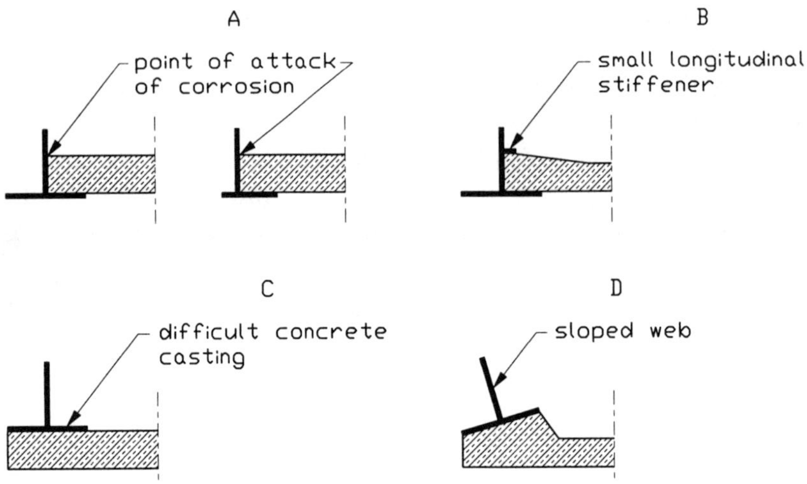

Figure 9.4 - Structural node between the steel girder and the bottom concrete slab.

Web Instability

The problem of instability is a basic issue in steel construction, and the plates are often designed to satisfy stability more than strength criteria.

From a general point of view, an I-beam subject to main flexural stresses can present three types of instability:

- the transverse deflection of the compression flange, which twists the girder and produces local deformations. In a prestressed composite section, this mode is prevented by the concrete slabs, which act as distributed bracing, and by the cross diaphragms, which avoid relative movements between the slabs;
- the torsional deflection of the compression flange, partially related to the previous mode and prevented by the slabs as well;
- the instability of web panels, along horizontal lines in the case of pure bending, and along sloped lines in the presence of shear forces. It can arise both in service and during launch, when the axial prestressing does not cooperate in reducing the shear diagram.

Besides, in an I-beam subject to in-plane concentrated loads, such as support reactions, two additional local phenomena related to instability can arise over the application point of the reaction:

- the local plasticization of the web, which is prevented by the load distribution created by the bottom slab;

- the creation of a local mechanism in the bottom flange, which is prevented by the slab as well, which increases the overall flexural stiffness of the steel flange.

Therefore, in a stiffened plate PCS, the problem of instability is limited to the buckling of the web panels.

It is well known that this type of instability does not represent a potentially dangerous situation like that due to the instability of slender compression members, since the panel edges are immobilized by the flanges and the vertical stiffeners, and the growth of out-of-plane elastic deformations is resisted by membrane stresses which develop a natural stiffening action of the panel. In other terms, after buckling, the steel plates develop a strength reserve because of the elastic work of membrane stresses. In the pre-critical flexural state, when stresses are low, the steel web deforms keeping its sections plane. At buckling, axial stresses begin to redistribute toward the stable zones of the panel, the slope of the tension diagonal changes, and a new equilibrium based on higher stresses arises.

Years ago, the problem of instability of a web panel was approached with manual methods. Once the plate was began to assume a transverse deformation, the loading was interrupted and the value of this "critical stress" was calculated. The present means of testing and calculation confirm that, over this "critical stress," the growth rate of the transverse deformation does not increase, but tends to decrease until the first plasticization occurs. Only at this point, the deformation begins to increase more quickly than the load and the plate collapses. Therefore, the critical stress represents only the border between two different states of elastic equilibrium, and does not constitute the limit value of the carrying capacity, which is much higher because of the behavior in the post-critical dominion.

The ratio of the ultimate stress to the critical stress increases with the plate slenderness, varies in the range 3 to 8, and for the usual slenderness of composite sections is in the range 5 to 6. In reality, the steel plates are not perfectly plane, and are often subject to internal stresses due to welding. Test models showed that, because of this, they adopt a portion of the post-critical dominion even in service conditions, and the actual safety factor is lower. Moreover, the post-critical behavior of a non-stiffened plate is different from that of a stiffened plate, due to the post-critical effects of the stiffeners.

Launch

In medium span (50 to $70\,m$) PCS bridges, front launch is decidedly simpler and more rational than either casting in-situ of the slabs over webs already launched, or construction on falsework.

Clearly, dead load during launch is larger than with a conventional composite section, axial prestressing is necessary to avoid or reduce edge cracking, and specific stresses arise because of the migration of the support reactions.

In the worst position of the support reaction with respect to the vertical stiffeners (figure 9.5), the vertical compressive stress in the webs is so high that steel plates must be sized for stability instead of strength, and their thickness increases.

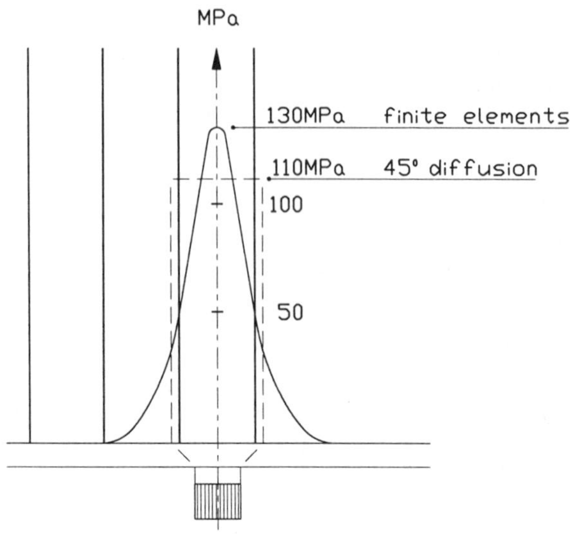

Figure 9.5 - In the worst position of the support reaction with respect to the vertical ribs, the vertical compressive stress in the steel girder can cause instability.

Figure 9.6 - Long launch supports are necessary under the steel girders of conventional composite bridges, in order to reduce the vertical compressive stress in the webs. Their articulation permits one to follow the flexural rotations of the deck support sections, and to guarantee a uniform contact pressure (author).

The same problem affects conventional composite sections, but is less prominent because of the lower dead load and greater thickness of the plates. In spite of this, long launching saddles are often necessary to reduce the average vertical compressive stress in the webs (figure 9.6). In turn, their length and the large flexural rotations of the support sections require articulation of these supports, in order to guarantee uniform contact pressure.

The launch of a stiffened plate PCS requires much heavier supports, and since mechanical articulation is no more convenient, long and stiff beams topped by stainless steel plates are placed over jacks fed by a common hydraulic plant so as to create an articulation. Even from this point of view, some types of lower node connections are more efficient than others. By placing the bottom steel flange over the concrete slab, it is possible to take advantage of the concrete thickness to improve the distribution of the support reaction. The average compressive stress in the web decreases, and the allowable stress increases thanks to the smaller depth of the plate.

When the spacing of the vertical web stiffeners is smaller than the pier cap width, movable bearings integral with the deck can be used during launch, placed every time under the next stiffener. But this requires lifting the deck to reposition them at the end of each stroke.

Additional problems arise from the alignment of the support reaction with the web axis[61]. If the concrete slab is cast over the steel flange, centering the flange and the web along a common axis as in figure 9.7 (scheme A) is not enough to center the support reaction. In fact, the pressure transferred by the neo-flon plates is distributed in the node according to the different stiffness, the one of a thick concrete slab on one side and the one of a thin steel plate on the other one. In turn, to widen the flange under the slab so that the support reaction is applied uniformly to the concrete (scheme B) creates a transverse flexural couple which is resisted practically only by the slab, the flexural stiffness of the web being much lower.

After some tests and a first small bridge (figure 9.9), the problems caused by migration of prestressing led to the adoption of folded-plate instead of stiffened plate PCS.

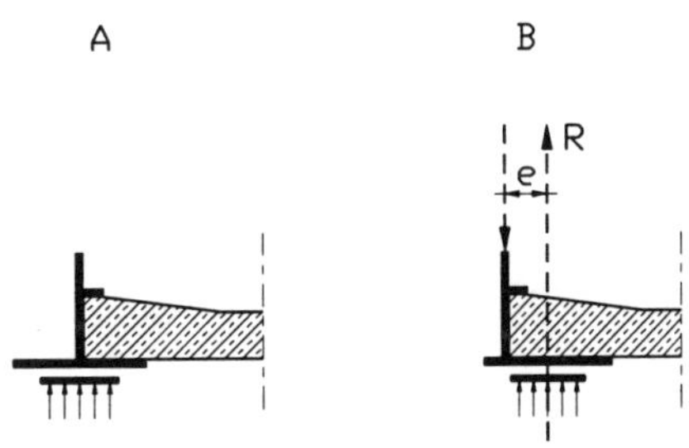

Figure 9.7 - When the concrete slab is cast over the steel flange, the contact pressure is distributed in the node according to the different stiffness, and overloads the concrete slab.

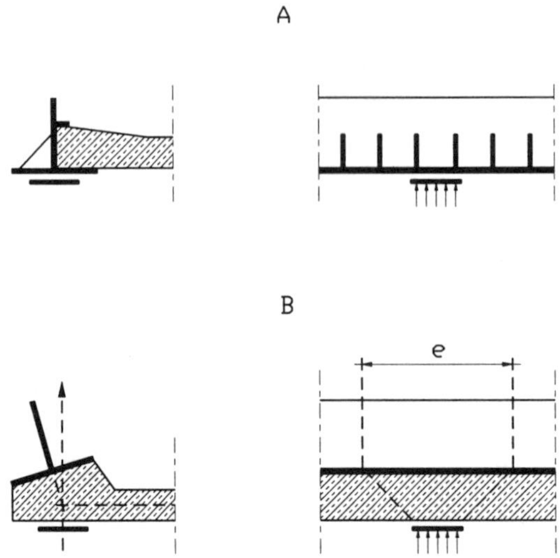

Figure 9.8 - By stiffening the cantilever portion of the steel flange it is possible to improve the transverse equilibrium of the node (scheme A). The solution is much simpler in the case of lower slab (scheme B).

Figure 9.9 - Ferté-Saint-Aubin Bridge, a stiffened plate PCS built by monolithic front launch on fixed supports (courtesy Virlogeux[61]).

Folded Plate Prestressed Composite Sections (PCS)

The worst defect of a stiffened plate PCS is the migration of prestressing toward the webs, which superimposes on the ideal situation of almost pure shear stresses.

The directional specialization of the orthotropic steel plates is commonly found in naval and aeronautical engineering, and several fuselages and wing structures have been built with undulated or folded plates. For the same reason, the concrete webs of a conventional box girder can be replaced with steel plates mechanically folded to create, at the same time, the axial deformability necessary to prevent migration of prestressing and the transverse stiffness necessary to prevent buckling without additional stiffeners (figure 9.10).

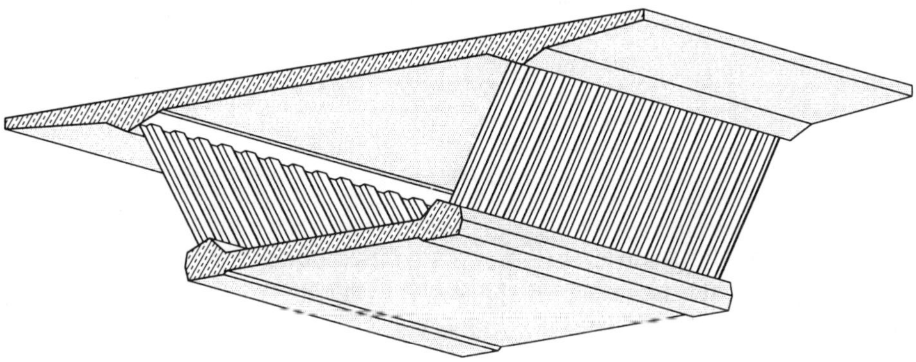

Figure 9.10 - Folded plate prestressed composite section. Steel plates are folded so as to create the axial deformability necessary to prevent migration of prestressing, and the transverse stiffness necessary to prevent buckling without additional stiffeners.

Finite element analysis, tests, and the first actual structures confirmed many advantages of the folded-plate PCS over the stiffened-plate PCS:

- the lowest axial stiffness of the webs avoids both migration of prestressing and internal stresses due to the deformation of the concrete slabs, and the stress state in the steel plates is in almost pure shear;
- consequently, the marked reduction of the plate thickness permits design according to strength criteria, while the shape of the folds can be chosen to avoid buckling;
- the transverse flexural stiffness of the folded plates does not require additional stiffeners, reduces the number of cross diaphragms, and permits savings in materials and labor generally competitive compared with the cost of folding;
- the sensitivity to premature buckling due to geometrical defects is decidedly lower. The defects of planeity must be compared with the amplitude of the folds

instead of the plate thickness, and their effects are *10* to *20* times lower. Local deformations due to welding are concentrated at the edges, which are fixed by the slabs. The internal stresses due to folding are uniform along the depth, and do not intervene in the elastic equilibrium. The steel plates act in their theoretical design dominion.

- the marked 3-D flexibility facilitates the construction of curved bridges and improves assembly conditions, avoiding any specific tolerance. In fact, the web panels are placed in contact, with an adequate overlap, and then fillet-welded on both sides.

Stress State in the Webs

The folded-plate webs are placed with vertical folds (figure 9.10). Set t_w, the plate thickness, and h_f, the transverse depth of the folds, the axial deformation modulus is:

$$E_x = E\left(\frac{t_w}{h_f}\right)^2 \qquad (9.4)$$

that is, about one thousandth of that of a stiffened plate.

This fact governs the behavior of the webs in terms both of elastic work and of instability, and since the vertical compressive stresses are generally very low as well, the state of stress in the webs is one of pure shear.

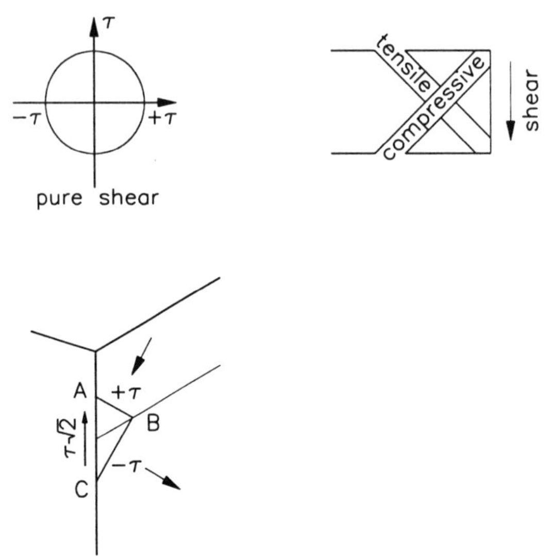

Figure 9.11 - Stress state in the folded plate girders.

In the web element of figure 9.11, setting $\sigma_x = 0$ and $\sigma_z = 0$, the two principal stresses $\sigma_1 = \tau$ and $\sigma_2 = -\tau$ are sloped at $45°$ and confirm the Mörsch lattice action of this type of plate. Along a fold, the forces, $F_{AB} = \tau t_w$ and $F_{BC} = -\tau t_w$, act on the sloped sides of the ABC element, with $AB = BC = 1$. Their longitudinal resultant is zero. The vertical resultant, $-\tau t_w \sqrt{2}$, resists the shear force, $\tau t_w \sqrt{2}$, along the AC fold. Transversely, the action of the compression fiber, which tends to push the fold outward, is balanced by the action of the tension fiber, which tends to pull it inward[125].

This behavior characterizes the central portion of the web, far from the slabs. In fact, in any type of PCS bridge, concrete shrinkage, the elastic shortening of the slabs at prestressing, and the additional deformations due to live loads and thermal gradients, generate longitudinal stresses at the web-slab interfaces. In folded-plate PCS, these stresses are transferred by the steel flanges which support the shear connectors. Therefore, their area should be as small as possible, and construction restraints on the width impose reduction in their thickness. Consequently, shear connectors must be very closely spaced to avoid buckling.

The fluctuations of these local stresses become smaller as the curvature of the plates becomes more constant. They do not seem to cause fatigue, as indicated by the rarity of problems in conventional composite sections which are subject to the same stresses.

Combined Compressive, Flexural, and Shear Stress

The application of bending moments and shear forces to a folded plate PCS mobilizes different actions of the materials. Because of the axial deformability of the webs, the bending moment is resisted only by the concrete slabs, while shear force is practically resisted only by the webs, as in conventional box girders.

The immediate consequences are the high cross-sectional efficiency (5.4) in the absence of bending-resistant material close to the centroid, and the behavior of each material in uniform instead of triangular stress patterns. Moreover, the force of a prestressing tendon placed in a slab remains concentrated in that slab without affecting the opposite slab and the webs, nor does elastic shortening induce tensile stresses in the slabs and compressive stresses in the steel plates.

Numerical analysis and data recording confirmed that the longitudinal stresses resisted by the webs are negligible.

On the contrary, they showed that the shear force resisted by the slabs, negligible in most of the deck, increases near the discontinuities of the shear force diagram. In correspondence with any discontinuity, ΔV, the related $\Delta\tau$ tend to deform the webs as a lozenge – that is, to produce a sudden angle of the deck axis equal to:

$$\Delta\gamma \cong \frac{\lambda_w}{2Gt_w H}\Delta V \qquad (9.5)$$

in which λ_w is the ratio of the plate length to its final length after folding (a_f is the longitudinal projection of the sloped strip and b_f the distance between two folds):

$$\lambda_w = \frac{2b_f}{a_f + b_f} \qquad (9.6)$$

t_w is the plate thickness, and H is the total depth of the cross section. On the contrary, the concrete slabs are subject to axial loads which do not produce any vertical deflection, and their flexural stiffness opposes sudden web rotation. Consequently, systems of orthogonal forces transfer the shear force necessary to bend the slabs until a common curvature arises at the web-slab interface (figure 9.12)[125]. These forces are not negligible, and require additional connectors in the tension zones of the interface. The slab reinforcement should be strengthened as well, since the flexural stiffness of the slabs is locally considerable and produces high tensile stresses.

Since this phenomenon is due to ΔV, it takes place near any vertical concentrated load – that is, near supports and tendon deviation points. It is not a peculiarity of folded-plate PCS, as similar effects take place in the stiffened-plate PCS and in the conventional composite sections as well, but in both cases, the greater thickness of the steel plates makes them less evident.

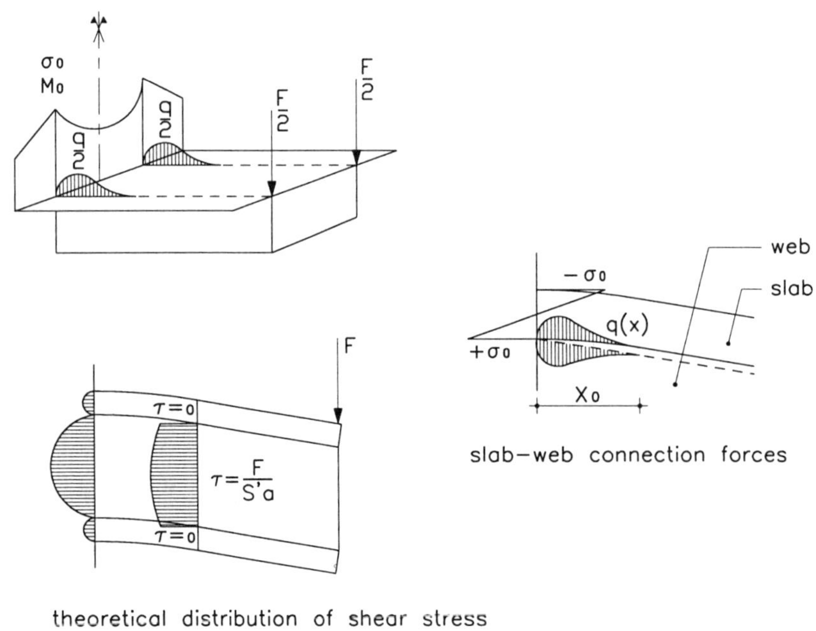

Figure 9.12 - Systems of internal forces arise near the discontinuities of the shear force diagram.

During launch, the migration of the support reactions, and the absence of deviated tendons which reduce the shear force due to dead load, generalize these problems to the whole deck. Therefore, at the end of launch, the mechanical properties of the folded plate PCS are generally already adequate for these local effects, and the deviation points of external prestressing may be spaced conventionally.

Deformability under Eccentric Loads

The application of an eccentric load to a conventional box girder produces torsion and distortion of the cross section in addition to bending moment and shear. Torsional effects are low due to the high stiffness. Distortion is rapidly absorbed by the interaction between the transverse frame action of the cross section and the in-plane flexural stiffness of its plates.

The behavior of a folded plate PCS is similar, but the effects are very different due to the different transverse and in-plane flexural strength of the individual plates.

The transverse flexural strength of the folded webs is much lower than that of concrete slabs. Transversely, the section can be schematized as hinged at the nodes, and the contribution of the cross-sectional frame action in (6.11) is completely lost. Moreover, the webs are unable to resist axial stresses – that is, in-plane bending moments, and C_d (6.11) decreases as well. Finally, shear deformability makes things worse. Consequently, distortion due to an eccentric load is mainly resisted by the opposite in-plane deflection of the concrete slabs.

Setting, F_h, as the hemisymmetrical component of the eccentric load, the shear force in each web due to distortion (with the symbols of figure 5.3) is:

$$V = F_h \frac{c}{h} \qquad (9.7)$$

It is higher than the Bredt's shear force due to a torsional moment equal to $F_h a$, and equal to the one due to a moment[56]:

$$T = F_h (a + b) \qquad (9.8)$$

Therefore, the stresses in the two slabs can be obtained by superimposing two opposite forces, equal respectively to $F_h a / h$ in the top slab and to $F_h b / h$ in the bottom one, to this torsional moment. Under these forces, the concrete slabs deflect in their plane according to their flexural stiffness and the constraint action of the diaphragms, and the edge stresses reach high values. Moreover, the hinged-rod action of the webs and the low torsional stiffness of the slabs superimpose large rotations on these transverse deflections (figure 9.13).

It shall be noticed that the shear forces in the slabs are similar (a is not very different from b) but the in-plane flexural stiffness is very different. Especially in the case of a bottom slab much narrower than the top one – that is, much less stiff, the webs are subject to wide transverse deflections. Although their stiffness is low, the lever arm of the folds is short as well, and the superficial stresses reach the same values as the tangential ones. These transverse bending moments propagate to the slabs, whose reinforcement is generally adequate, since the top slab resists the service loads as simply supported on the webs, and the bottom slab must resist the transverse launch stresses.

For all these reasons, distortion should be limited by frequent steel bracing or diaphragms which integrate the action of support and tendon deviation diaphragms. Due to the fundamental role of prestressing in reducing shear force, the spacing of the main diaphragms may be defined for optimum tendon deviation, and distortion can be reduced by means of additional devices.

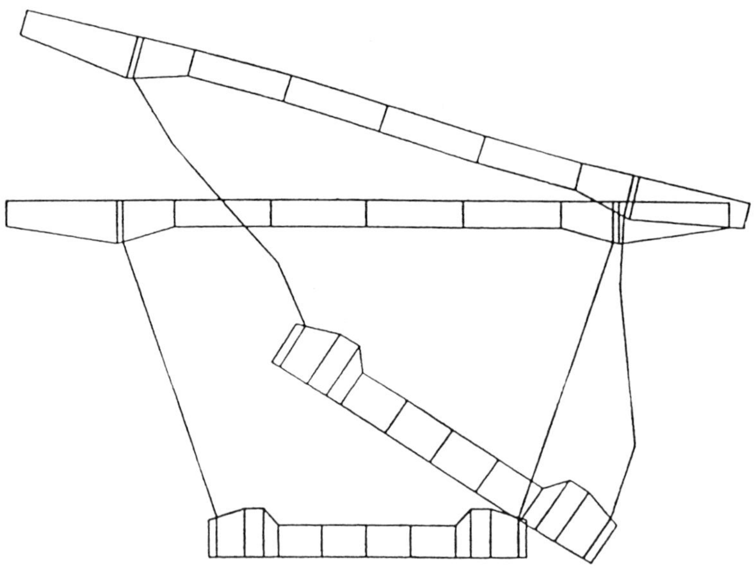

Figure 9.13 - In a folded plate PCS, distortion is mainly resisted by the in-plane deflection of the concrete slabs. Hinged-rod action of the webs, along with the low torsional stiffness of the slabs, produces large transverse rotations.

Web Instability

Unlike stiffened-plate webs, whose design (plate thickness and spacing of stiffeners) is generally based on stability instead of on strength criteria, folded webs make the most of material. In fact, their thickness can be defined based on the allowable tangential stress (prevention of corrosion apart). Then, the shape and the amplitude of the undulations is defined to prevent buckling. This allows reaching the desired safety factor with marginal costs, by lengthening the initial plate (9.6).

Folded plates show three different buckling modes, related to the ratio of the depth of the undulation to its width:

- a local buckling mode, which corresponds to the instability of a steel strip simply supported between two folds, is critical only for deep folds, and is defined by Timoshenko's formula. Setting Poisson's ratio, $\nu = 0.30$, in a panel of indefinite length, the critical shear stress is[125]:

$$\tau_{Tim} = 4{,}826\, E \left(\frac{t_w}{b_f} \right)^2 \qquad (9.9)$$

and since the web panel is hinged to elastic constraints, it can be assumed, for instance:

$$\tau_{crL} \cong 0.88 \tau_{Tim} = 4{,}247\, E \left(\frac{t_w}{b_f} \right)^2 \qquad (9.10)$$

- a general buckling mode, characterized by a regular wide wave-length deformation, progressive in development if in absence of local buckling, and similar to the one of an orthotropic plate. It is critical in webs with shallow folds. Once the transverse flexural stiffness of a unit length of web is defined (h_f is the fold depth):

$$D_x = E \frac{t_w b_f h_f^2}{3(a_f + b_f)} \qquad (9.11)$$

and the longitudinal flexural stiffness[125], reduced by (9.6):

$$D_z = \frac{E t_w^3}{12(1-\nu^2)\lambda_w} = 0.0916\, E \frac{t_w^3}{\lambda_w} \qquad (9.12)$$

the critical shear stress is defined by the Easley's formula:

$$\tau_{crG} = K \frac{D_x^{3/4} D_z^{1/4}}{t_w h_p^2} \qquad (9.13)$$

in which h_p is the net depth of the web panel. K varies from *36.0* for a hinged panel to *68.4* for a perfectly fixed one according to Easley[125], and from *32.4* for a panel fixed to elastic slabs to *60.4* for stiff unsettling slabs according to Bergfeld;
- an intermediate mode, identified in tests and characterized by a sudden irreversible steel plasticization along the folds. It is critical in webs with folds of medium depth, and is still the object of research[155].

Once the depth of the web panel and the spacing of the folds are defined, the critical tangential stress evolves as shown in figure 9.14.

The usual safety factors for composite bridges may be used, as they already take into account the possible interaction between different buckling modes, and therefore the intermediate instability as well. But once buckling has been reached, the particular post-critical dominion reduces the shear strength of the plates suddenly, and they have to assume the large deformations required to stretch the folds before being able to transfer diagonal tension. Therefore, the safety factor should be prudently high, even when the design standards allow for the post-critical behavior of stiffened plates, and therefore permit a lower factor.

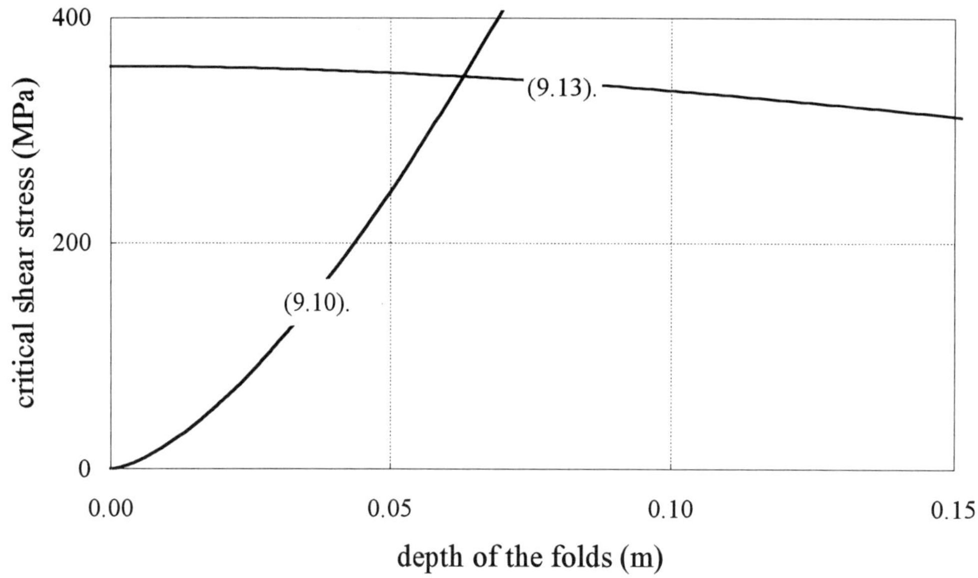

Figure 9.14 - Evolution of the critical tangential stress for $t_w = 8\ mm$, $a_f = b_f = 0.40\ m$, and constant wave-length undulation. In the case of deep folds, it is convenient to maintain the strip width constant, and this reduces the wave-length $(a_f < b_f)$.

Since the only orthogonal stresses the web can resist are vertical, the critical compressive stress is, for instance:

$$\sigma_{crL} = \frac{4\pi^2 E t_w^2}{12 - b_f^2 (1 - \nu^2)} \qquad (9.14)$$

and taking into account the shear-compressive stress interaction at the ultimate limit state, it should be, for instance:

$$\left(\frac{\sigma_u}{\sigma_{crL}}\right)^2 + \left(\frac{\tau_u}{\tau_{crL}}\right)^2 \leq 1.8 \qquad (9.15)$$

Finally, local buckling of the connection flanges will be avoided by adjusting their thickness to the width, or by reducing the spacing of shear connectors.

Launch

Folded-plate PCS are particularly suited for cantilever construction. Their construction on falsework is generally expensive, and incremental launch, although competitive, is affected by the migration of the support reactions. In this last case, the use of the bottom node scheme with the steel flange over the concrete slab is practically unavoidable, since the small thickness of the steel plates imposes a careful distribution of the concentrated actions, obtained by increasing the slab thickness locally. In turn, the undulated shape of the steel plates does not allow centering the support reaction on the web axis[61], and the slab has to resist transverse stresses as well (figure 9.15).

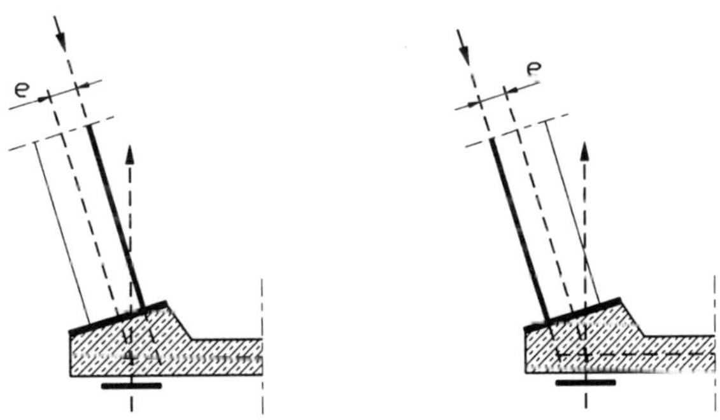

Figure 9.15 - Bottom concrete slab is thickened to better distribute the support reaction and the transverse flexural stresses due to steel plate undulation.

In spite of distribution, launch causes high local stresses in the plates[61].

Elastic stability is better than in a stiffened plate PCS (figure 9.5), since the distance between the folds is much shorter than the spacing of the vertical stiffeners. Local instability must be verified, (9.14), and if the distribution produced by the launch saddles and the concrete slab is insufficient, it is necessary to lengthen the saddles. For this reason, in the Charolles Viaduct they reached a length of *2.5 m*.

In this case, the launch saddles must be stiff, and able to follow the large rotations of the deck support sections to guarantee a uniform contact pressure. Therefore, it is convenient to place them over jacks fed by a common hydraulic plant, so as to create an articulation.

Case Studies - Folded Plate PCS Bridges

Within a research program in the early eighties, the French Direction des Routes decided to build some small folded-plate PCS bridges.

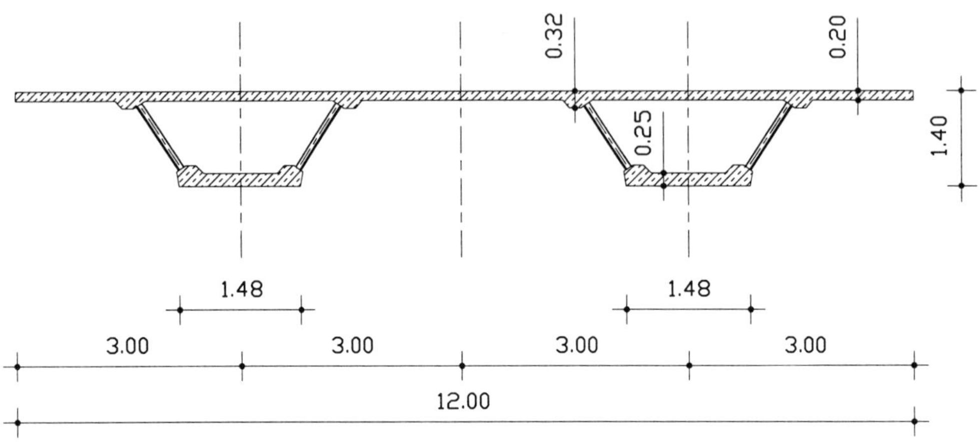

Figure 9.16 - Cross section of a preliminary project for a highway overpass.

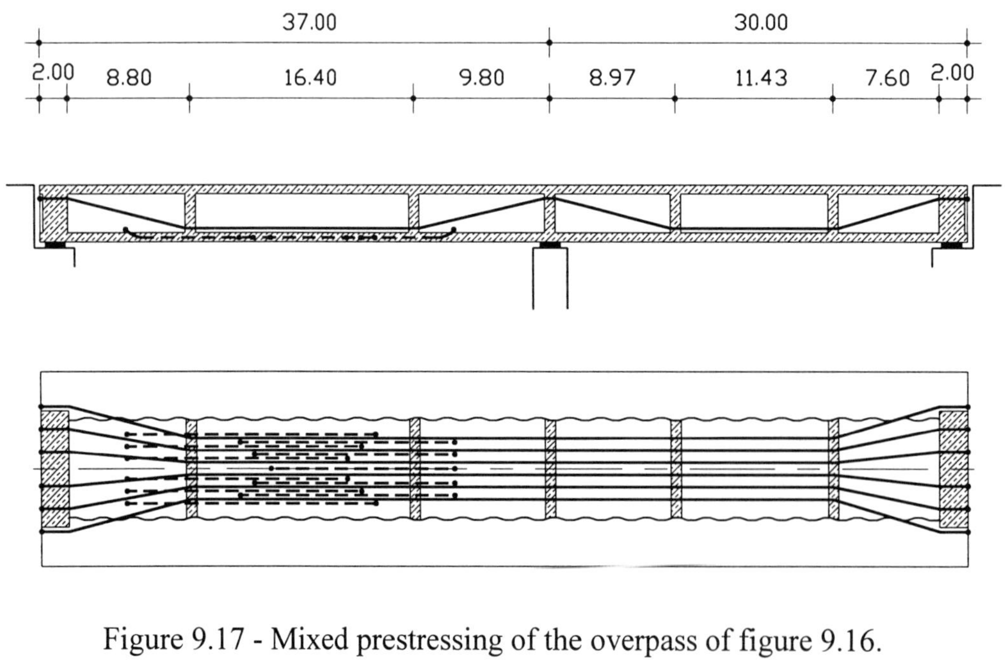

Figure 9.17 - Mixed prestressing of the overpass of figure 9.16.

A preliminary project for an overpass over the A86 Highway, never built, defined the section of figure 9.16 for the two span bridge of figure 9.17. The webs, figure 9.18, composed of *8 mm*-thick steel plates with $h_p = 1.1\ m$, were to be connected to the slabs by steel flanges *12 mm* thick. Dead load would have been $85\ kN/m$, and the average concrete thickness, *0.28 m*. Construction was foreseen on falsework, and allowing minimum tensile stresses in the bottom slab under exceptional loads, the mixed prestressing of figure 9.17 would have required concrete deviation diaphragms.

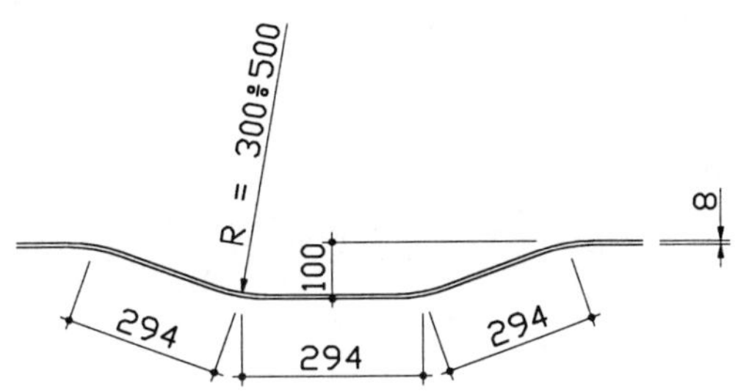

Figure 9.18 - Plate folding for the overpass of figure 9.16.

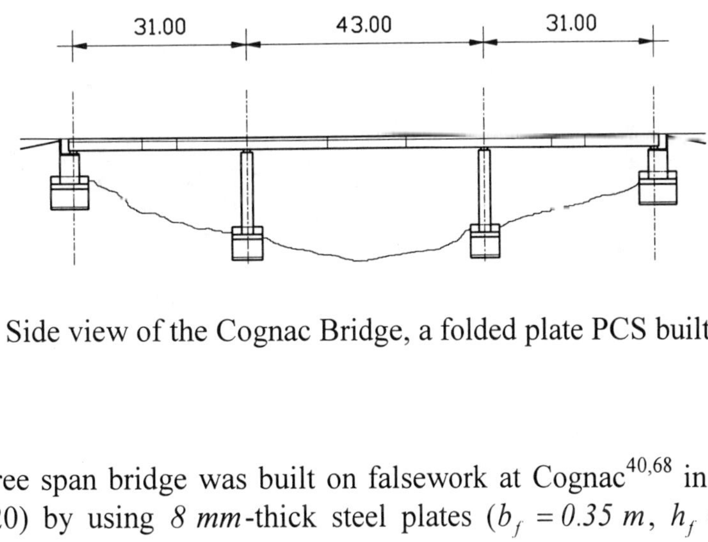

Figure 9.19 - Side view of the Cognac Bridge, a folded plate PCS built on falsework.

A first three span bridge was built on falsework at Cognac[40,68] in 1986 (figures 9.19 and 9.20) by using *8 mm*-thick steel plates ($b_f = 0.35\ m$, $h_f = 0.15\ m$, and $h_p = 1.77\ m$) with *10 mm*-thick connection flanges (figure 9.21). The tangential stress in the webs reached $73\ MPa$, and the vertical compressive stress over the supports, $170\ MPa$[56]. Dead load was $102\ kN/m$, and the average concrete thickness, *0.35 m* (figure 9.22). Totally external prestressing (eight *19T15* tendons) required support and mid-span concrete deviation diaphragms (figure 9.23).

The construction of the Charolles Bridge[30,41,42,56,129] gave researchers the opportunity to compare different types of PCS. The tender design proposed two

Figure 9.20 - Cognac Bridge (reproduced with permission, Campenon Bernard SGE Photothèque).

Figure 9.21 - Cognac Bridge. Detail of the folded plate web (courtesy Campenon Bernard SGE Photothèque[68]).

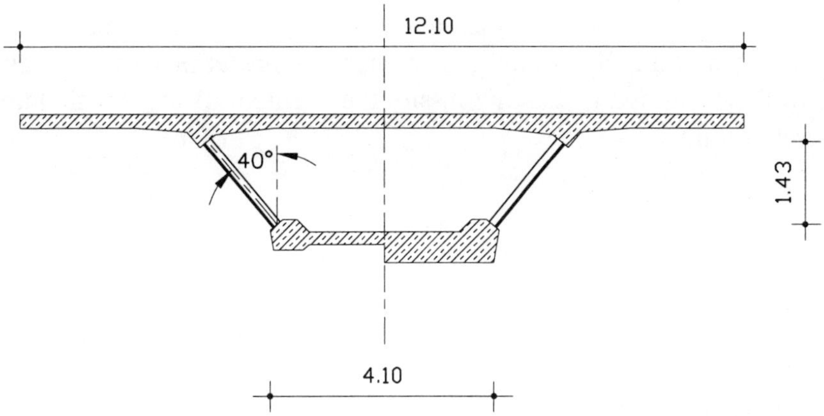

Figure 9.22 - Cross section of the Cognac Bridge.

Figure 9.23 - Internal view of the Cognac Bridge (reproduced with permission, Campenon Bernard SGE Photothèque).

different PCS (a stiffened plate PCS built by incremental launch and a truss PCS built by cantilevering), but the final choice was left to the bidders.

Among several solutions, the folded plate PCS of figure 9.24 was preferred, for a seven span continuous beam with a *53.6 m* main span and a total length of *327 m*. In this cross section, a steel pipe filled with reinforced concrete replaced the bottom concrete slab.

The contractors planned incremental launching without temporary piers. Each deck segment was built by welding a segment of prefabricated steel girder to the back end of the deck, and by casting in-situ the segment of the top slab and the steel pipe concrete filling.

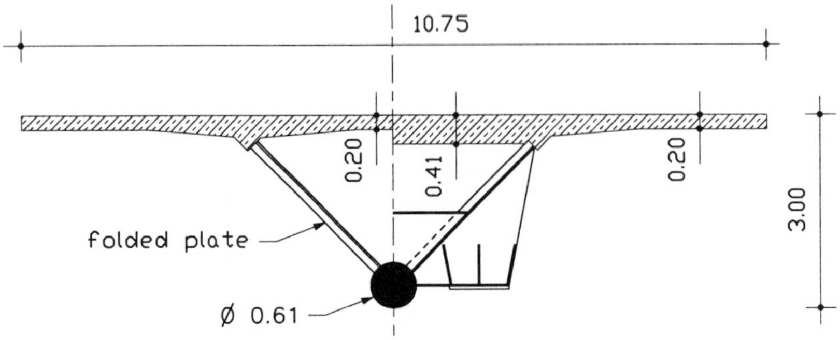

Figure 9.24 - Cross section of the Charolles Bridge.

Figure 9.25 - Detail of the pier cap and the steel support diaphragm of the deck (reproduced with permission, Campenon Bernard SGE Photothèque).

Figure 9.26 - The front span top slab was built only at the end of launch, to use this lighter deck portion as a launching nose. Its tip was equipped with a hydraulic device to recover elastic deflection (reproduced with permission, Campenon Bernard SGE Photothèque).

The triangular cross section is very stiff transversely, but its moment of inertia is very small, more similar to that of a conventional composite section than to that of a PCS.

Launch prestressing was obtained with final horizontal external tendons placed near the top slab and anchored to the support diaphragms (figure 9.25). The top slab of the front span was built only upon completion of launch, so that this lighter front deck portion could be used as a launching nose (figure 9.26). At the end of launch (figure 9.27), prestressing was completed with external tendons anchored to the end diaphragms, and deviated by steel diaphragms over the piers and steel saddles, welded to the bottom pipe, in mid-span. Their length required a final regulation of their stress, obtained by forcing the deviation diaphragm upward over the central support.

The use of a composite pipe instead of a bottom concrete slab guaranteed a good distribution of the support reaction and avoided secondary transverse stresses. But it posed new problems due to the long launch saddles, the transverse stability of the

Figure 9.27 - At the end of launch, the back end diaphragm of the deck is reaching the abutment (reproduced with permission, Campenon Bernard SGE Photothèque).

deck during launch, and the need to guarantee a perfect connection between the steel pipe and its concrete filling.

Truss Prestressed Composite Sections (PCS)

Truss schemes were widely tested in the nineteenth century, grand bridges were built, and some of them are still operating nowadays. Almost all of them were composed of only one material, steel or cast iron, since the use of concrete or masonry would have increased dead load excessively.

In these last decades, prestressing permitted the use of reinforced concrete in tension, overcoming limits in resisting tensile stresses. This is the reason for the exceptional advance of prestressed concrete, which made heavier structures advantageous thanks to the lower cost of materials.

Even though it improves the efficiency of concrete, prestressing does not succeed in making the most of it, due to the Navier's triangular orthogonal stresses which unload the material near the neutral axis – that is, the deck webs.

Recent advances in external prestressing freed designers from the need to contain tendons inside the webs, and brought prestressing, even visually, back to a system of external forces rather than of coactions. This permitted conception of new truss schemes in which different materials cooperate on the basis of their mechanical and economical properties.

The webs of a conventional box girder, subject mainly to shear stresses, can be replaced with trusses which concentrate the Mörsch lattice action in their members. Since instability penalizes compression steel more than concrete, the material composing their members does not have a decisive influence on the overall efficiency.

Similarly to the steel plate PCS, truss PCS also use prefabricated web panels placed directly inside the formwork. Their assembly is simpler, since longitudinal joints may be avoided by joining the deck segments only through the slabs.

The unusual site problems and the particular attentions required affect the span and bridge dimensions which represent the economic threshold of these innovative structural schemes. Only under certain conditions the lightening – that is, the savings in materials, can be competitive in spite of an undeniable technological complication.

Truss PCS are not very fit for front launch due to the migration of the support reactions, although the use of movable launching bearings placed under the bottom nodes of the truss could reduce launch problems. On the contrary, truss PCS are suitable for cantilevered construction (with both fully precast segments and concrete slabs cast in-situ in sliding formworks hung on the cantilevered trusses), since construction does not involve static schemes excessively different from the final one. In variable-depth cross sections, the deck lightening permits longer cantilevers and reduces the problems of longitudinal balancing.

Prestressing can be conceived in different ways. In the most intuitive scheme, concrete tension slabs are compressed by internal tendons anchored at the truss nodes, following the discrete variations of tension. Concrete tension diagonals are prestressed by internal tendons; steel tension diagonals are helped by external tendons, as the higher efficiency of prestressing steel still encourages the combined use of the two materials.

Since weak point of this scheme is the need to design the members and their connections for the whole shear force, it is convenient to introduce additional tension diagonals, which reduce the shear force resisted by the webs. For this purpose, external tendons are convenient due to both their high efficiency and the possibility of shaping them following the permanent load funicular diagram, to avoid or reduce edge decompression in the slabs. Therefore, the contribution of the individual structural components in resisting shear forces can be adjusted as in folded-plate PCS, sizing the tendons to resist permanent loads and only one half of live loads.

These additional diagonals, along with the low shear forces resisted by the webs, modify the design criteria for the trusses. The optimum truss scheme deviates from the conventional Pratt one, with tension diagonals and compression verticals, since each member is compressed by the shear fluctuation, and slab deformations dictate use of more symmetrical schemes.

As with the folded-plate PCS, the advantages over conventional box girders in terms of structural efficiency can be significant. And the interesting aspects of this type of PCS are not limited to the savings in materials. Many qualitative aspects are also involved:

- most of conventional box girders have only two webs, due to the complicated formwork and the higher dead load of multicellular sections. Consequently, in wide decks, axial stresses in the slabs are rather different from the average values given by elastic calculation. Even though truss PCS can have only two webs as well, their number can easily be increased to create a space-frame PCS, which is elastically much more reliable. The uniform transverse deflection of space-frame PCS makes the assumption of an unidimensional beam correct, whereas distortion is not a negligible phenomenon in a conventional box girder;
- in a conventional box girder, the different thicknesses of the webs and the slabs cause stress gradients at the cross sectional nodes due to differential shrinkage. In a space-frame PCS, this phenomenon is reduced by the similar thickness of the structural elements;
- load distribution occurs through several paths, and in the case of collapse of a member the stresses are redistributed toward alternative paths thanks to the higher grade of indeterminateness.

Truss Scheme and Secondary Stresses

Even a preliminary truss analysis shows that some truss types interfere more than others with slab deformations due to prestressing, main bending moments, thermal gradients, and time-dependent phenomena, and suffer a progressive amplification of secondary stresses along the deck[169].

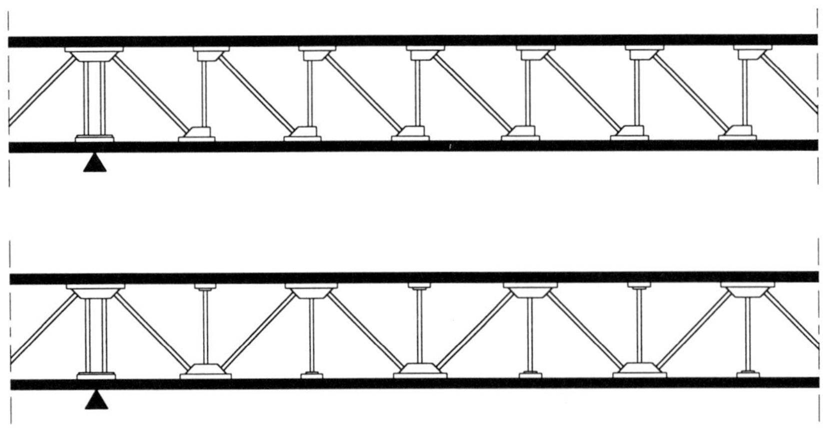

Figure 9.28 - The optimum scheme for a truss PCS deviates from the Pratt design (above), as the slab deformations suggest more symmetrical schemes, such as the Warren (under) and the X scheme.

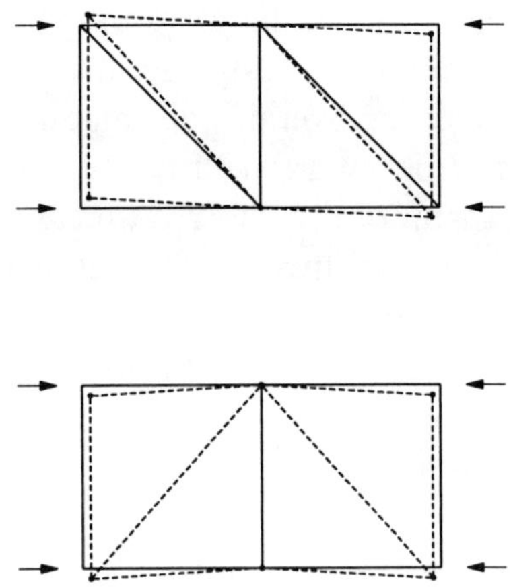

Figure 9.29 - Hinged node schemes. Due to the slab deformations, each truss panel deforms as a parallelogram. In the Warren truss, the rectilinear layout of the deck gravity axis is not compromised.

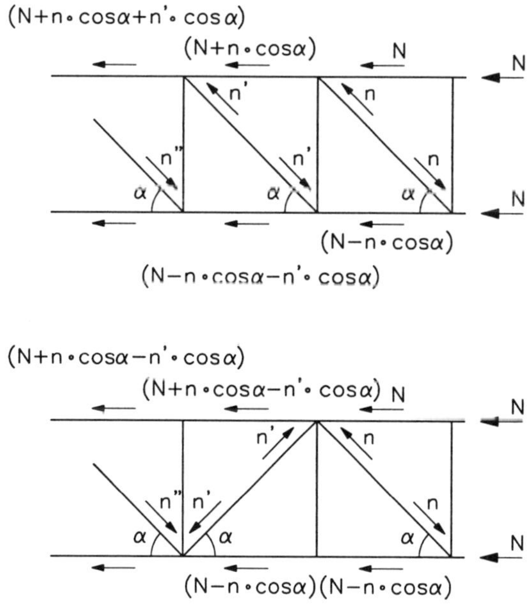

Figure 9.30 - The symmetrical truss scheme (under) is more efficient even taking into account the member flexural stiffness, since in each node the diagonal forces are balanced.

Truss behavior can be studied with the hinged node scheme (in which the axial components of prestressing forces are resisted only by the slabs, as in the folded-plate PCS), while the stresses due to nodal continuity can be evaluated later.

In this scheme, the main difference between the symmetric truss types, as in the Warren or the X design, and the non-symmetric types, as in the Pratt design (figure

9.28), is due to the consequences of the slab deformations on the gravity axis of the deck (figure 9.29). In a uniformly compressed Pratt truss, each panel deforms as a parallelogram, and this produces an overall deformation similar to the effects of a shear force. In contrast, in the Warren truss, each double panel undergoes deformations which do not compromise the rectilinear layout of the deck gravity axis, and therefore do not induce any stress accumulation due to their restraint[61].

Even taking the flexural member stiffness into account, the symmetrical truss schemes are more efficient (figure 9.30). In the Pratt truss, the longitudinal components of the diagonal forces add to each other proceeding toward the supports, and progressively overload one slab and discharge the other one. In contrast, in each node of the Warren truss the diagonal forces are balanced, and similar deformations of the two slabs reduce secondary flexural stresses[61].

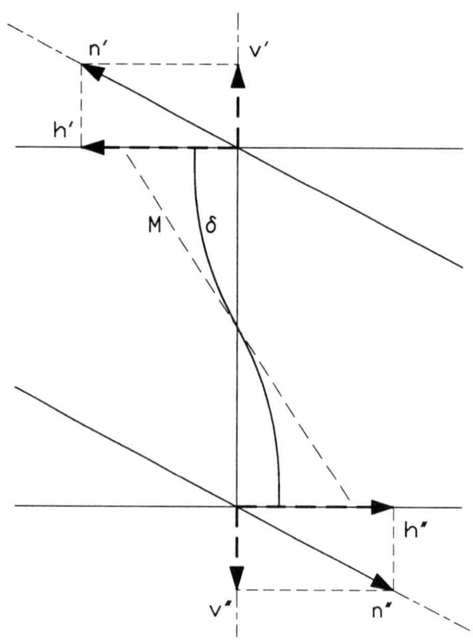

Figure 9.31 - The different slab deformations due to thermal gradients produce secondary stresses both in the diagonals and in the verticals.

Finally, the two truss types act similarly under thermal gradients[61], as the different deformations of the slabs cause secondary stresses both in the diagonals and in the vertical members (figure 9.31).

Since all these secondary stresses depend on the flexural stiffness of the members, stiffening them locally cannot solve the problcm.

High local flexural stresses arise in the members and the slabs near the discontinuities of the shear force diagram, for the same reasons already mentioned for the folded-plate PCS. The flexural stiffness of the slabs and the members which meet in the node reduces the nodal displacements of the truss, diagonals can resist shear forces only partially, and these migrate towards the slabs. In spite of the high compressive stress due to prestressing and bending moment, these local effects can decompress the slab edges. In this case, the actual secondary moments at the application points of concentrated loads are lower than the values of the elastic calculation, as if continuity is only partial.

The local phenomenon of the "shear curvature," more uniform in the folded-plate PCS and more concentrated in the truss ones, is a common characteristic of all these shear-deformable schemes, whose behavior becomes more effective the less the shear force varies or prevails over bending.

Especially in the Pratt scheme, the truss should be reinforced at the deviation points of external tendons (figure 9.32). At these points, the concentrated force transferred to the deck is inclined and eccentric with respect to the point of convergence of the truss members. This causes local flexural stresses both in the slabs and in the trusses, which add to the effects of the discontinuity of the shear force diagram.

For this reason, the truss scheme is generally reinforced on both sides – that is, two crosses are introduced. Aesthetic reasons often suggest to widen this scheme to the whole central part of the span, or to the whole span in the case of very deviated tendon layouts.

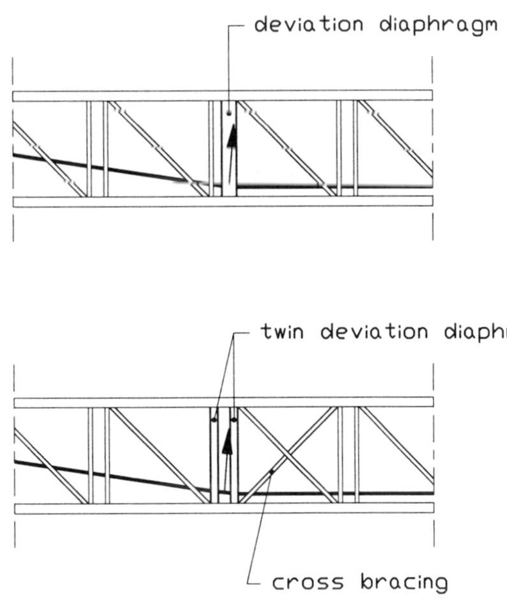

Figure 9.32 - In the Pratt scheme, the truss should be reinforced at the deviation points of external tendons.

Even from this point of view, the Warren truss is preferable. The presence of two diagonals meeting in the node improves the distribution of the concentrated load and reduces the local effects, even though it slightly increases the compressive stresses in one member.

As already mentioned, the tendon deviation reduces the shear force resisted by the webs, and with it the permanent deformation of the deck. This is an important advantage because of the difficulty in creating cambers. It reduces the secondary stresses as well, and their contribution in resisting external loads is negligible. Consequently, they are a problem instead of an advantage. Their calculation is relatively simple, although non-linear effects affect the flexural stiffness of the members. Considering a generic member, $i-j$, of length, l, and constant EI, its flexural stiffness (bending moment to apply to its end, i, to cause its unit rotation when the other end, j, is fixed) is:

$$M_i = C_1 \frac{EI}{l} \qquad (9.16)$$

$$M_j = C_2 M_i \qquad (9.17)$$

In the absence of axial force, the two coefficients, C_1 and C_2, are respectively equal to 4 and 0.5. In the presence of an axial force, N, they depend on both the force and its sign. In a compression member they are[134]:

$$C_1 = \frac{\alpha(1-2\alpha \cot g2\alpha)}{tg\alpha - \alpha} \qquad (9.18)$$

$$C_2 = \frac{2\alpha - sin2\alpha}{sin2\alpha - 2\alpha \cos 2\alpha} \qquad (9.19)$$

and in a tension member[134]:

$$C_1 = \frac{\alpha(1-2\alpha \cot gh2\alpha)}{tgh\alpha - \alpha} \qquad (9.20)$$

$$C_2 = \frac{2\alpha - sinh2\alpha}{sinh2\alpha - 2\alpha \cosh 2\alpha} \qquad (9.21)$$

with:

$$\alpha = \frac{\pi}{2}\sqrt{\frac{N}{N_E}} \qquad (9.22)$$

where N_E is the Euler's critical load of the member (8.10).

The problem is non-linear (the stiffness matrix depends on the loads) and may be solved by iteration. The correction converges rapidly, and it is usually sufficient to define the axial force distribution in the hinged node truss and to solve the frame with the member properties so obtained.

As with folded-plate PCS, particular attention should be paid to bracing systems, necessary to make the deck behavior as close as possible to the theoretical rigid cross section. Since concrete slabs guarantee the best result in their own plane, the problem is limited to the distribution of the eccentric loads toward the webs – that is, to the positioning and sizing of the cross diaphragms.

While in folded-plate PCS, diaphragms cause undesired local effects, in truss PCS the presence of the lower nodes facilitates the use of steel or concrete struts which support the top slab transversely and transfer loads directly to the bottom tension chord. These struts may be placed in the cross plane, not to interact with the truss scheme of the main girders. As an alternative, they can be arranged in 3-D schemes, and the truss PCS evolves towards the space-frame PCS.

Thus, the deck can be very wide, since the transverse truss action permits one to decrease the thickness of the top slab and prevents the relative movements of the slabs, cooperating with the tendon deviation diaphragms in limiting distortion.

The design of a space-frame PCS is rather complex, as the magnitude of the flexural and shear deflection is often similar, and the shear stiffness is much more difficult to evaluate than the moment of inertia. The superposition of several finite-element schemes may be useful. For instance, three specific schemes may be adopted[123]:

- a 3-D model of the typical span, composed of truss members and plates connecting nodes with six degrees of freedom. The boundary conditions are limited to the vertical constraint of the support nodes at the end sections. It permits one to analyze any load condition, and the stresses in the members when the load position and the continuity moments vary;
- a unidimensional model of the continuous beam, which takes shear strain into account and permits one to define the support reactions and the continuity moments to apply to the other models. Both the moment of inertia and the shear stiffness are defined with the 3-D model, by imposing that the rotations and the deflections under unit loads are the same;
- a 2-D model of the typical span, composed of a simply supported truss with the same mechanical properties as the unidimensional model and subject to the same forces and continuity moments. It permits rapid definition of the longitudinal forces and the bending moments in the slabs.

Steel Truss Members

In truss PCS, the steel members are practically as efficient as the concrete ones, since instability reduces steel's compression efficiency to the same values as those of concrete. Since cast-in-situ concrete truss members would be very expensive and a precasting plant requires a large number of members to be amortized, in the case of short bridges, or of variable depth decks, the use of steel members is preferable (figure 9.33).

In steel truss PCS, tendon deviation reduces the stresses in the members to values comparable to those due to live loads only, and commercial H-beams are generally adequate. Welded beams are more expensive, and tubular members are penalized by lower efficiency in the principal plane and the higher cost of connections.

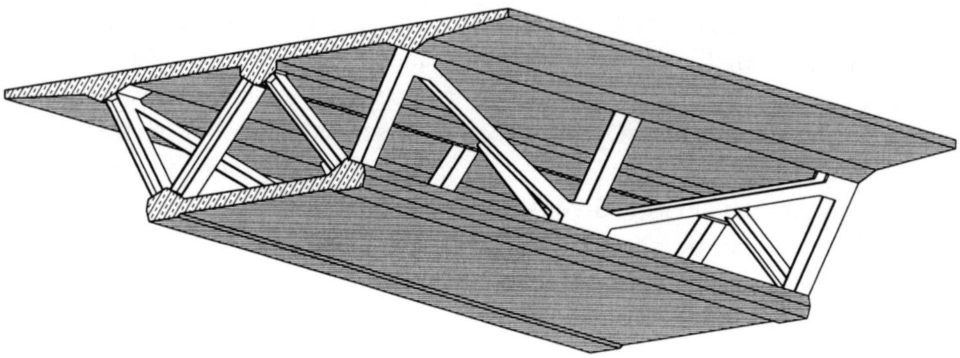

Figure 9.33 - Steel truss PCS.

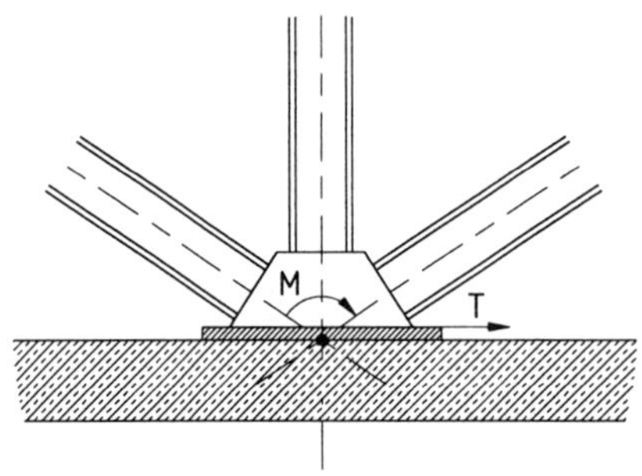

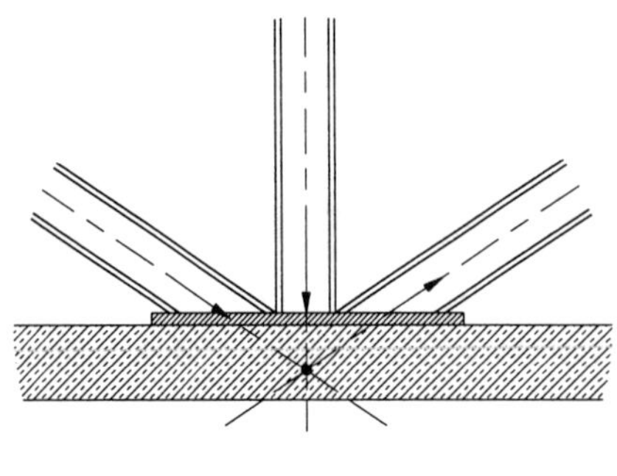

Figure 9.34 - Schemes of the node between the truss members and the concrete slab.

The nodes between truss members and the concrete slabs are rigid, since the cost of mechanical articulations would be prohibitive. A first scheme is with a steel node integral with the truss members; a second scheme is with single gusset plates, sometimes common to several members (figure 9.34).

In the second scheme, the force in the tension diagonal is transferred to the concrete slab, which deviates it towards the vertical and the compression diagonal. In contrast, in the first scheme, the components of the diagonal forces orthogonal to the contact surface balance inside the node, and only tangential stresses are transferred to the concrete slabs, as in a conventional composite section. In reality, secondary orthogonal contact stresses are required for nodal equilibrium, but the connectors[7] are much less stressed that in the previous solution. The node is a balanced system in itself, mechanically simple, built in a workshop, and homogeneous in material.

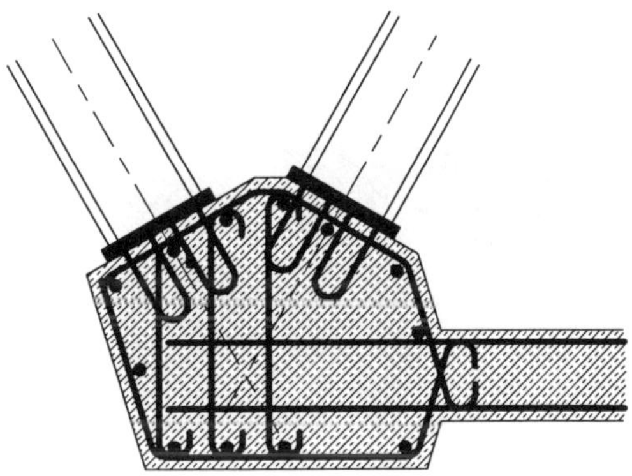

Figure 9.35 - Lower node reinforcement.

In both schemes, stress concentration imposes careful design and construction of the nodes, in terms of both connectors and slab reinforcement. Even concrete casting and vibration should be carefully studied, because of the small thickness and the dense reinforcement, especially in the lower node (figure 9.35). If needed, the steel nodes can be completed with precast concrete blocks joined by means of bolts or ordinary connectors, and equipped with emerging reinforcement to embed in the cast in-situ concrete slab. In turn, these precast blocks can be the deviation saddles of external tendons.

The use of steel nodes does not eliminate all the problems of structural schematization, since the point of geometrical convergence of the member centerlines still needs to be defined[61].

The hinged node truss scheme suggests convergence at the mean fiber of the slab (figure 9.36/B), but its position is not well defined. By taking into account the whole slab portion connected to the web, the result is different than by using only the slab area of the anchorage zone, generally stiffened and with a more internal centroid. In both cases, a flexural moment arises at the gusset plate.

To reduce this moment, the centerlines can converge at the gusset (figure 9.36/A), and a flexural moment equal to the shear force multiplied by the distance of the gusset from the slab mean fiber arises in the slab. This solution permits reduction in the dimensions and the cost of the node and avoidance of bending moments in the connection, makes the stress distribution in the gusset plate uniform, and simplifies connectors.

An intermediate solution (figure 9.36/C) generates secondary moments both in the truss members and in the slab, but these are lower than the maximum values reached in the two previous schemes.

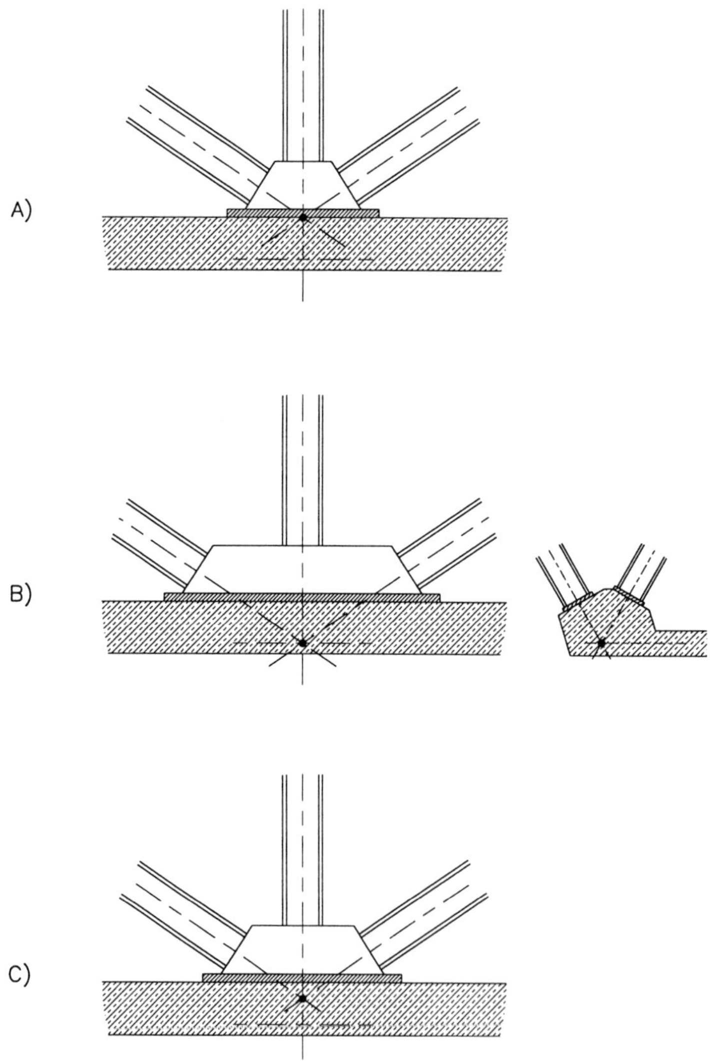

Figure 9.36 - Geometrical convergence of the member centerlines.

The choice of the point of convergence is a rather complex problem, since the related stresses add to secondary stresses due to the flexural stiffness of the members and the nodes. The optimum equilibrium between the secondary and the connection moments must be searched case by case.

Finally, whichever the point of convergence may be, it is possible to include a tendon anchorage in the node (figure 9.37), with some complication in the design of both the node and the connectors, but with an optimum mechanical continuity between the tendon and the tension diagonal[61].

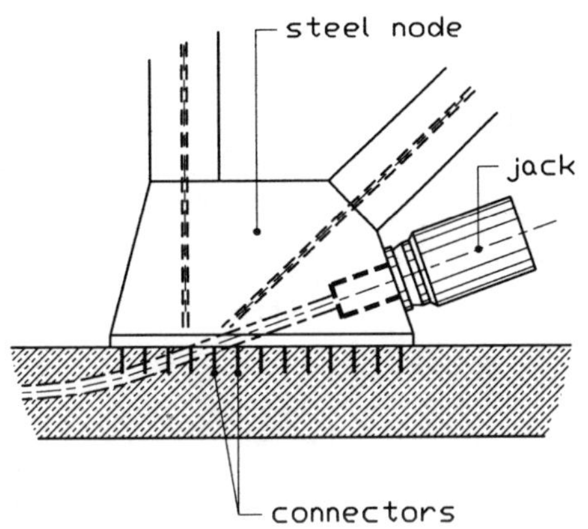

Figure 9.37 - Internal tendon anchored at the node.

The particular stresses transferred through the gusset plate require a careful evaluation of the behavior of the connectors.

In a conventional composite section, the connection stresses are mainly shear stresses. Low orthogonal stresses arise at the application points of concentrated loads, and have only a local significance. In folded-plate PCS the situation is similar, in spite of the larger number of sections affected by vertical loads, and the higher stresses due to the “shear curvature” because of the lower thickness of the steel plates. Therefore, in both cases, conventional connectors may be used.

In truss PCS, the connection problems are specific, and strictly related to the type of node.

In a connection with individual gussets, the stresses are transferred from the truss members directly to the concrete slab, and are balanced by the reinforced concrete. Connection can be made by embedding steel bolts in the concrete slab, or by introducing a local prestressing which resists the transferred tensile stresses, but in both cases the high value of the tensile force, especially at the ultimate limit state, creates some difficulties.

In prefabricated nodes, the contact force may be decomposed into a tangential force, an orthogonal force (generally low) and a secondary bending moment whose value depends on the eccentricity of the point of convergence of the member centerlines from the connection plate. Therefore, connectors must resist both tensile and shear stresses, and their action is substantially similar to that of reinforcement in a reinforced concrete cross section. Tests on hook connectors sized to resist the combined compressive and bending forces, added to conventional connectors sized to resist the shear force (figure 9.38) confirmed good performance of the connection under variable shear force, and in terms of ductility as well[61].

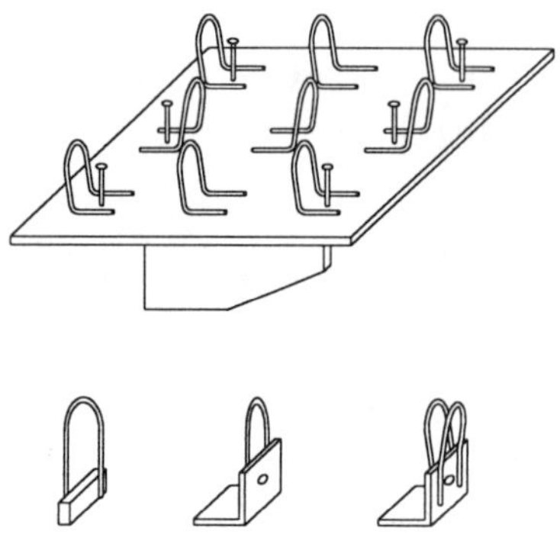

Figure 9.38 - Combined hook and conventional connectors.

Figure 9.39 - Arbois Bridge. The steel members are placed into the formwork (courtesy Virlogeux[61]).

An interesting steel truss PCS structure is the Arbois Bridge, a three span, *100 m* long continuous beam (spans of *29.85 m*, *40.40 m* and *29.85 m*) built on formwork (figure 9.40). The consumption of structural materials was $0.49\ m^3/m^2$ of concrete (*0.36 m* for the current section and *0.13 m* for diaphragms and tendon deviators), $69.3\ Kg/m^2$ of reinforcement, $11.7\ Kg/m^2$ of prestressing steel, and $57.1\ Kg/m^2$ of commercial HEB-E36-4 steel girders (figure 9.39).

Figure 9.40 - Arbois Bridge (courtesy Virlogeux-ACI[33]).

Precast Concrete Truss Members

Prestressed concrete trusses have been tested for decades in industrial ceilings. However, they are just beginning to be used in highway bridges, although they are the most representative PCS ever built.

The first concrete trusses were affected by several construction problems due to the high amount of reinforcement, the complex formworks, and the difficult concrete casting and vibration. This made these structural schemes not competitive, and the use of concrete was limited to solid webs, in spite of the higher dead load.

Since the main limitation of concrete trusses lies in the complexity of casting in-situ, greater economy may be found in schemes composed of a large number of

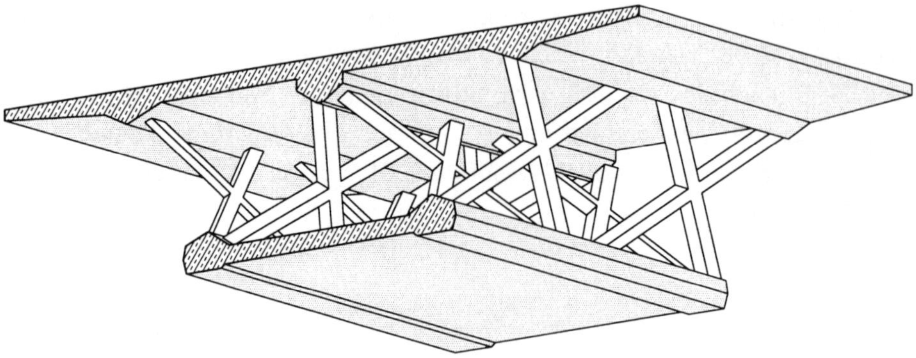

Figure 9.41 - Concrete space frame PCS.

Figure 9.42 - Bubiyan Bridge (reproduced with permission, Bouygues).

equal elements precast in simple formworks. In turn, this requires the use of a prestressing layout compatible with reduced member dimensions.

The recent advances of external prestressing encouraged the use of space-frame PCS composed of precast concrete members and cast in-situ concrete slabs (figure 9.41). Truss members are positioned in the bottom-slab formwork, and the top-slab formwork is placed over them. The cast in-situ slabs guarantee a perfect structural continuity and avoid problems of tolerances. Once the slabs have cured enough, external prestressing guarantees a perfect clamping of the precast elements.

Figure 9.43 - The Glacières Bridge was built with precast HPC segments assembled by balanced cantilevering (courtesy Bouygues[213]).

Figure 9.44 - Sylans Bridge (courtesy Bouygues[213]).

Grand examples of concrete space-frame bridges are the Bubiyan Bridge[47,118,123] in Kuwait, built with precast segments hung on a launching gantry with a typical span of *40.1 m* and a total length of *2503 m* (figure 9.42), and the Sylans and Glacières Bridges[31,32,213], built by balanced cantilevering assisted by a service launching truss with a typical span of *60 m* and a total length of *1266 m* (figures 9.43 and 9.44). A simpler concrete truss has been used in the Rip Bridge, in Australia.

As in steel truss PCS, most of the longitudinal prestressing is external, and only local tendons are contained in the slabs. External tendons are shaped so as to cover the diagram of moment capacity (partially or fully, according to the type of joint), and to resist the shear force due to dead loads plus a portion of live loads.

Further improvements may avoid truss member decompression under service loads – that is, the need to prestress them. As an alternative, tensile stresses may be allowed, and member reinforcement is sized to limit the crack width to *0.1 mm* or *0.2 mm*. This is generally sufficient to guarantee an adequate durability, since the members resist tensile stresses only under live loads, and remain compressed during most of their life.

Secondary stresses are generally low, since under permanent loads the deflection is relatively small, and with it the secondary flexural moments. They are limited to live load effects, and are resisted by member reinforcement.

On the other hand, stresses in the exterior trusses under eccentric loads are high, since a concrete space-frame is generally less stiff than a conventional box girder in longitudinal bending, and much less stiff torsionally.

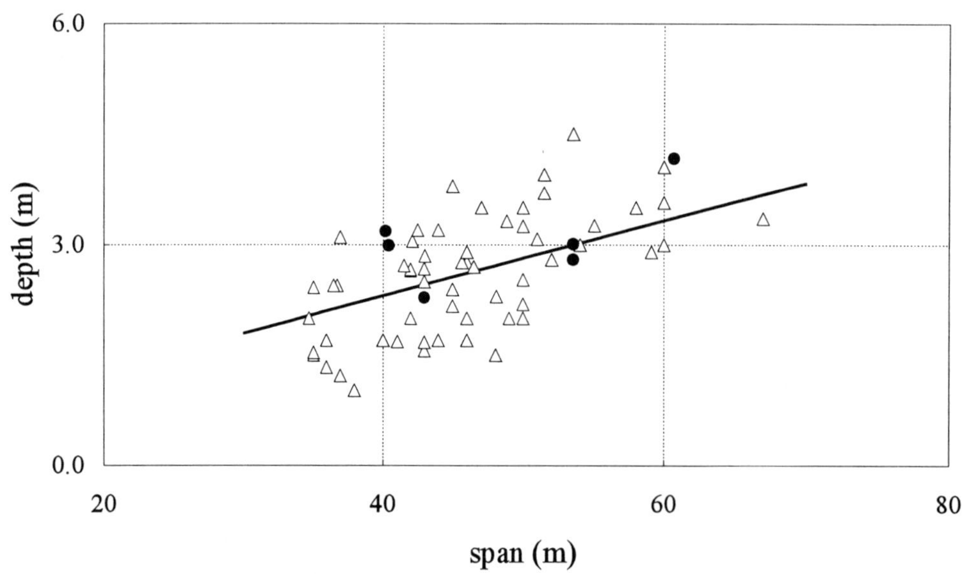

Figure 9.45 - Prestressed composite sections (PCS, black dots) are substantially as slender as the conventional box girders (CBG, white triangles).

In terms of durability, the larger surface exposed to the atmospheric agents does not require particular care, since the cross section is ventilated and does not suffer problems of condensation. Therefore, crack width limitation is often sufficient.

Conclusions

To compare the efficiency of prestressed composite sections (PCS) and conventional box girders (CBG), a database of highway bridges with a continuous beam static scheme, constant depth monocellular cross section, internal prestressing, and spans ranging from *35 m* to *100 m*, has been analyzed[169].

In the database, seventy bridges are CBG, and six bridges are PCS (no other PCS has been found in the literature). Clearly, the statistical analysis of so few PCS has a debatable significance, but by replacing the concrete webs of an actual CBG with light webs it is possible to simulate the behavior of a PCS, and the few PCS ever built can be used to verify the reliability of the data so obtained.

In figure 9.45, the relationship between depth and span confirms that the actual PCS (black dots) are substantially as slender as the CBG (white triangles) – that is, their higher efficiency was not obtained by increasing the depth excessively.

The structural efficiency, ρ_s (5.4), of the CBG was calculated, and represented with white triangles along with their regression line in figure 9.46.

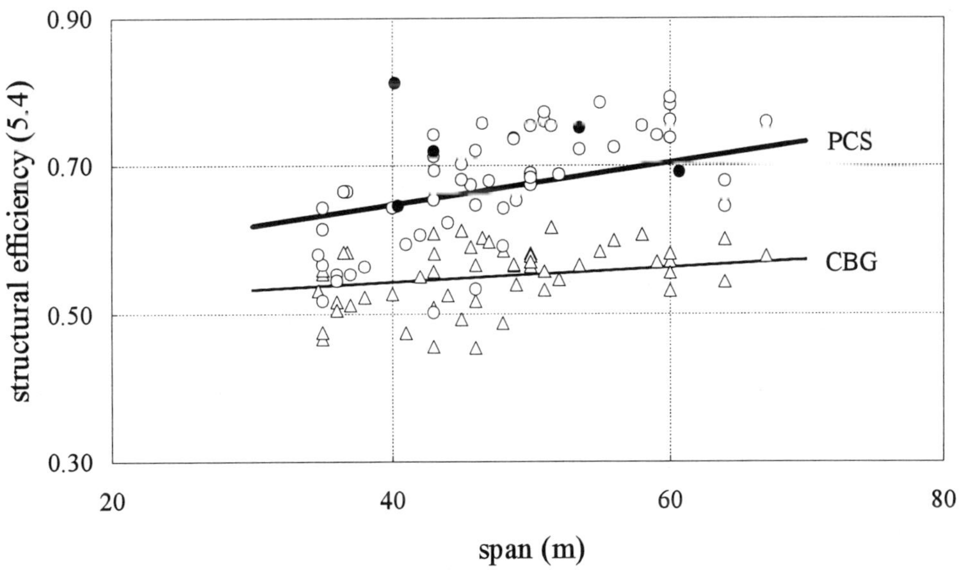

Figure 9.46 - The structural efficiency, ρ_s (5.4), of the CBG is represented with white triangles. White dots represent theoretical cross sections obtained from the actual CBG by neglecting the contribution of the webs. The few actual PCS (black dots) distributed conservatively among them and have the same static behavior.

The efficiency of each CBG was calculated a second time by neglecting the contribution of the concrete webs both in the moment of inertia and in the area. Then, the area of the slabs so obtained was increased by *8%* to take the web dead load of the actual PCS into account (steel webs weigh about *5%* of the slabs, and the concrete space-frames about *13%*), and the results are represented with white dots in figure 9.46.

The few actual PCS distribute conservatively among them, and have the same static behavior. This seems to validate the use of a common regression line, which shows an average increase of the structural efficiency of about *22%*, which improves with the span.

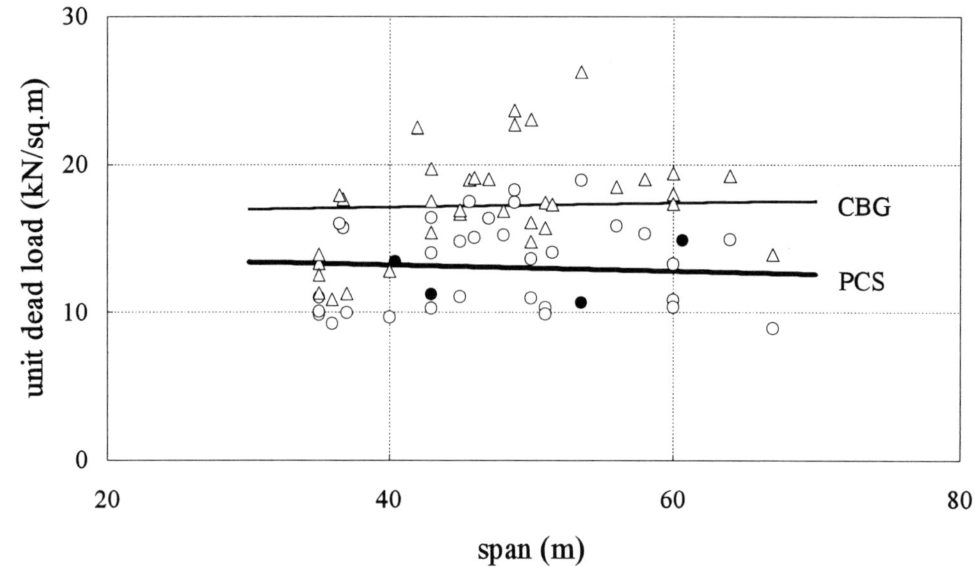

Figure 9.47 - White triangles represent the actual CBG, white dots the theoretical PCS, and black dots the actual ones.

In figure 9.47, the white triangles represent the dead load of the actual CBG (including diaphragms and for the database average width of *12.6m*) and the white dots the theoretical dead load of the PCS. Between them, the black dots of the actual PCS confirm the regression line. The saving in dead load is significant, and remains practically constant with the span.

The distributed service loads (live and superimposed dead) are represented along with their mean value and the dead load in figure 9.48, and finally the general efficiency, ρ_g (5.1), is drawn in figure 9.49. The improvement is *15%* on a *30 m* span, and *27%* on a *70 m* span.

This better performance opens new perspectives in bridge construction. The prime cost is not dramatically different, since the higher cost of the steel webs is balanced by the lower costs of concrete, reinforcement, and prestressing. Nor does

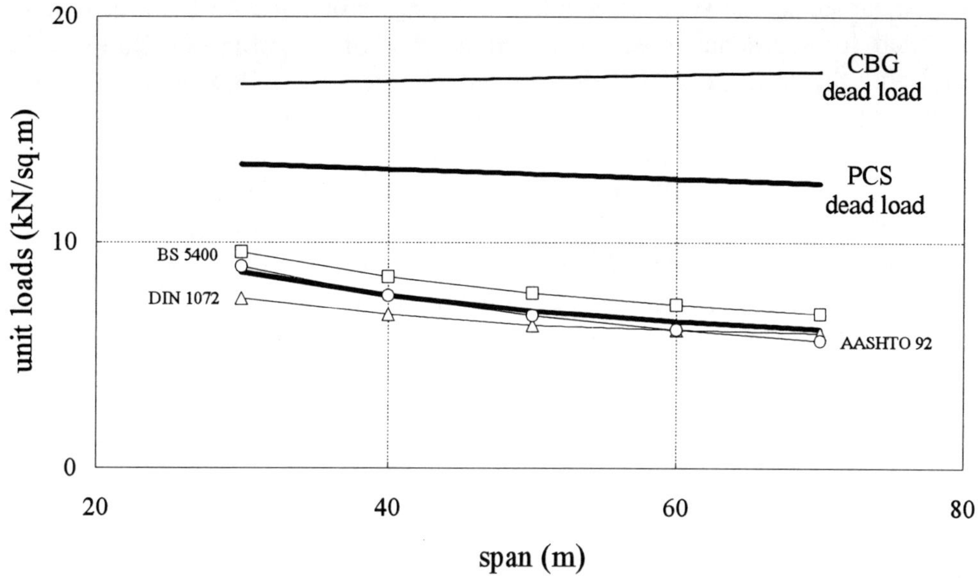

Figure 9.48 - Service loads (AASHTO Standard 92, BS 5400, and DIN 1072) are always definitely lower than dead load.

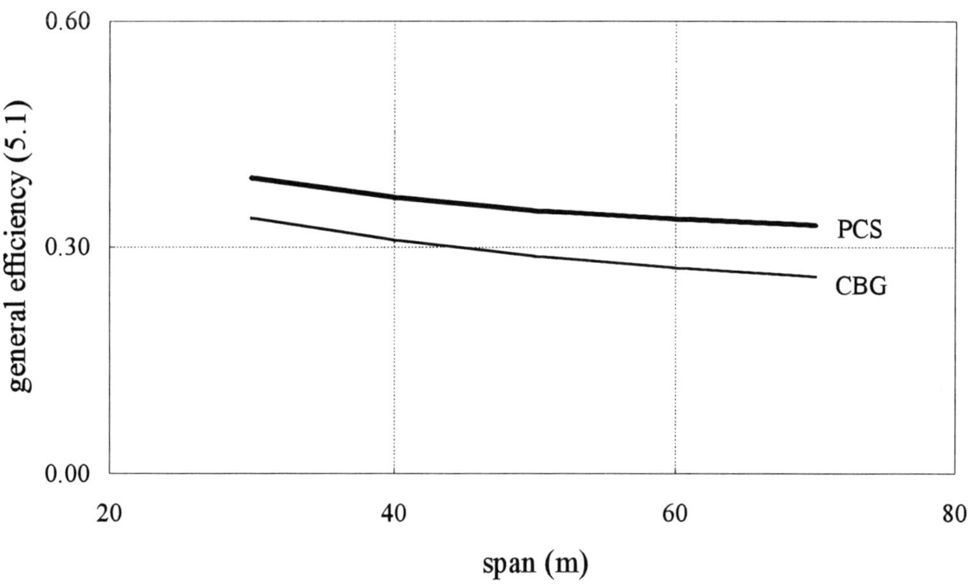

Figure 9.49 - General efficiency, ρ_g (5.1).

construction involve particular costs, as savings in equipment balance the risks of inexperience.

Of course, the economic thresholds must be verified, and construction methods must be adapted to the specific requirements. In reality, truss PCS are particularly well suited for segmental precasting, but the use of movable launching bearings, integral with the deck, permits their front launch. Steel plate PCS are more suited for launching. The Charolles Bridge was built by incremental launch, and the Ferté-Saint-Aubin Bridge by monolithic launch, both with encouraging results.

The horizons of the PCS are even wider for variable depth bridges[190] built by rotation, and for cable-stayed bridges launched over temporary piers and then stayed. Beyond certain spans, the higher efficiency of the PCS opens new perspectives in launched bridges, and the possibility of overcoming longer spans may increase the upper limit of economics of the constant depth bridges.

Chapter 10

MONOLITHIC LAUNCH

INTRODUCTION

The balanced casting construction method permits one to build two-span "hammer" bridges and three-span continuous beams with a central span of length double the side ones (figure 1.13). Moreover, it allows the building of slender variable-depth bridges with only one apparent central span, the counterweight spans being hidden in the embankments (figure 10.1).

Its most frequent adoption is for the three-span continuous beam. In this structural scheme, long spans are necessary to amortize the cost of the equipment needed for conventional free cantilevering construction, and alternative methods are soon competitive.

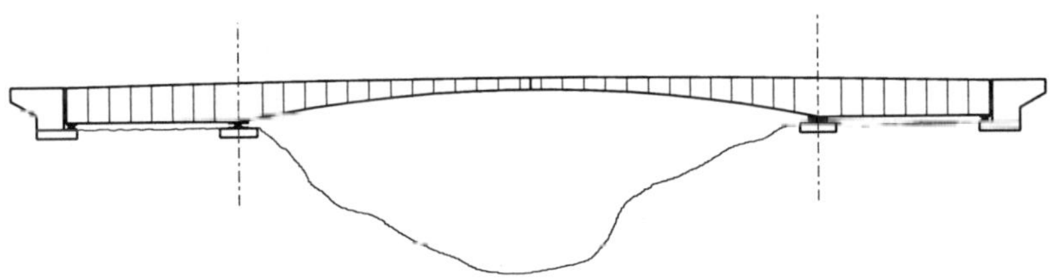

Figure 10.1 - Variable depth bridge with a slender central span and counterweight spans hidden in the embankments.

The technological limits are very few.

With launch by rotation it is not possible to build two parallel bridges, as the presence of the first one prevents the construction of the second one. Two yards parallel to the obstacle to overpass are necessary, and the interior piers must be adequately distant from the obstacle – that is, the main central span increases.

On the other hand, with symmetrical launch, two yards as long as half the deck are necessary on both banks, but parallel bridges are possible, and the interior piers can be closer to the obstacle.

There are several advantages:

- the construction and the launch requires simple and inexpensive equipment;
- the construction method is perfectly suited to several structural types, such as cable-stayed bridges, constant or variable depth bridges, box girders or open sections, conventional box girders or prestressed composite sections;

- design geometry is totally free;
- regular traffic on the obstacle to overpass (in case of roadways, railways, and channels) is absolutely safe during each construction stage;
- the bridge is built under optimum conditions, since concrete casting occurs in a formwork supported on the ground as in incremental launching;

Consequently, it is possible to obtain high quality results, simple site organization, and savings at the same time.

The main span often ranges from *30 m* to *60 m*, but much longer spans have been reached. In the Danube Bridge at Woerth[74], for instance, the three spans are *108 m*, *168 m*, and *128 m* long.

Short highway overpasses apart, the deck is generally a monocellular box girder. Compared with a ribbed slab, this scheme offers higher structural efficiency, a lower position of centroid (which improves the absorption of negative moments, critical in these construction techniques), a higher torsional stiffness, easier inspections, and the same consumption of materials.

The deck depth may be constant or variable. Constant depth bridges are generally more simple to build, but convenient in a limited range of spans, up to *60* to *70 m*, due to the greater amount of prestressing which reduces their competitivity. Beyond *70 m*, variable depth bridges are generally less expensive and aesthetically more pleasing.

The law of variation of the depth is conceived with the same criteria as for free cantilevering. The most frequent is the parabolic law, which better evokes ancient bridges, but linear, parabolic, or cubic laws near the internal supports along with constant depth mid-span sections yield valid results as well.

However, the construction method is rather different from free cantilevering, since the formwork is supported on the ground instead of hung from the deck. Moreover, the deck is often supported during its construction, and reaches the cantilever configuration immediately before launch.

In most of the constant depth bridges, and always in the variable depth ones, each half-deck is built symmetrically from a central segment. In the case of front launch, this segment is placed at a distance from the internal pier equal to the length of launch. In the case of rotation, it is placed directly over the interior pier, which acts as a pivot during launch.

In front launch, the launch bearings may be placed following two different schemes: fixed supports, and movable supports.

Fixed supports keep the deck statically determinate during launch, but the bending moment diagram changes due to the varying length of the cantilevers. This requires a temporary launch prestressing, removed at the end of launch. This scheme is generally adopted in the case of foundations so subject to settlement as to require piles under the temporary bearings.

On the other hand, movable supports are integral with the deck during launch. They require stiff foundation beams, oversized with respect to the loads transferred by the formwork and the deck support struts, but avoid the need for temporary deck prestressing.

Front launch with fixed bearings is similar to incremental launch, and is applicable only for constant depth bridges. On the other hand, both in front launch with movable bearings and in rotation, the deck is always supported at the central segment, which in the first case moves toward the internal pier, and in the second case is already placed over this. The deck is supported at a second balancing point as well, but neither support shifts along the deck. Therefore, cylindrical geometry is no longer necessary, and the deck depth can vary.

Based on these general schemes, several options adapt the bridge design to the nature of the obstacle to overpass and to geometrical restraints. The safety factor for tilting about the front support may be corrected with appropriate combinations of different materials, such as conventional and lightweight concrete, or steel and concrete, or by creating temporary or permanent back counterweights.

Whatever the launch scheme may be, the lack of balance of the cantilever masses and the accidental loads require a careful longitudinal and transverse balancing of the half-bridge. This is aimed at reaching an adequate safety factor to lifting at the less loaded support and, in the case of rotation with the pivot fixed to the deck, also at avoiding excessive stresses in this device.

To guarantee an adequate safety factor at overturning, several load combinations should be verified, as in the ordinary free cantilevering construction[28]:

$$\begin{cases} 0.9\,(Q_{max}+Q_{min}) + 1.25\,P_{c+w} \\ 1.1\,(Q_{max}+Q_{min}) + 1.25\,P_{c+w} \\ 0.9\,(Q_{max}+Q_{min}) + P_a + P_{c+w} \\ 1.1\,(Q_{max}+Q_{min}) + P_a + P_{c+w} \end{cases} \tag{10.1}$$

in which:

- Q_{max} and Q_{min} are the dead loads of the two cantilevers (the first increased by *2%* with respect to the average design value, and the second reduced by the same quantity, to take geometrical tolerances into account). They are chosen so as to obtain the most unfavorable condition in relation to the considered action;
- P_{c+w} is a load that takes into account the construction loads and the relieving wind action. It can be evaluated by placing a uniform load of $0.20\ kN/m^2$ on one of the two cantilevers (always chosen in the less favorable way) and a concentrated load of $(50+5B)\ kN$ at the tip of the cantilever, where B is the top slab width in meters;
- P_a is the load due to the fall of the closure joint formwork hung on the tip of the cantilever. It can be evaluated with a dynamic amplification of its self-weight equal to *2* – that is, with an upward force equal to the formwork dead load.

The forces which participate in the longitudinal stability of the deck are very high, and their moment arms very long. Therefore, the fluctuations of the eccentricity of the resultant vertical force from the central support axis are very wide. It is not convenient to rely only on the strength of the support device to resist them, and the use of a second movable support is generally advisable.

Therefore, during launch the deck is no longer a double balanced cantilever, but a simply supported beam with a long front cantilever. The main support is placed at its center and resists most of the vertical load. The balancing support is placed at the back end, and resists a low vertical force, intentionally created or controlled according to the extreme stability conditions. In the case of rotation, the back support slides on an arc of circle whose center is the pivot – that is, the interior pier. In the case of front launch, it slides along the launch rails of the deck.

In the final static scheme of the deck, the optimum side span, l_s, is:

$$\frac{l_s}{l} = 0.58 \text{ to } 0.60 \qquad (10.2)$$

for variable depth bridges and:

$$\frac{l_s}{l} = 0.68 \text{ to } 0.70 \qquad (10.3)$$

for constant depth structures, where l is the central span[28]. Since the end supports generally have a diaphragm while the mid-span cross section is much lighter, the longitudinal balance is often satisfied by a general design aimed at avoiding negative support reactions at the abutments under live loads. Otherwise, to achieve a balanced launch and avoid decompression of the end supports, the relative self-weight of the side spans may be increased using lightweight concrete in the main span and conventional concrete in the side ones, or by filling the box cell with concrete or sand, or by oversizing its cross section. As an alternative, external counterweights may be used during launch and removed once the bridge has been closed by completing the mid-span segment. In this case, lifting the end supports and/or lowering the internal ones may modify the support reactions, although creep of the concrete progressively reduces the effects.

At the end of launch, the deck is generally hinged at the piers, and conventional bearings are introduced after the deck lifting and the removal of the launch bearings.

As already mentioned, with these construction methods, no restraints affect the design of variable depth bridges, and the cross section may be dimensioned according to the statistical formulae for bridges built by free cantilevering.

The present trend is towards monocellular box girders even for the widest decks, and, with B the deck width and a the distance between the web axes at the level of the upper nodes, a correct distribution of the transverse flexural stresses may be obtained with[28]:

$$a \cong \frac{B}{2} \qquad (10.4)$$

The position of the drainage pipes should be verified based on the internal tendon layout, and the width of the side cantilever should not be such to shade the whole web, since the aesthetic results are rarely the best.

The webs may be inclined, especially in wide decks, to improve the aesthetic aspect and reduce the transverse width of the piers. Their vertical slope should not be excessive, to avoid casting problems, and it may be convenient not to exceed $12°$.

With H_s the deck depth at supports and H_m at mid-span, in a highway bridge correct dimensions may be obtained with[28]:

$$\frac{l}{H_s} = 16 + 0.25\left(\frac{l}{100}\right)^4 \qquad (10.5)$$

$$\frac{l}{H_m} = 0.21l + 25 - 7.5\left(\frac{l}{50}\right)^{-3} \qquad (10.6)$$

where l is the central span. In any case, H_m should not be lower than $2.3\ m$, to facilitate working inside the box cell.

In case of constant depth bridges, the slenderness of the main span may be[28]:

$$\frac{H}{l} = 20 \text{ to } 22 \qquad (10.7)$$

The top slab is sized to resist transverse bending and concentrated loads. With t_t its minimum thickness, in the absence of transverse ribs[28]:

$$\frac{a}{t_t} = 25 \text{ to } 30 \qquad (10.8)$$

The web thickness, t_w, does not depend on the width of the tendon anchorages, and may be pre-sized with[28]:

$$t_w = 0.26 + \frac{l}{500} \qquad (10.9)$$

which is statistically correct for $0 \le l \le 170$ and width up to $15\ m$. Beyond this value, it should be slightly increased.

In the case of external continuity prestressing, when the internal cantilever tendons are anchored at the cross sectional upper nodes the web thickness may be reduced, depending only on the shear strength and the case of casting. It may be pre-sized with[28]:

$$t_w = \frac{l}{275} + 1.25\frac{B}{l} - 0.125 \qquad (10.10)$$

The web thickness is generally constant along the depth, although with cantilever tendons anchored at the upper nodes, the shear stress diagram can suggest a variable thickness.

The thickness of the bottom slab depends on the transverse flexural stresses. In the case of internal continuity prestressing, some tendons will be arranged in the bottom slab, and to avoid excessive deformations of the cross section, its minimum thickness is[28]:

$$t_b \geq Max\left\{0.18; 3\phi; \frac{t_w}{3}\right\} \qquad (10.11)$$

where ϕ is the sheath diameter. The bottom slab must resist the longitudinal flexural compressive stress at piers, where its thickness may be about *2.5* times the mid-span thickness[28].

Finally, for a central span $60 \leq l \leq 120$, the average concrete consumption (equivalent thickness) may be estimated with[28]:

$$h_c = 0.4 + 0.0035\, l \qquad (10.12)$$

In free cantilevering[28], the average consumption of reinforcement is of 110 to $130\ Kg/m^3$, and of prestressing of 45 to $50\ Kg/m^3$. Since sliding formworks need not be supported, these values may be slightly reduced.

YARD ORGANIZATION

Generally, the yard is subjected to many constraints, since these construction methods are mainly used to span over navigable channels, railways, and highways, which all have specific safety clearances.

The arrangement of the yard with respect to the obstacle depends on the launch technique. In front launch it is transverse, along the final layout of the bridge, while in rotation it is parallel. In both cases, the half-bridge is launched only once finished, and the yard must contain it in its entirety.

In the case of launch by rotation, adequate safety clearances should be kept from the obstacle, to avoid the falling of materials. In the case of bridges across channels, this clearance is required to avoid building the bridge over the banks, generally subject to settlement (figure 10.2).

The equipment is used only once, and therefore is rather rudimentary. Temporary foundations are necessary only in the case of front launch with movable bearings, and are limited to a central base and a pair of longitudinal beams which support the sliding formworks.

Figure 10.2 - La Flèche Bridge. When overpassing a channel by rotation, the interior piers should be moved away from the banks, subject to settlement (courtesy Virlogeux[104]).

Constant-depth bridges may be composed of box girders or ribbed slabs. Box girders are built with a bottom formwork placed directly over the foundation beams, while ribbed slabs are cast in one stage only in an even simpler formwork, composed of a central tunnel and two sides, as shown in figure 3.7. In both cases, the webs are match-cast against the foundation beams, and no dismantling device is necessary. The segmental casting sequence may be symmetrical, moving from the central support segment towards the ends (figure 10.3) or in longitudinal progression. In both cases, prestressing is introduced only once the whole half-bridge has been built.

In the more efficient variable-depth bridges, the deck is generally a box girder, and the bottom formwork, supported on modular scaffolding, rises progressively as shown in figures 10.4 and 10.5. The construction sequence is necessarily symmetrical.

Figure 10.3 - The La Flèche Bridge was built symmetrically, moving from the central support segment towards the ends (courtesy Virlogeux[104]).

The symmetrical cantilever static scheme may be used during construction, but it slows down the operations due to the curing time needed to introduce prestressing. Therefore, it may be convenient to introduce distributed struts at the segment dismantling, and to apply prestressing directly to the finished structure, as in figure 1.17. To reach a correct final geometry of the bridge, it is necessary to correct its profile during construction, taking into account the immediate and time-dependent deformations due to dead load and prestressing at the strut removal. Time-dependent phenomena are less important than in free cantilevering, thanks to the longer curing time both at prestressing and at the mid-span closure.

The deck segments are generally *4* to *7 m* long – that is, longer than in free cantilevering[28], since formworks are supported on the ground with no restraints of weight and deformability. The mid-span closure joint is about *1 m* long.

PRESTRESSING

With these construction methods, each half-bridge is prestressed based on the same criteria used for free cantilevering, but without the need to anchor tendons at each joint and to tension them as construction proceeds. Prestressing may be conventional, with all internal tendons, or mixed, as shown in figure 10.6.

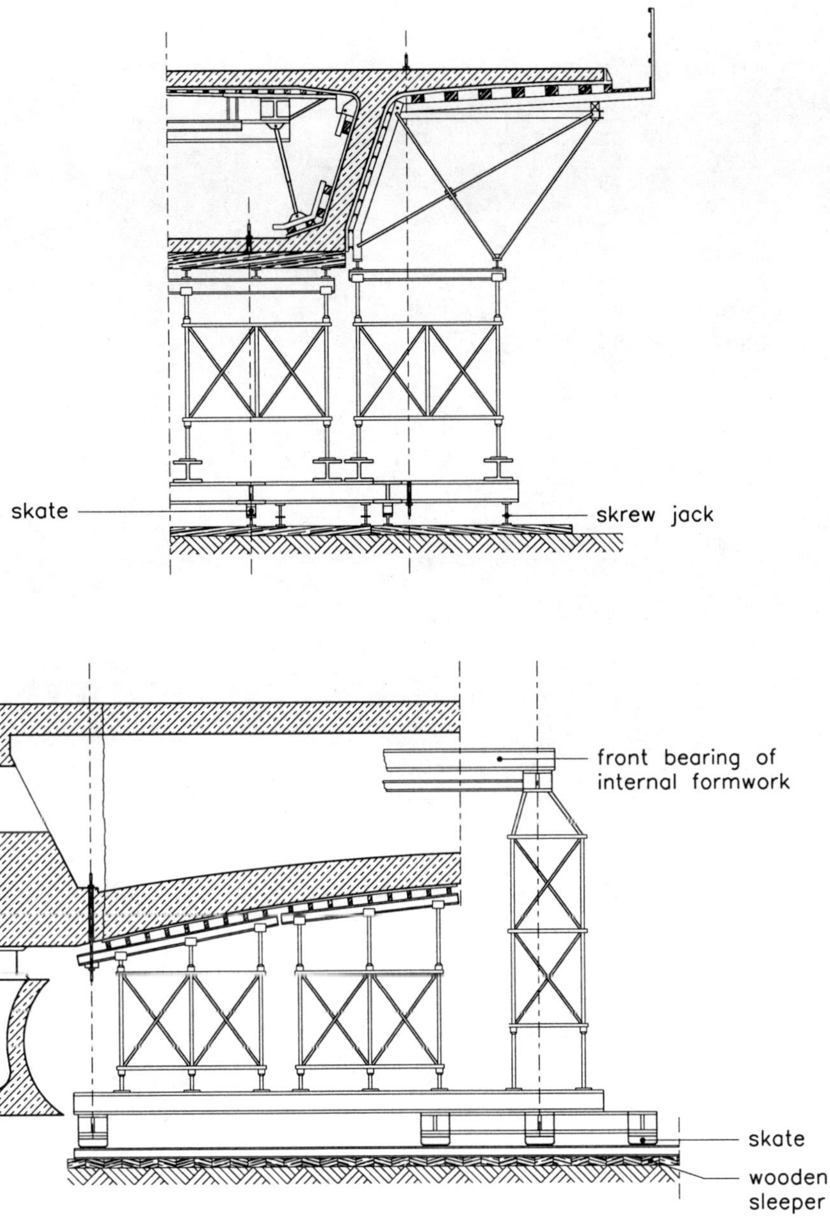

Figure 10.4 - Adjustable formwork placed on a movable platform.

The main cantilever tendons are arranged in the top slab and sized for the negative moment due to dead load in launch configuration. They may be:

- horizontal, with anchorages placed near the upper nodes of the cross section, as in figure 10.6. In constant depth bridges, the non-reduction of the shear force may require vertical prestressing of the web zones near supports, which is expensive, cumbersome, and of doubtful effectiveness;

Figure 10.5 - Detail of the first cantilever segment of the La Flèche Bridge (courtesy Virlogeux[104]).

– lowered in the webs (allowing the positioning of the stressing jacks) with a better shear efficiency, but with sharp plan and profile curvatures that produce losses of prestress.

Continuity tendons may be internal or external. External tendons may be anchored at either the support or the end diaphragms, to make the positioning of the stressing jacks easier, and to reduce the number of anchorages. Due to the presence of cantilever tendons, continuity tendons need to resist only thermal effects, stress redistribution due to time-dependent phenomena, and service loads. If sized for the mid-span section, continuity tendons tend to be oversized at the support sections. The use of some temporary cantilever tendons to relieve at the end of launch may be useful. Otherwise, continuity tendons may be sized only to increase the moment capacity at supports. In this case the mid-span closure prestressing, originally aimed at joining the two half-bridges and resisting the effects of the moment redistribution due to creep, may be sized also for a portion of the service stresses. It is composed of short and large local tendons, which allow acceleration of the completion of the bridge and guarantee early resistance to thermal effects.

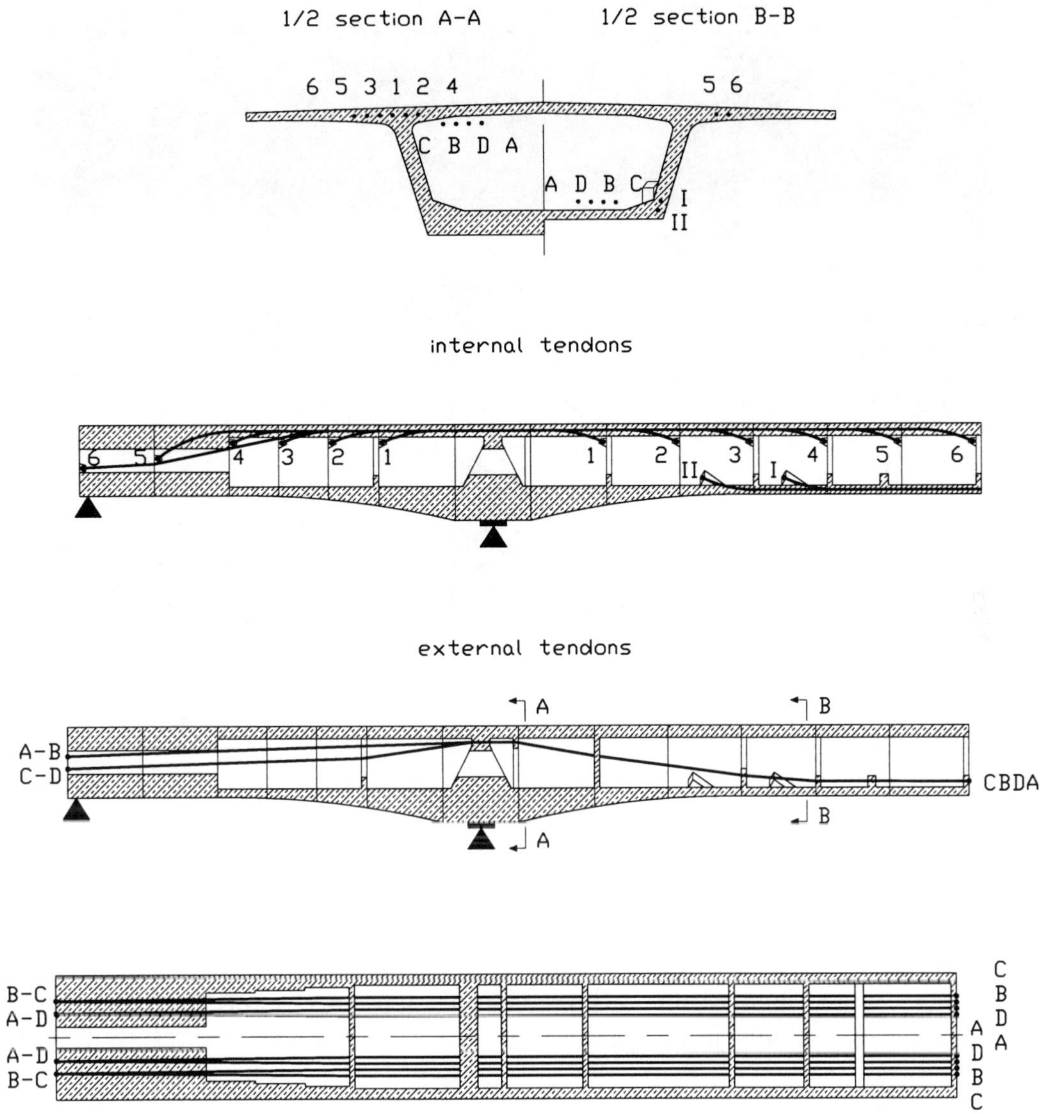

Figure 10.6 - Mixed prestressing with internal cantilever and closure tendons and external continuity tendons anchored at the end diaphragms.

In constant-depth bridges, the optimum mid-span position of the continuity tendons is inside the bottom slab. The anchorages may be placed at the cross sectional lower nodes, or may be raised in the webs, or placed directly over the deck extrados, even though this increases the risk of infiltration. On the other hand, in variable-depth bridges, continuity tendons should be placed in the webs, since in the bottom slab their lifting action is locally resisted by the curved layout of the slab, and this tends to crack the latter.

Figure 10.7 - External continuity tendons (courtesy Virlogeux[104]).

Even the use of external continuity tendons (figure 10.7) has some particularities. Using two deviation diaphragms in each span, as in figure 10.8/a, they would fall in low depth deck zones, and the tendon slope – that is, the shear force reduction, would be low. Introducing two additional deviation diaphragms, they would fall where the higher depth of the deck generates a high shear strength, while the webs need to be helped only in the intermediate zones of the span. Therefore, it may be advantageous to use polycentric laws of variation of the bridge depth, or parabolic near supports and straight in mid-span to obtain a longer section with constant depth.

No unique criterion has been defined for deciding the portion of prestressing to produce with external tendons. Some designers use the minimum number of internal tendons compatible with a good joint between the two half-bridges. Other designers size external prestressing to resist only the moments due to live loads and temperature.

Finally, front launch over fixed bearings requires an appropriate prestressing scheme, as shown in figure 10.9:

- internal cantilever tendons sized for dead load and anchored at the upper nodes of the cross section, since the action of tendons descending in the webs would be totally incoherent during launch, and would require an expensive and cumbersome corrective temporary vertical prestressing;
- temporary internal or external lower tendons, symmetrical with respect to the main support, of length *0.5* to *0.65 l* (where *l* is the main span), and with a maximum strength of about one third of the strength of the upper tendons[28]. This

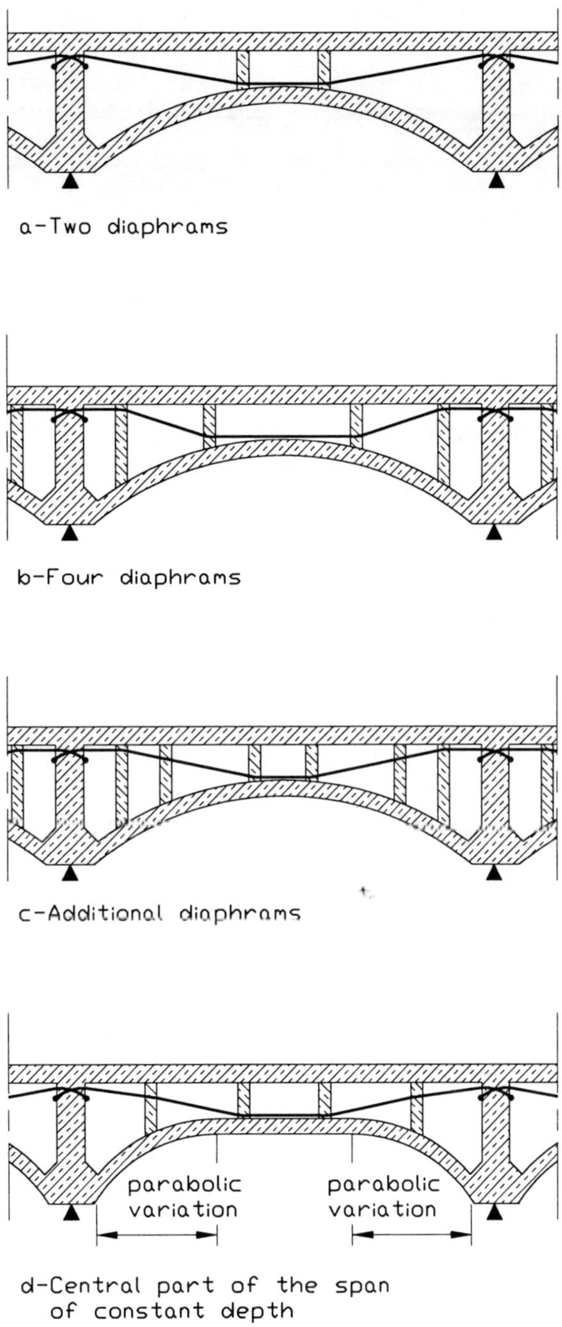

Figure 10.8 - External continuity prestressing.

corrective prestressing is aimed at balancing the eccentricity of the upper tendons and reducing the elastic deflection of the deck, which compromises the slope of the soffit necessary for a correct launch;

- parabolic internal tendons or polygonal external ones introduced once the mid-span segment is cast to correct the shear force diagram;

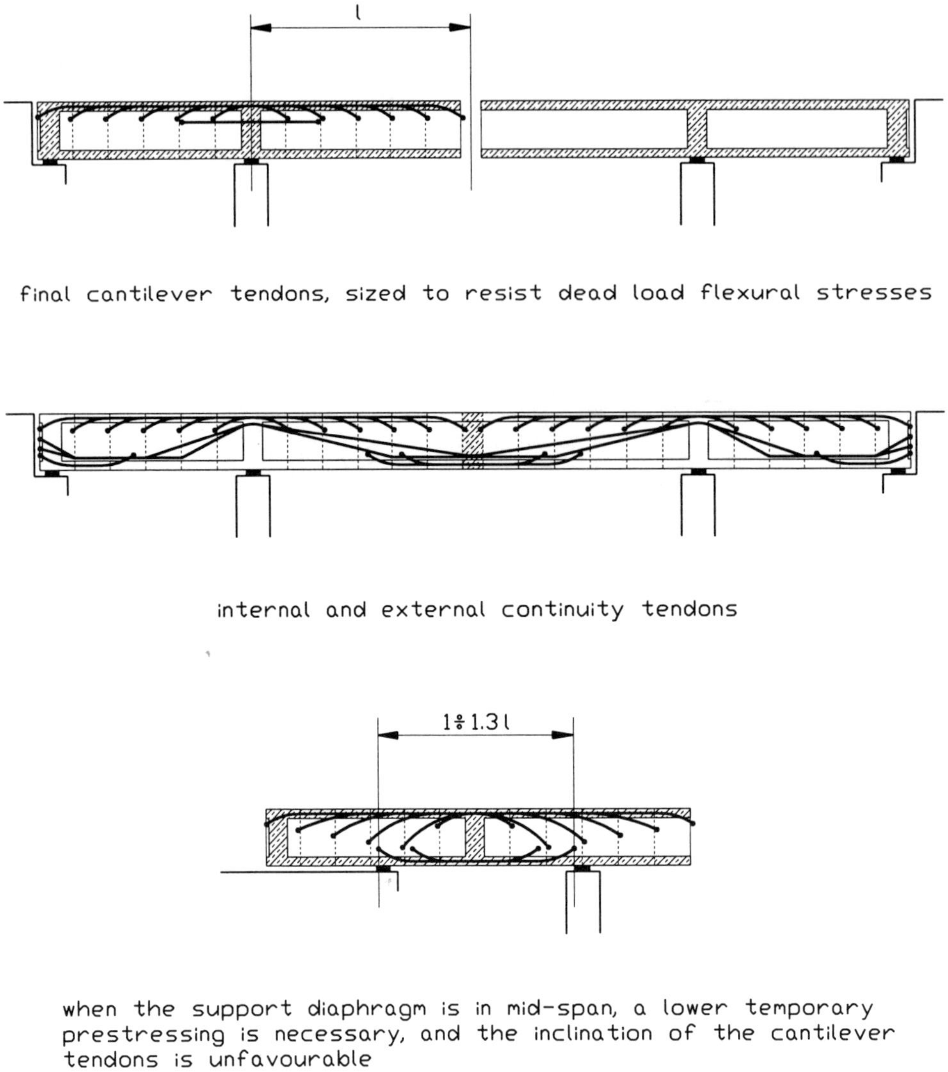

Figure 10.9 - Prestressing scheme for front launch over fixed bearings.

- if needed, additional local tendons aimed at covering the bending diagram completely.

LAUNCH

Monolithic front launch is very similar to incremental launching, and requires the same equipment. The main difference is that the deck has to be lifted before launch, to introducc the movable bearings or to load the fixed ones. For this reason, the foundation beams have niches which contain the lifting jacks. During the deck casting, these are protected by precast concrete planks (figure 3.26).

On the contrary, rotation involves specific problems, due to the stability of the half-bridge in the temporary configurations.

The rotation of the half-bridge occurs about the vertical axis of the internal support, which resists practically the whole dead load, except for the low balancing reaction. This support may be designed according to different schemes, which depend on the role of the mobile balancing support in longitudinal and transverse stability.

The choice is not easy, since longitudinal stability involves high forces and suggests long lever arms, while the transverse forces are lower (besides the same longitudinal loads, wind thrust and in some cases the curved alignment) but the lever arm is short as well. The design depends on the torsional stiffness of the deck and on its capacity to resist concentrated moments.

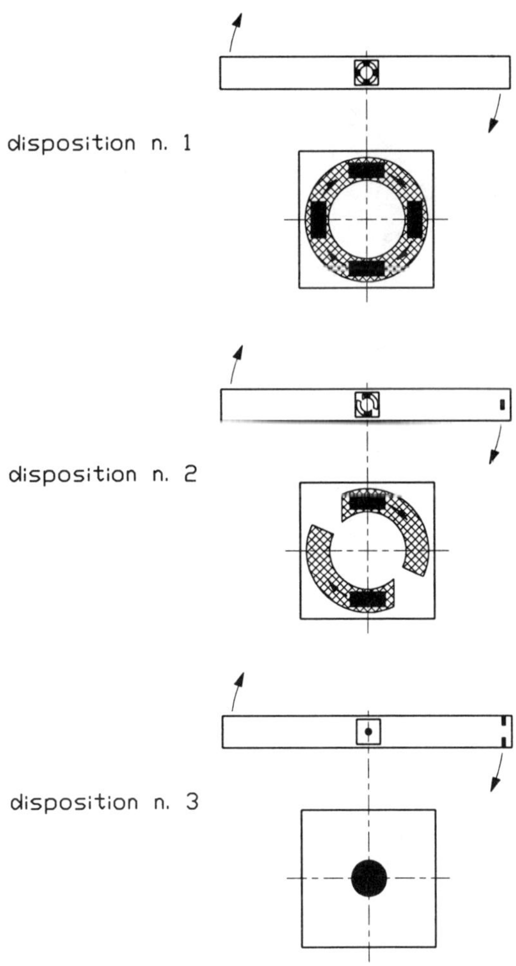

Figure 10.10 - Support schemes for bridges launched by rotation.

If the torsional stiffness is high, as in the box girders, transverse stability may be totally entrusted to the movable balancing support. In this case, this device must be rigidly connected to the deck, and is composed of two distant verticals which resist the torsional moment. In the longitudinal static scheme of the half-bridge, the support is placed at a location such as to obtain a high reaction. On the other hand, if torsional stiffness is low, as in the ribbed slabs, transverse stability should be ensured by fixing the deck at the pivot.

Three different general schemes are possible (figure 10.10).

A first scheme (disposition no. 1) entrusts both longitudinal and transverse stability to the pivot. It is used rarely in prestressed concrete bridges due to the high forces deriving from the short lever arm of the supports, while it is not unusual in steel bridges. It is adopted when the surrounding areas are inaccessible (a pier inside a river, or between several roads or railway tracks, as in figure 1.17) and the "hammer" should be as symmetrical as possible.

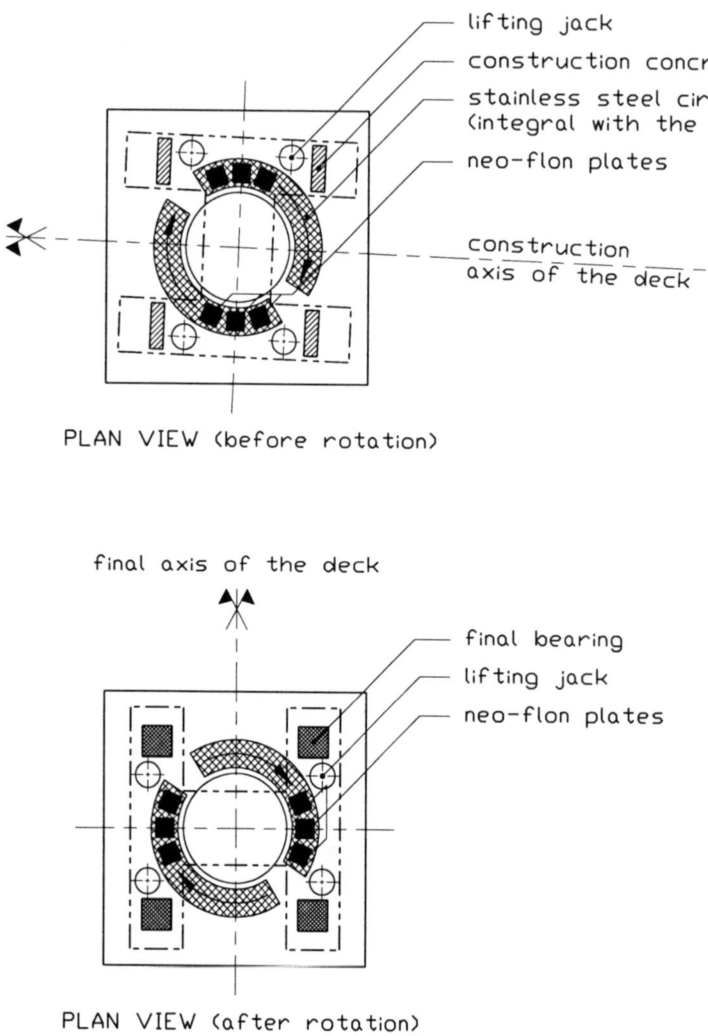

Figure 10.11 - Detail of the rotating support.

A second scheme (disposition no. 2) entrusts longitudinal stability to a pair of independent supports (the pivot and the back balancing support) and the transverse stability to the pivot. This optimizes the launch static scheme by fixing the deck torsionally at the most loaded support. No torsional couple is resisted by the back support, which cooperates only for the longitudinal equilibrium and may be a simple axial strut. Thus, a support triangle is obtained, and this avoids indeterminate effects during launch. This scheme requires high torsional stiffness of the deck. Otherwise, even the back balancing support should provide torsional constraint, with distant verticals.

In both these schemes, the rotating support is composed of neo-flon plates which slide over stainless steel circular crowns (figure 10.11). If the loads are low, steel crowns may be embedded in the plinth of foundation. If the loads are high, they are embedded in the deck, and slide over neo-flon plate circular crowns placed directly over the smooth concrete of the foundation to increase the contact surface.

In both cases, a guiding device is necessary to keep the deck centered with respect to the pier – that is, to create the pivot. A steel pipe integral with the pier may be used. It is removed upon completion of launch, and acts inside a second steel pipe, integral with the deck, through neo-flon rings which reduce friction and avoid relative movements.

The rotating support should be designed and built carefully, since the removal of defective elements is very difficult. It may be convenient to cast the deck segment placed over the rotating support in the position it will assume at the end of launch, and then to rotate it in the opposite direction until reaching the position of construction of the deck. Thus, it is possible to check the proper functioning of the rotating bearing without the weight of the half-bridge. Moreover, this prevents sharp movements at the beginning of launch due to the presence of cement residue.

Finally, if the plan dimensions of the main support are so small as to make the insertion of the circular crowns and the service jacks impossible, a compact rotating support may be adopted (disposition no. 3 of figure 10.10). This third support scheme entrusts the longitudinal stability of the half-deck to a pair of independent supports and the transverse stability to the balancing support, and again creates a support triangle. It requires a high torsional stiffness of the deck, simplifies the pivot support, and permits design of the pier for the final position of the deck, leaving rotation out of consideration.

A first type of compact rotating bearing is composed of a neo-flon plate placed over a stainless steel plate, both crossed by a removable vertical pin. Similar bearings were used in the Danube Bridge at Vienna, and in the La Flèche Bridge[48]. As an alternative, a hydraulic jack with a rotating piston may be used, as in the Fontenelle Bridge and the Trith Saint-Lèger Bridge. This scheme assures several functions at the same time (deck lifting, rotation, centering, lowering over the permanent bearings, dampening of vibration, reduction of friction), but requires a particularly careful construction of the jack, which must rotate under a heavy load.

The balancing support is composed of a pair of struts which slide over a circular steel rail by means of lubricated steel sledges (figure 10.12). In the case of high loads, the struts are equipped with stainless steel saddles which slide over neo-flon plates arranged along a concrete beam. Thus, the support is more flexible, and the launch movement more regular.

When deck torsional stiffness is sufficient, this third constraint scheme is the most efficient and economic, although it needs a very rigid pier diaphragm which transfers the whole support reaction during launch, applied as a concentrated load at the deck axis instead of under the webs.

Whichever the support scheme may be, the tangential force necessary to rotate the bridge is applied as far as possible from the pivot, to make the most of the lever arm. In the second and third support scheme, it is applied directly to the balancing support. Since low friction at the pivot is necessary to avoid secondary stresses and sharp movements, the launch force is generally low, and may be obtained with small self-clamping jacks, whose action is more uniform than that of pulling bars or strands.

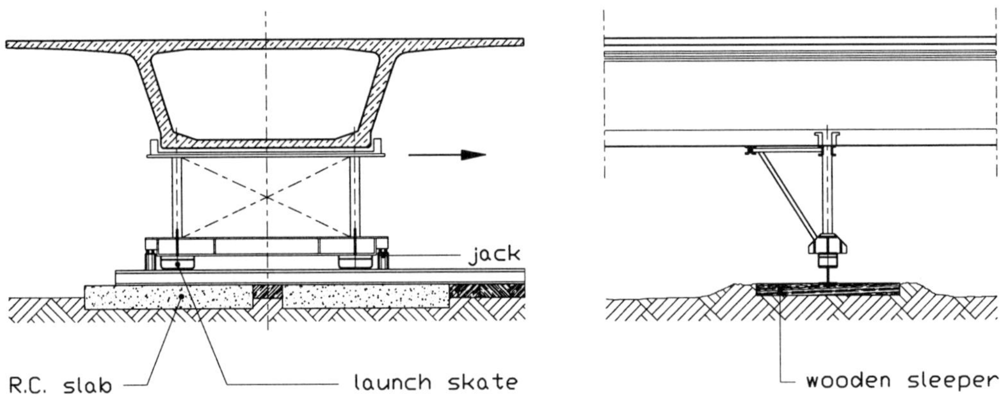

Figure 10.12 - Back balancing support.

SIDE LAUNCH

In construction by side launch, the bridge deck is built on one side of its final position, and then moved transversely.

In the narrow field of utilization of this construction method, advantages are several:

- different from both front launch and rotation, the deck static scheme is the same during launch and in service, and no temporary stresses arise during launch;
- in the case of parallel bridges, it is possible to use the same falsework for both of them, especially when it is an expensive one (high, long, placed over highways, railways or channels, or supported on piles due to settling soils) and its disassembly and reassembly would be excessively expensive;
- it is possible to place the yard on one side of the final position of the bridge. When the new structure replaces an existing structure still in service, the

interference between the yard and the old structure is reduced, and safety is improved.

The disadvantages are the need to build a falsework anyway, and the deck movement along transfer rails which must be designed for the same vertical loads as the final supports, and can be expensive in case of settling soils.

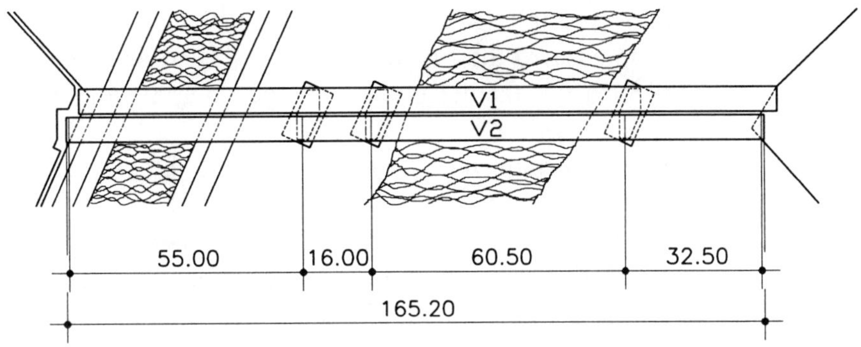

Figure 10.13 - The Bridge over the Armancon and Bourgogne Channel was built by side launch.

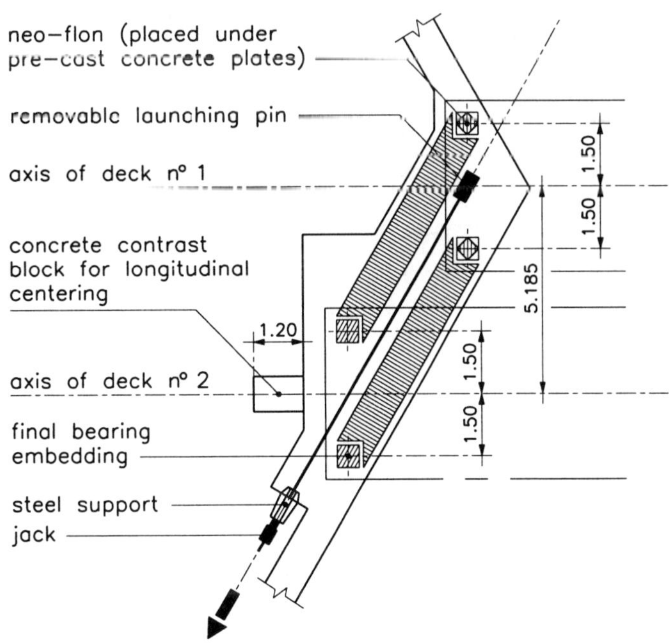

Figure 10.14 - Support and launch devices at an abutment.

In the presence of two parallel bridges, side launch can be combined with incremental launching, to make the most of the yard. The first bridge is launched into the final position of the second one, and then shifted transversely until reaching its final position. Then, the launch of the second bridge can begin. This technique has been used in the Donau Woerth Bridge[74].

An example of double utilization of the same falsework is the Armancon-Bourgogne-Channel railway Bridge[185] of the French SNCF-TGV. Two variable depth continuous beams were built on a fixed falsework placed under the first lane (figure 10.13). Upon completion of the first deck, it was shifted into its final position under the second lane, releasing the falsework for the construction of the second deck.

Launch occurred by supporting the deck on concrete blocks equipped with bottom stainless steel plates sliding over neo-flon plates. The launch force was applied by means of prestressing bars at the opposite abutments (figure 10.14). A careful monitoring of the deck horizontal deflection was necessary, since the actual in-plane bending moment was dependent on the launch forces at the abutments and on the friction forces at the internal supports. Some final corrections were required (figure 10.15).

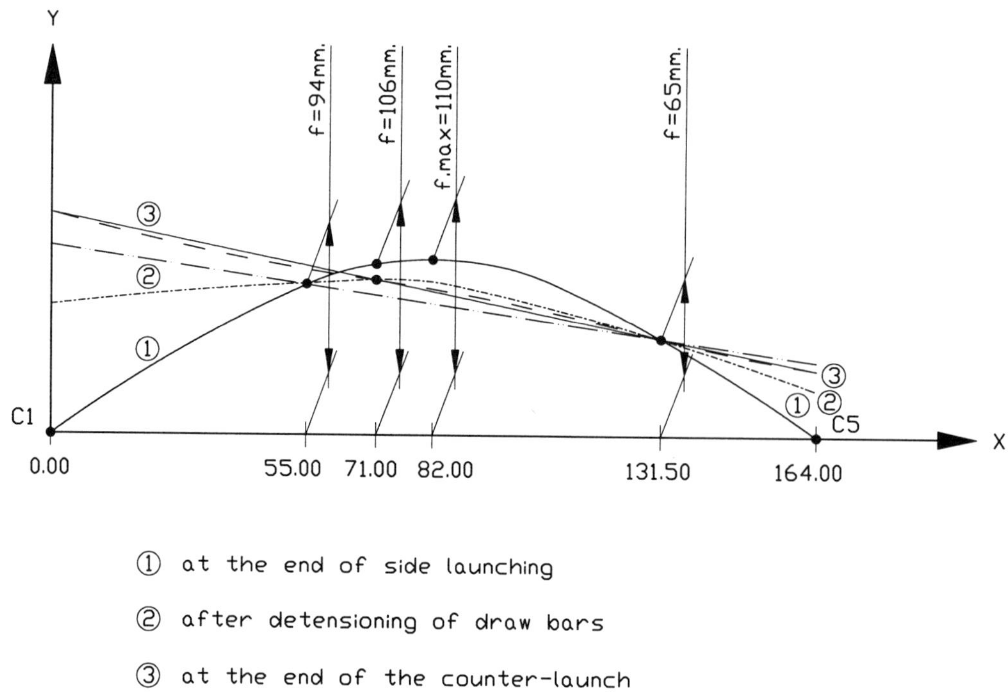

Figure 10.15 - Final correction of the in-plane deck deflection due to side launch.

An example of a bridge built near a highway in service is the new Azergues Bridge[156] of the A6 Highway, and a bridge built near a railway is the RN 504 Bridge[4] at Rossillon, both in France. In the first case, two new decks were built with conventional techniques parallel to a highway bridge, one on each side. With traffic on only one lane of the existing bridge, the superstructure of the second lane was demolished and replaced with the new deck. With traffic on this new portion, the second half of the existing bridge was demolished and replaced as well.

In the second case, the bridge of figure 10.16 was built as two half-bridges parallel to the railway, rotated until alignment with pivots at the abutments. Although geometrically it is a rotated bridge, it presents all the problems of the side launch, since the sliding supports resist high loads and the rotating supports only low loads, contrarily to the typical load distribution of the rotated bridges.

Sliding occurred over temporary reinforced concrete walls covered by stainless steel plates, which were demolished at the end of launch.

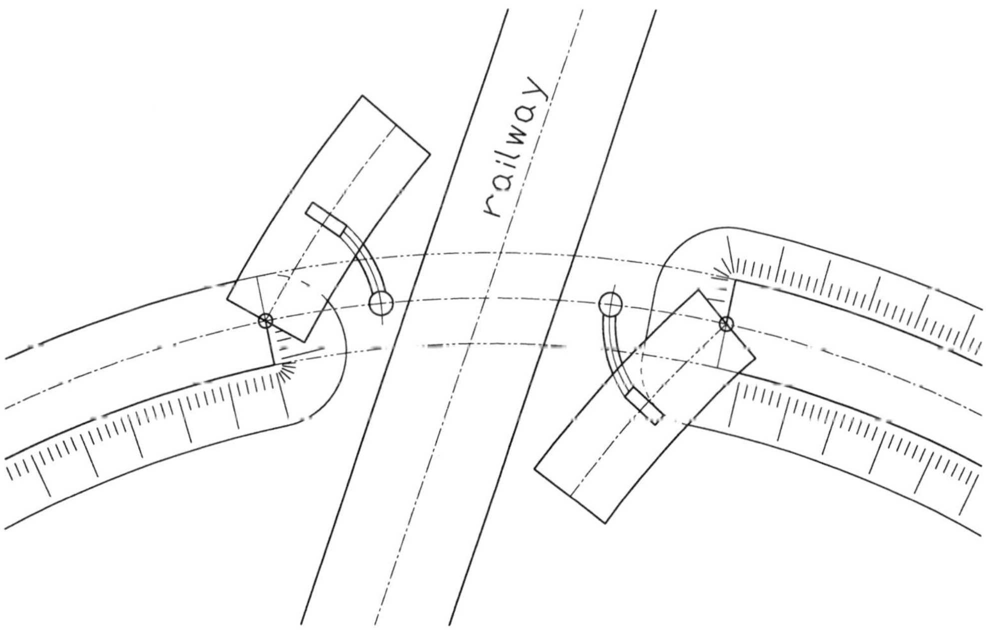

Figure 10.16 - Bridge over SNCF at Rossillon.

BIBLIOGRAPHY

References are in alphabetic order, and the contribution of the individual Authors according to chronology.

1 Abdel Samad, S.R.; Robinson, A.R.; Wright, R.N.. "BEF analogy for analysis of box girders." *ASCE - Journal of the structural division*, July 1968.

2 *ACI Fall Convention Atlanta, GA, 1974* – Symposium on segmental bridge design and construction.

3 *ACI monography n. 10.* "Concrete box girder bridges."

4 Amat, M.; Andre, M.; Carsignol, M.. "Le pont de la R.N. 504 a Rossillon." *Travaux* - July-August 1983.

5 Arcangeli, A.. "Il viadotto sopraelevato Hamburg ad Amburgo." *L' Industria Italiana del Cemento* - January 1990.

6 Arcangeli, A.. "Viadotto sulla vallata dell' Our a Steinebruck, Belgio." *L' Industria Italiana del Cemento* - May 1990.

7 Aribert, J.M.. "Connexion au cisaillement des poutres fatigue - determination experimentale de la resistance." *Annales ITBTP* - September 1988.

8 Aschrafi, M.; Lustgerten, P.; Rinne, K.D.; Saul, R.. "Verbundbrücke mit rekordspannweite uber den Rio Caroni / Venezuela." *Der Stahlbau* - January 1992.

9 Aswad, A.; Aswad, G.G.. "Rational prediction of bridge girder reinforcing and strength." *PCI Journal* - May-June 1991.

10 Attal, A.; Calgaro, J.; Causse, G.; Gesta, P.; Lacroix, R.; Marchetti, M.; Mauboussin, J.; Moreau, Ph.; Placidi, M.; Thivans, P.; Virlogeux, M.. "New composites structures. Concrete and steel." *AIPC - IABSE Congress Vancouver 1984* - AFPC Publication.

11 Attal, A.; Calgaro, J.; Causse, G.; Gesta, P.; Lacroix, R.; Marchetti, M.; Mauboussin, J.; Moreau, Ph.; Placidi, M.; Thivans, P.; Virlogeux, M.. "Nouvelles structures mixtes associant le metal et le beton." *AIPC - IABSE Congress Vancouver 1984* - Final Report.

12 Barfoot, J.. "Incremental launching at Storeham bridge." *Concrete* - November 1986.

13 Bartels, H.J.; Berz, P.; Jaschke, W.; Schwarz, J.; Meyer. H.. "Die neue Weserbrücke Bodenwerder." *Der Bauingenieur* - February 1987.

14 Bartoli, C.; Rosignoli, M.. "Il Ponte sul fiume Serio a Bergamo." *L' Industria Italiana del Cemento* - in publication

15 Battermann, W.. "Elastomer-Verformungsgleitlager mit lastabhaengiger volumenverdraengungs-Schmierung." *Die Bautechnik* - January 1987.

16 Baum, P.; Chassagnette, P.; Launay, P.; Stouf, H.. "Une nouvelle méthode de construction des ouvrages de franchissement en béton précontraint: le poussage." *Annales ITBTP* - May 1970.

17 Baumann, H.; Schwichow, V.. "Vom Bau der Steinaggertalbrücke." *FIP Congress Stockholm 1982* Final Report.

18 Baur, W.; Leonhardt, F.. "Neue Verfahren zur Herstellung weitgesspannter, mehrfeldriger Balkenbrücken aus Spannbeton." *Beton und Stahlbetonbau* – 1962.
19 Baur, W.; Leonhardt, F.. "Die Agerbrücke." *Die Bautechnik* – 1963.
20 Baur, W.; Leonhardt, F.; Trah, W.. "Brücke uber den Rio Caroni, Venezuela." *Beton und Stahlbetonbau* – 1966.
21 Baur, W.; Leonhardt, F.. "Erfahrungen mit dem Taktschiebeverfahren im brücken und hochbau." *Beton und Stahlbetonbau* - July 1971.
22 Baur, W.. "Bridge erection by launching is fast, safe and efficient." *ASCE - Civil Engineering* - March 1977.
23 Baur, W.; Leonhardt, F.. "Taktschiebeverfahren." *Stuttgard 1978.*
24 Begusch, W.. "Grossbrücken der Wechselautobahn." *Osterreichische Ingenieur und Architekten Zeitschrift* - May 1987.
25 Begusch, W.. "Ponte sulla vallata di Edlitz per l' Autostrada del Sud "A2" in Austria." *L' Industria Italiana del Cemento* - June 1995.
26 Bellero, C.; Colazingari, C.; Gozzi, V.. "Ponte Posadas-Encarnacion sul fiume Paranà tra Argentina e Paraguay." *L' Industria Italiana del Cemento* - February 1995.
27 Belluzzi, O.. "Scienza delle costruzioni" *Zanichelli editore.*
28 Bernard-Gély, A.; Calgaro, J.A.. "Conception des Ponts" *Presses de l' Ecole Nationale des Ponts et Chaussées - Paris 1994.*
29 Best, K.H.; Kingston, R.H.; Whatley, M.J.. "Incremental launching at Shepherds House Bridge." *Proceedings of the Institution of Civil Engineers* - February 1978.
30 Bonnet, M.; Bouvy, B.; Causse, G.; Chaussat, J.C.; Combault, J.; Duviard, M.; Guichard, M.; Guiet, G.; de Matteis, G.; Ponchaud, B.; Porcherel, H.; Thibonnet, J.L.; Thivans, P.; Virlogeux, M.. "Viaduc du vallon de Maupré, à Charolles (Saone-et-Loire)." *Travaux* - October 1988.
31 Boudot, J.; Radiguet, B.; Thao, P.X.. "The Sylans and Glacières Viaducts. Contribution of the French Group" *AIPC - IABSE Congress Paris-Versailles 1987* - AFPC Publication.
32 Buodot, J.; Radiguet, B.; Thao, P.X.. "Les viaducs de Sylans et des Glacieres." *AIPC - IABSE Congress Paris-Versailles 1987* – Report.
33 Breen, J.; Naaman, A.. "External prestressing in bridges" *ACI SP 120.*
34 Bruneau, J.; Causse, G.; Radiguet, B.; Raspaud, B.. "Test loading of a concrete truss. Contribution of the French group." *AIPC - IABSE Congress Vancouver 1984* - AFPC Publication.
35 Bruneau, J.; Causse, G.; Radiguet, B.; Raspaud, B.. "Expérimentation d' une structure spatiale en trellins." *AIPC - IABSE Congress Vancouver 1984* - Final Report.
36 Campbell, D.; Wedgwood, J.L.. "Need for diaphragms in concrete box girders." *ASCE - Journal of the structural division* - March 1971 (discussions on December 1971).
37 Campenon Bernard. "Due viadotti per l' autostrada Esterel - Costa Azzurra." *L' Industria Italiana del Cemento* - April 1986.

38 Capitanio, S.. "The King Taskin Bridge in Bangkok." *L' Industria Italiana del Cemento* - September 1985.

39 Capra, R.; Martinez y Cabrera, F.. "Tecnica della prefabbricazione per la realizzazione di sezioni strutturali a cassone." *L' edilizia e l' industrializzazione* - January-February 1988.

40 Cassiau, A.; Combault, J.; Duviard, M.; Fontaine, J.F.; Gauche, J.F.; Guichard, M.; Lecroq, Ph.; Thibonnet, J.L.; Thivans, P.; Virlogeux, M.. "Pont sur la Charente à Cognac." *Travaux* - October 1988.

41 Causse, G.; Duviard, M.. "Maupre viaduct near Charolles. A bridge deck with corrugated steel plate webs constructed by the incremental launching technique." *AIPC - IABSE Congress Paris-Versailles 1987* - AFPC Publication.

42 Causse, G.; Duviard, M.. "Ouvrage a ames plissees mis en place par poussage, Charolles." *AIPC - IABSE Congress Paris-Versailles 1987* – Report.

43 *CEB-FIP Bulletin d' Information 222* "Application of High Performance Concrete."

44 Ceccherelli, D. "Il ponte sul fiume Brembo della SS 342 Briantea." *Le Strade* - January-February 1996.

45 Cestelli-Guidi, C.. "Cemento Armato precompresso" *Hoepli editore.*

46 Cézard, C.; Servant, C.I.. "Charix Viaduct. Contribution of the French Group." *AIPC - IABSE Congress Helsinki 1988* - AFPC Publication.

47 Cocomello, S.. "A spatial lattice in prestressed concrete for the deck of the Bubiyan bridge in Kuwait." *L' Industria Italiana del Cemento* - August 1984.

48 Colas, M.; Fresnais, P.; Hirsch, D.; Lacoste, G.; Mossot, J.; Placidi, M.; Virlogeux, M.. "Le nouveau pont sur le Loir à La Fleche." *Travaux* - July-August 1983.

49 Collepardi, M.; Coppola, L.; Troli, R.. "Materiali cementizi innovativi: dagli HPC verso gli RPC." *L' Industria Italiana del Cemento* - March 1995.

50 Collepardi, M.; Coppola, L.; Monosi, S.. "Il calcestruzzo leggero." *Il nuovo cantiere* - October 1995.

51 Collepardi, M.. "Scienza e tecnologia del calcestruzzo." *HOEPLI editore.*

52 Colombo, A.; Gentile, R.; Martinez y Cabrera, F.; Rosignoli, M.; Rossi, S.. "Il ponte strallato di Via Palizzi a Milano." *Giornate AITEC 1997.*

53 Combault, J.. "Evolution et developpement des ponts modernes a precontrainte totalement exterieure au beton." *Travaux* – January 1986.

54 Combault, J.; Leveille, A.; Neron, P.; Thibonnet, J.L.. "Incrementally launched bridges with total external prestressing. Contribution of the French Group." *AIPC - IABSE Congress Paris-Versailles 1987* - AFPC Publication.

55 Combault, J.; Leveille, A.; Neron, P.; Thibonnet, J.L.. "Ponts poussés à précontrainte totalement extérieure." *AIPC - IABSE Congress Paris-Versailles 1987* – Report.

56 Combault, J.. "The Maupre Viaduct near Charolles." *AISC NEC/COP 1988 National steel construction conference, Miami Beach, Florida.*

57 Curbach, M.; Seitz, G.. "Unusual applications of the incremental launching method." *IABSE Proceedings.*

58 Deinhard, J.; Schafer, H.; Schlosser, H.. “Hochstrasse Ratsweg in Frankfurt a.M. im zuge der BAB 661.” *Beton und Stahlbetonbau* - November 1987, January and February 1988.

59 De Vriend, H.; Oud, H.J.C.. “Unieke brug - tunnelverbinding in ringweg om Amsterdam.” *Cement* - August 1987.

60 Dezi, L.; Menditto, G.; Rosignoli, M.. “In tema di predimensionamento dei ponti a cassone in calcestruzzo armato ed in calcestruzzo armato precompresso.” *L' Industria Italiana del Cemento* - June 1982.

61 Different Authors. “Innovation dans le domaine des structures mixtes: ponts mixtes metal-beton precontraint” *Annales ITBTP.* - October and November 1987.

62 Dimitrov, D.G.. “Spannungsverteilung querbelasteter Stegbleche mit abgestuften Dicken unter einseitiger begrenzter gleichmassiger last.” *Der Stahlbau* - August 1986.

63 Dimitrov, D.G.. “Zum taktschiebeverfahren bei Verbundbrücken.” *Der Stahlbau* - October 1986.

64 Dorner, M.; Franzi, J.. “Der bau des Taluberganges Enzenreith.” *Osterreichische Ingenieur und Architekten Zeitschrift* - March 1986.

65 Dorton, R.A.. “Safety considerations for the Burlington Skyway project.” *AIPC - IABSE Congress Tokio 1986* – Report.

66 Dureil, P.; Placidi, M.. “Le pont sur le Drac a Grenoble.” *Annales des Ponts et Chaussees* - second quarter 1982.

67 Dutoit, D.; Muller, J.. “Projects Marta CS 360 ed CN 480. Atlanta mass transit: Marta CS and CN 480. Railways and mass transit systems bridges. French experience.” *AIPC - IABSE Congress Vancouver 1984* - AFPC Publication.

68 Duviard, M.. “Le premier pont à ames plissées à Cognac.” *AIPC - IABSE Congress Paris-Versailles 1987* – Report.

69 Falempin, M.; Glasson, J.P.; Lacroix, A.; Servant, C.; Simon, J.E.; Villette, P.. “Le pont sur la voie 3 de Clergy Pontoise (Val d' Oise).” *Travaux* - October 1988.

70 Fernandez, A.. “Construction d' un pont sur le Caroni (Venezuela).” *Travaux* - July-August 1965.

71 Ferretti Torricelli, L.; Furlanetto, G.. “Il viadotto Lijak presso Nova Gorica nella Repubblica Slovena.” *L' Industria Italiana del Cemento* - June 1997.

72 Flugel, K.. “Der bau der Maintalbrücke Veitshochheim, ein neuer Weltrekord im taktschieben.” *Tiefbau - Berufsgenossenschaft* - August 1987.

73 Freystein, W.; Wallas, H.. “Planung der Eisenbahnbrücke uber die Werra mit uberbau in der verbundbauart.” *Beton und Stahlbetonbau* - April 1987.

74 Fritsch, R.; Scheidler, J.. ”Donaubrücke Woerth - Taktschiebeverfahren mit Querverschub.” *Der Bauingenieur* - May 1980.

75 Fritsch, P.. “Schragkabel und taktschiebebrücken in Korea und Taiwan.” *Osterreichische Ingenieur und Architekten Zeitschrift* - September-October 1988.

76 Frizzi, D.; Giovannini, B.. "Ponte sul Po a Casale Monferrato per l' Autostrada dei Trafori." *L' Industria Italiana del Cemento* - April 1977.

77 Froehlich, H.. "Neubau der Donnersberger Brücke in München unter Anwendung weiterentwickelter Lager bei der Herstellung in Taktschiebeverfahren." *Beton und Stahlbetonbau* - January 1984.

78 Gillet, G.; Jacquet, Ph. "L' ouvrage 33 sur l' autoroute du Littoral à Marseille." *Annales ITBTP* - October 1988.

79 Giovannini, B.. "Un impalcato estruso: il viadotto in Val Restel." *L' industria delle costruzioni* - September-October 1972.

80 Gisi, R.. "Die Birsbrucke Rutihard bei Münchenstein. - Prestressed concrete in Switzerland, 1982 - 1986" *FIP Congress New Delhi 1986* – Report.

81 Giuliani, G.. "Il nuovo ponte per l' oleodotto dell' Europa Centrale, realizzato sul fiume Po con il sistema dell' autovaro." *L' Industria Italiana del Cemento* – March 1970.

82 Gobel, W.; Mauch, A.; Muller, B.; Schuldt, O.. "Brücken Hetzdorf - Die bautechnologie fur das Taktschiebeverfahren und ihr einfluss auf die bautechnische losung." *Strasse* - February 1989.

83 Godart, B.; Lenoir, B.; Léveillé, A.; Neron, P.; Tonnoir, B.. "Viaduc d' Amiens. Conception et realisation de l' ouvrage. Suivi experimental sur site, pendant la construction." *Annales ITBTP* - March-April 1989.

84 Grant, A.. "Incremental launching of concrete structures." *ACI Journal*- August 1975.

85 Greisch. "Il ponte di Ben-Ahin sulla Mosa in Belgio." *L' Industria Italiana del Cemento* - April 1991.

86 Greisch. "Il ponte strallato di Wandre sulla Mosa in Belgio." *L' Industria Italiana del Cemento* - May 1993.

87 Grennerat, Y. "Le viaduct de Poncin sur l' autoroute A40 (Ain)." *Travaux* - October 1988.

88 Hanel, D.; Kusche, R.. "Sonderlosung zur Aufnahme grosser Horizontalkrafte bei Eisenbahnbrücken - Beispiel Hetzdorfer Viadukt." *Die Bautechnik* - January 1993.

89 Harries, H.; Kinkel, H.; Kleinmanns, R.; Petri, H. "Zwei trogbrüken uber die Sieg bei Hennef." *Beton und Stahlbetonbau* - September 1986.

90 Harries, H.; Kinkel, H.; Petri, H.. "Sliding formwork launching gantries. Transferring the principle of incremental launching to formwork launching gantries in bridge construction. - Spannbetonbau in der Bundesrepublik Deutschland 1983 - 1986" *FIP Congress New Delhi 1986* – Report.

91 Hetenyi, H.. "Beams on elastic foundation, theory with application in the fields of civil and mechanical engineering" *Ann Arbor - The University of Michigan 1971.*

92 Heywood, P.. "Corrugated box-girder web lowers bridge weight, cost." *Engineering News - Record* - 17 December 1987.

93 Hurd, M.K.. "Segmental box-girder bridge construction." *Concrete International* - September 1986.

94 Hurd, M.K.; Watson, S.C.. "Esthetics in concrete bridge design" *ACI MP 1.*

95 Isiguro, W.; Murata, Y.; Sugo, T.; Uehira, K.. "Stability of Prestressed Concrete Bridge with Corrugated Steel Web." *AIPC - IABSE Congress Innsbruck 1997* – Report.

96 Klein, W.. "Montage der Sauertalbrücke im taktschiebeverfahren." *Der Stahlbau* - December 1986.

97 Koncz, T.. "Vorschub-Freivorbau bei einer hohen talbrücke, dem Semorile-Viadukt." *Beton und Stahlbetonbau* – 1968.

98 Korn, H.. "Bridge construction by extrusion sliding." *Concrete* - May 1975.

99 Krautwald, W.; Schutt, K.; Thormahlen, U.. "Berbke Valley Bridge, incremental launching and external prestressing. - Spannbetonbau in der Bundesrepublik Deutschland 1987" *FIP Congress Hamburg 1990* – Report.

100 Krebs, F.; Schambeck, H.. "Grossdoordamm flyover in Hamburg - Incremental launching under difficult geometrical conditions - Spannbetonbau in der Bundesrepublik Deutschland 1983 - 1986" *FIP Congress New Delhi 1986* – Report.

101 Kroppen, H.; Schambeck, H.. "Die Donaubrücke Metten." *Beton und Stahlbetonbau* – 1982.

102 Kusche, R.; Verch, W.. "Brückenbau in der VDR Jemen." *Bauplanung, bautechnik* - January 1986.

103 Lacroix, R.. "La méthode des matrices-transfert" *Annales ITBTP* - March-April 1967.

104 Lavigne, C.; Placidi, M.; Virlogeux, M.. "The new bridge over the Loir river at La Flèche." *L' Industria Italiana del Cemento* - April 1985.

105 Lehmann, W.. "Ein "Schieber" mit vielen Trumpfen. Viadukt Lowenberg n. 1." *Bau* - April 1992.

106 Leonhardt, F.. "Prestressed Concrete." *Wilhelm Ernst & Sohn, Berlin 1964.*

107 Leonhardt, F.. "Abminderung der Tragfahigkeit des betons involge strabformiger, rechtwinklig zur Druckrichtung angeordneter Einlagen." *Stahlbetonbau* – 1969.

108 Leonhardt, F.. "Procédé de construction par cycles de bétonnage en coffrage fixe et cycles de poussage." *Annales ITBTP* - January 1973.

109 Leonhardt, F., Andra & Partners. "Il ponte ferroviario sul Fiume Meno a Gemunden." *L' Industria Italiana del Cemento* - December 1987.

110 Leonhardt, F.; Andra & Partner. "Ponte ferroviario sulla valle del Meno a Veitshochheim" *L' Industria Italiana del Cemento.* - February 1991.

111 Leonhardt, F.; Andra & Partners. "Il viadotto ferroviario sul fiume Enz presso Enzweihingen (Germania)." *L' Industria Italiana del Cemento* - October 1994.

112 Leonhardt, F.. "C.A. e C.A.P. calcolo di progetto e tecniche costruttive - Vol. 4 - I ponti, dimensionamento, tipologia, costruzione" *Edizioni di scienza e tecnica.*

113 Libby, J.R.. "Modern prestressed concrete, design principles and construction methods." - *Ed. Van Nostrand Reinhold, New York 1990.*

114 Losinger Ltd. "The incremental launching method in prestressed bridge construction." April 1977.

115 Mabounga, D.; Sakala, L.; Servant, C.; Sibide, Y.; Tran Vo Nhiem, Zumaglini, M.. "Pont sur le Kouilou." *Travaux* - January 1986.

116 Maddison, M.. "Crossing the cutting with segments at Sonning." *Concrete* - February 1978.

117 Marchetti, M.. "Specific design problems related to bridges built using the incremental launching method." *2nd Conference on Structural Engineering, Bangkok, december 1982.*

118 Martin, R.A.; Raspaud, B.; Richard, P.. "Le pont de Bubiyan au Koweit." *Travaux* - November 1982.

119 Martinez Y Cabrera, F.; Menni, C.. "I ponti a cassone monocellulare a profilo deformabile." *Technical Note n. 33 ISTC May 1974 - Politecnico di Milano - Tamburini Editore.*

120 Mathivat, J.. "L' evolution recente des ponts en beton precontraint." *Ingenieurs et Architectes Suisses* - March 1987.

121 Mathivat, J.; Virlogeux, M.. "Le pont en arc sur la Rance - la technique francaise du beton precontraint." *FIP Congress Hamburg 1990.*

122 Matildi, P.; Matildi, G.. "Ponti metallici, esperienze vissute" *Cimolai SpA.*

123 Mireles, A.A.; Podonly, W.. "Kuwait's Bubiyan bridge. A 3.D. precast segmental space frame." *PCI Journal* - January-February 1983.

124 Moreau, Ph.; Thivans, P.. "Composite structures Steel-Prestressed Concrete." *IABSE Proceedings* - February 1983.

125 Moreau, Ph.; Thivans. P.. "Structures composites acier-beton precontraint." *Annales ITBTP* - January 1984.

126 Moschner, H.. "Der bau der Landlbrücke im Zuge der B 25 Erlauftal. - Bundesstrasse" *Zement und beton* - February 1986.

127 Muller, J.M., Podolny, W. Jr.. "Construction and design of prestressed concrete segmental bridges." *John Wiley & Sons.*

128 Munz, H.T.. "Zwei brücken aus einem guss vermessungsarbciten beim bau von gekoppelten taktschiebebrücken fur die B 224 n bei Wulfrath." *Der Vermessungsingenieur* - January 1987.

129 Nadal, P.. "Le Viaduct de Maupré à Charolles." *Chantiers de France* - November 1987.

130 Nadal, P.. "L' Echangeur A1/Parc Astérix." *Chantiers de France* - September 1988.

131 Nam Shiu, K.; Russell, H.G.. "Knowledge gained from instrumentation of the Kishwaukee river bridge." *PCI Journal* - September-October 1983.

132 O' Connor, C. "Design of bridge superstructures." *John Wiley & Sons.*

133 Peters, H.. "Vom Bau der "Pont de Normandie" bei Le Havre." *Beton und Stahlbetonbau* - January 1993.

134 Petrangeli, M.P.. "Progettazione e costruzione di ponti." *Masson editoriale ESA.*

135 Placidi, M.. "La construction d' ouvrages d' art avec mise en place par rotation." *FIP Congress Stockholm 1982.*

136 Placidi, M.. "La construction au sol et la mise en place par deplacement." *Annales ITBTP* - December 1983.

137 Placidi, M.. "Evolution et tendances des méthodes de construction des tabliers de ponts poussés." *Annales ITBTP* - October 1988.
138 Placidi, M.. "La mise en place par rotation." *L' Ingenieur Constructeur* - June-July 1990.
139 Placidi, M.; Virlogeux, M.. "Il ponte di Trellins: un arco in c.a. sul fiume Isere." *L' Industria Italiana del Cemento* - October 1991.
140 Poier, P.. "Eisenbahnbrückenbau im Iran." *Zement und beton* - March 1986.
141 Poppy, W.. "Neue technische verfahren: strassen, brücken und tunnelbau." *Osterreichische Bau-Zeitung* - n. 20 1989.
142 *Portland Cement Association* "Wabash River Bridge, Covington, Indiana." Bridge Report SR 201.01E, 1978, Skokie, Ill.
143 *Post Tensioning Institute, Phoenix, Arizona* "First incrementally launched post tensioned box girder bridge to be built in the United States." Bridge Report - December 1976.
144 Pozzati, P.. "Teoria e tecnica delle strutture." *UTET.*
145 Rademacher, C.H.. "Die montage der neuen Hammer Eisenbahnbrücke." *Eisenbahningenieur* - November 1986.
146 Raspaud, B.. "Les couvertures de l' aerogare n. 2 de l' aeroport Charles de Gaulle." *AIPC - IABSE Congress Wien 1990.*
147 Reichelt, A.. "Brückenbau in der VDR Jemen – Bauausführung." *Strasse* - January 1989.
148 Reifegerste, L.; Zbynek, P.. "Externe Vorspannung im Brückenbau." *Die Stasse* - December 1990.
149 Reinbeck, U.; Voss, W.. "Entwurf und Bau der Strassenbrücke uber den Bahnhof und die Ruhr in Meschede - Anwendung des Taktschiebeverfahrens in einer Wendelinie. - Spannbetonbau in der Bundesrepublik Deutschland 1983 - 1986" *FIP Congress New Delhi 1986* – Report.
150 Richard, P.. "Structures triangulées en béton." *Annales ITBTP* - January 1984.
151 Richter, K.; Schmackpfeffer, H.. "Herstellung von LP - Blechen und deren Verwendung im Brückenbau." *Der Stahlbau* - February 1988.
152 Riedl, R.. "Das Taktschiebeverfahren im Brückenbau." *Zement und beton* – 1970.
153 Riedmann, E.; Schewior, E.. "Talbrücke Marktbreit - Taktschiebebrücke von zwei Widerlagern aus eingeschoben. - Spannbetonbau in der Bundesrepublik Deutschland 1983 - 1986" *FIP Congress New Delhi 1986* – Report.
154 Riedmann, E.. "Mainbrücke Oberhaid Bundesautobahn A 770 Schweinfurt – Bamberg." *Tiefbau - Berufsgenossenschaft* - July 1991.
155 Reinhard, J.M.. "Pont de la Corniche à Dole." *Ouvrages d' Art* - n. 19 - November 1994.
156 Rimboeuf, M.; Salzmann, C.. "Ouvrage de franchissement de l' Azergues." *AIPC - IABSE Congress Wien 1990.*
157 Rosignoli, M.. "Ponti a cassone: la distorsione della sezione trasversale secondo la analogia della trave su mezzo elastico." *University of Ancona – 1981.*

158 Rosignoli, M.. "Incremental launching method in P.C. bridge construction." *Italian building and construction* - n. 54/55, July-December 1993.
159 Rosignoli, M.. "Ponti stradali in c.a.p. per getto cadenzato con varo di punta." *Le Strade* - n. 1296 - September 1993.
160 Rosignoli, M.. "L' interazione avambecco - impalcato nei ponti in c.a.p. realizzati per varo frontale." *L' Industria Italiana del Cemento* - March 1995.
161 Rosignoli, M.. "Evoluzione di una tecnica." *Le Strade* - October 1995.
162 Rosignoli, M.. "Il ponte spinto contro la piena." *Le Strade* - June 1996.
163 Rosignoli, M.. "Una specie ibrida." *Le Strade* - November-December 1996.
164 Rosignoli, M.. "Sul dimensionamento degli impalcati da ponte in c.a.p. realizzati per varo frontale progressivo." *L' Industria Italiana del Cemento* - December 1996.
165 Rosignoli, M.. "Evolution of a technique." *Concrete International* - February 1997.
166 Rosignoli, M.. "Influences of the incremental launching construction method on the sizing of prestressed concrete bridge decks." *Proceedings of the Institution of Civil Engineers - Structures and Buildings* - August 1997.
167 Rosignoli, M.. "Tolleranze di posizionamento degli appoggi di varo nei ponti spinti in c.a.p." *L' Industria Italiana del Cemento* - September 1997.
168 Rosignoli, M.. "Solution of the continuous beam in launched bridges." *Proceedings of the Institution of Civil Engineers - Structures and Buildings* - November 1997.
169 Rosignoli, M.. "Prestressed Composite Box Girders." *IABSE Structural Engineering International* - November 1997.
170 Rosignoli, M.. "Nose-deck interaction in launched prestressed concrete bridges." *American Society of Civil Engineers - Journal of Bridge Engineering* – February 1998.
171 Rosignoli, M.. "Serio River Bridge, Creep and Incremental Launching." *Proceedings of the Institution of Civil Engineers - Structures and Buildings* – February 1998.
172 Rosignoli, M.. "Milano, Via Palizzi." *Le Strade* - in publication.
173 Rosignoli, M.. "La risoluzione della trave continua nei ponti realizzati per varo frontale." *L' Industria Italiana del Cemento* - in publication.
174 Rosignoli, M.. "Il nodo di Via Palizzi a Milano." *L' Industria Italiana del Cemento* - in publication.
175 Rosignoli, M. "Serio River Bridge." *Concrete International.*
176 Rosignoli, M.. "Milan-Palizzi road interchange." *Concrete International* - in publication.
177 Rosignoli, M.. "Ponti ad arco a sezione tubolare composita." *L' Industria delle Costruzioni* – in publication.
178 Roth, C.. "Neue Wege im Eisenbahnbrückenbau." *Beton* - June 1984.
179 Rowley, F.N.. "Incremental launch bridges: UK practice and some foreign comparison." *The structural engineer (GB)* - April 1993.
180 Schauch, G.. "Von Lyon nach Valence mit rasender Geschwindigkeit." *Stassen und tiefbau* - October 1992.

181 Scheidler, J.. "Bauverfahren und ihre kritischen Montagezustande bei Grossbrücken." *Tiefbau - Berufsgenossenschaft* - May 1990.
182 Scheller, A.. "Die Donaubrücke Fischerdorf." *Tiefbau - Berufsgenossenschaft* - January 1992.
183 Schrader, R.. "Erneuerung der Schwarzbachtalbrücke bei Dusseldorf unter Aufrechterhaltung des Verkehrs auf der BAB A 3 - Spannbetonbau in der Bundesrepublik Deutschland 1983 – 1986." *FIP Congress New Delhi 1986* – Report.
184 Seifried, G.; Wittfoht, H.. "Die Brücke über den Shatt-al Arab in Basrah (IRAQ)." *Beton und Stahlbetonbau* - April 1979.
185 Servant, C.. "Le pont sur l' Armancon ed le canal de Bourgogne - Ligne T.G.V." *FIP Congress Stockholm 1982*.
186 Spielmann, A.. "Il viadotto di Roquebilliere a Cahors." *L' Industria Italiana del Cemento* - October 1988.
187 Stich, W.. "Rekonstruktion der Stadtbahnbogen in Berlin, Hauptstadt der DDR." *Strasse* - July 1986.
188 Svensson, H.; Zellner, W.. "Incremental launching of structures." *ASCE - Journal of Structural Engineering* - February 1983.
189 Theiner, J.. "Weltrekord im Taktschiebeverfahren. Maintalbrücke Veitshoechheim fertiggestellt." *Beton* - January 1987.
190 Treffot, G.. "Le viaduc de la Corniche à Dole." *Ouvrages d' Art* - n. 19 - September 1995.
191 Twiehaus, W.. "Eisenbahnbrücke uber den Rhein zwischen Dusseldorf und Neuss." *Baktuell* - January 1990.
192 Twiehoff, R.. "Stahlverbundbrücke uber den Guterbahnhof Dortmund." *Der Vermessungsingenieur* - June 1991.
193 Unidentified Author. "Noch nie zuvoe so weit im Takt geschoben. Strabag bau AG techn. federführung und WTB erzielen Weltrekord beim bau der Maintalbrücke Veitshochheim." *Bauwirtschaft* - n. 49/50 1986.
194 Unidentified Author. "Box Girder will be Launched in the U.S." *Engineering News - Record* - 27 January 1977.
195 Unidentified Author. "Extruded Span Cut Bridge Costs in Half - Mainbrücke Mainflingen." *Engineering News - Record* - 17 November 1977.
196 Unidentified Author. "Segmental box girder bridges make the big time in U.S." *Engineering News - Record* - 2 March 1978.
197 Unidentified Author "Arch Slipformer Shuns Ground Support to Cross Valley." *Engineering News - Record* - 1 June 1978.
198 Unidentified Authors. "Monographies d' ouvrages." *Travaux* - January 1986.
199 Unidentified Author. "Sovrappasso ferroviario in cemento armato lungo la linea Wuppertal - Colonia nella Germania Federale." *L' Industria Italiana del Cemento* - May 1986
200 Unidentified Author. "Remarkable structures 1983 - 1986 - Spannbetonbau in der Bundesrepublik Deutschland 1983 – 1986." *FIP Congress New Delhi 1986* – Report.

201 Unidentified Author. "Three - dimensionally curved prestressed concrete bridge built by the incremental launching method for railway traffic - Spannbetonbau in der Bundesrepublik Deutschland 1983 – 1986." *FIP Congress New Delhi 1986* – Report.

202 Unidentified Author. "Prestressed concrete in Switzerland, 1982 – 1986." *FIP Congress New Delhi 1986* – Report.

203 Unidentified Author. "Bau einer raumlich gekrummten spannbetonbrücke fur den Eisenbahnverkehr im taktschiebeverfahren." *Tiefbau - Berufsgenossenschaft* - May 1988.

204 Unidentified Author. "Le passage superieur n. 8 sur l' autoroute A 71." *Travaux* - October 1988.

205 Unidentified Author. "Evolution recente des ponts en beton precontraint." *IABSE Proceedings* - November 1988.

206 Unidentified Author. "Sviluppi recenti nel campo dei ponti in c.a.p.." *L' Industria Italiana del Cemento* - June 1989.

207 Unidentified Author. "Un sovrappasso autostradale di concezione innovativa." *L' Industria Italiana del Cemento* - October 1989.

208 Unidentified Author. "Autobahnbrücke, Frankreich." *Der Bauingenieur* - November 1989.

209 Unidentified Author. "Il ponte sull' estuario di Mandurah. Primo ponte australiano realizzato con il procedimento a spinta." *L' Industria Italiana del Cemento* - February 1990.

210 Unidentified Author. "Il viadotto Lirone per il tratto S.S. 229 - Gattico dell' Autostrada dei Trafori a Borgomanero, Novara." *L' Industria Italiana del Cemento* - April 1990.

211 Unidentified Author. "Entwicklungen im spannbetonbau. Ruckschau und ausblik." *Betonwerk und Fertigteil-technik* - April 1990.

212 Unidentified Author. "Ponti a conci prefabbricati in c.a. Ultime realizzazioni e prospettive in Florida." *L' Industria Italiana del Cemento* - May 1990.

213 Unidentified Author. "I viadotti di Sylans e Glacieres con struttura a traliccio spaziale in c.a.p. realizzati per l' Autostrada A40 nel massiccio del Jura." *L' Industria Italiana del Cemento* - June 1990.

214 Unidentified Author. "Eine brücke uber die Donau." *Der Baukasten* - September 1990.

215 Unidentified Author. "Baustelle eines Weltrekords. Die pylone der Pont de Normandie." *Bauzeitung* - November 1992.

216 Verch, W.. "Taktschiebeverfahren im Spannbetonbrückenbau der DDR." *Bauplanung, bautechnik* - April 1990.

217 Virlogeux, M.. "Développement de l' innovation dans le domaine des grands ouvrages" *Travaux* - December 1981.

218 Virlogeux, M.. "Evolution de la conception des ponts de portée moyenne." *FIP Congress Stockholm 1982* - Final Report.

219 Virlogeux, M.. "La précontrainte extérieure." *Annales ITBTP* - December 1983.

220 Virlogeux, M.. “Bilan de la politique d' innovation dans le domaine des ouvrages d' art” *Travaux* - March 1985.
221 Virlogeux, M.. “La precontrainte exterieure au beton.” *FIP Congress New Delhi 1986* - Final Report.
222 Virlogeux, M.. “Die externe vorspannung.” *Beton und Stahlbetonbau* - May 1988.
223 Virlogeux, M.. “Les ponts de portee moyenne.” *L' Ingenieur Constructeur* - June-July 1990.
224 Virlogeux, M.. “La conception et la construction des ponts à précontrainte extérieure au béton.” *Annales ITBTP* - November and December 1991.
225 Volke, E.. “Funktions und ausfuhrungsgerechte planung von stahlbrücken.” *Eisenbahningenieur* - August 1992.
226 Walter, S.. “10.000 mq Fertigteile über den Bahnhof in Basel.” *Bauwirtschaft* - 12 May 1997.
227 Wittfoht, H.. “Wohin entwickelt sich der spannbeton-brückenbau? Analysen und tendenzen.” *Beton und Stahlbetonbau* - February 1986.
228 Wittfoht, H.. “Building Bridges” *Beton Verlag, Dusseldorf 1984.*
229 Wolff, H.J.. “Spannbetonbrücke im Taktschiebeverfahren” *Beton* - May 1981.
230 Zilch, K.. “Maintalbrücke Veitshochheim - Maschinen und rusttechnik fur eine ungewohnliche bauaufgabe.” *BMT Baumaschine und bautechnik* - January 1987.

INDEX